Contributions from Biology Education Research

Contributions from Biology Education Research (CBER) is the international book series of the European Researchers in Didactics of Biology (ERIDOB). The series includes edited collections of state-of-the-art papers presented at the ERIDOB international conferences, and monographs or edited collections of chapters by leading international scholars of the domain. The aim of the series is to shed light on global issues and trends in the teaching and learning of biology by gathering cutting edge research findings, theoretical views, and implications or concrete suggestions for everyday school practice regarding biology. The books may serve as resources for (a) getting informed about the most recent findings of biology education research to possibly integrate them in new personal research, and (b) studying about the teaching and learning of biology as a pre-service or in-service biology teacher. So, the main audiences for the series range from senior to early career biology education researchers and pre- or in-service biology teachers working at all educational levels. Book proposals for this series may be submitted to the Publishing Editor: Claudia Acuna E-mail: Claudia.Acuna@springer.com

Olivier Morin · Catherine Bruguière ·
Marcus Hammann
Editors

Challenges of a Changing World in Biology Education

Selected Papers from the ERIDOB 2024 Conference

eridob
european researchers
in didactics of biology

Springer

Editors
Olivier Morin
Claude Bernard University Lyon 1
Villeurbanne, France

Catherine Bruguière
Claude Bernard University Lyon 1
Villeurbanne, France

Marcus Hammann
Münster University
Münster, Germany

ISSN 2662-2319 ISSN 2662-2327 (electronic)
Contributions from Biology Education Research
ISBN 978-3-032-05345-9 ISBN 978-3-032-05346-6 (eBook)
https://doi.org/10.1007/978-3-032-05346-6

This work was supported by Universität Münster, Zentrum für Didaktik der Biologie, Schlossplatz 34, 48143 Münster, Germany.
Université Claude Bernard Lyon 1, S2HEP - Sciences et Société; Historicité, Education et Pratiques UR 4148, Campus de la DOUA, 38 Boulevard Niels Bohr, 69622 Villeurbanne, France.

This Springer imprint is published by the registered company Springer Nature Switzerland AG
The registered company address is: Gewerbestrasse 11, 6330 Cham, Switzerland

Preface

This book of selected papers is a collection of 22 original research contributions presented at the 14th conference of European Researchers in Didactics of Biology (ERIDOB), which took place at the University Claude Bernard Lyon 1, France, from July 1 to 5, 2024. ERIDOB contributes to shaping the future of research in biology education, providing a unique platform for researchers in biology education to discuss the theoretical and conceptual underpinnings of biology education, as well as empirical aspects of research on teaching biology and biology education.

In the current period, marked by both a strong expectation for science to meet the challenges of our era and by the discrediting of scientific results, the theme of the ERIDOB 2024 Conference was "Biology education to bridge the gap between science and society." Under this banner, the contributions addressed a wide range of different research topics related to the teaching of biology, united by the conviction that the specificities and the nature of the knowledge to be taught are the basis for the construction of the meaning of learning.

Following the conference, the submitted chapters have been selected after passing a peer-review process by experienced reviewers of the ERIDOB academic community, followed by a monitoring of revisions by a member of the ERIDOB 2026 Academic Committee. We believe that this edited volume has managed to capture the diversity and quality of current research being carried out across Europe and beyond. The chapters presented in this book are organized in four parts, representing recent trends in biology education research and reflecting the main research strands of the conference.

Important aims for biology education include equipping young people with the tools and resources needed to become scientifically literate, fostering critical thinking, and promoting the ability to actively engage in and solve socioscientific issues. Therefore, we chose the following title for this book of selected papers from the ERIDOB conference: *Challenges of a Changing World in Biology Education.*

The first group of chapters (Part I: *Socioscientific Issues, Environmental Issues and Climate Changes*) deals with the ways researchers, students, and teachers might respond to the new realities of the world they are facing. The section begins with two special chapters by authors who kindly agreed to hold plenary keynote lectures at the ERIDOB conference. Looking back to inspire the future of biology education, the authors of the first two chapters provided the community with their analyses of the evolution of their shared field of research on socioscientific issues in biology education aimed at developing students' critical and responsible citizenship and their ability to become agents of change.

Marie-Christine Knippels and Laurence Simonneaux (Chap. 1) give an overview of the research related to the links of education for action competence and environmental citizenship with the diverse visions of science education. They explore the educational steps that can help design socioscientific inquiry-based learning, and they argue for significant changes in teacher education, curriculum design, and assessment methods, particularly the consideration of action-taking as part of the learning process and the required development of a hope-based education.

The chapter by Justin Dillon (Chap. 2) notes that the current education system has not been successful in preparing young people to cope with wicked problems. He addresses the difficult integration of such problems in science curricula and argues for radical change and for hope in the role of biology education to make society a better place.

After these research reviews, the following seven chapters focus on various issues related to current concerns about sustainability. The chapter by Maren Skjelstad Fredagsvik, Marthe Lønnum, and Ragnhild Lyngved Staberg (Chap. 3) deals with Norwegian student's decision-making related to a local and authentic sustainability dilemma on environmental management. By analyzing the articles written by the students and characterizing students' justifications in arguments, the authors found that students have difficulties justifying their claims with factual information and scientific knowledge. Furthermore, the authors stress the importance of valuing emotions as an important dimension that affects the students' decisions. In sum, this study highlights that dilemmas are a suitable method for students to develop the ability to identify different perspectives and to assess these perspectives against each other in order to make a decision.

The contribution of Theodora Boubonari and Athanasios Mogias (Chap. 4) focuses on Greek primary students' knowledge of ocean acidification, before and after an intervention including 8 hours of inquiry-based learning and knowledge-integration activities. The results serve as evidence that the intervention had a positive effect on students' understanding of the subject matter. Furthermore, it supports the suggestion that the carbon cycle and ocean acidification can be taught in primary education.

Georgios Ampatzidis and Konstantinos Korfiatis (Chap. 5) investigated the ways in which environmental problems, their causes and proposed solutions are presented in Greek secondary education biology textbooks. They found that biology textbooks address some popular issues such as biodiversity loss, pollution, overgrazing and eutrophication, but their main result consists in the lack of a substantial discussion of currently debated and practiced pro-environmental actions.

Regarding the actions to be taken in response to environmental issues such as the decline in biodiversity, the chapter by Dorothee Benkowitz and Holger Weitzel (Chap. 6) reports on findings from an evaluation study involving German students in a hands-on conservation program with tangible experiences. The results show the relevance of fieldwork and how the intervention significantly increased the participants' knowledge, motivation and appreciation of nature.

The focus of the chapter by Anna Pshenichny-Mamo, Maggie Demarse, Roberta Howard Hunter, and Dina Tsybulsky (Chap. 7) is an exploration of the educational potential of climate change exhibits in natural history museums. The authors' analyses of the exhibition signages and of 15 semi-structured interviews with museums staff members reveal which aspects of the Nature of Science are integrated into two Climate Change exhibitions, one in Israel, the second in Michigan (USA). The authors describe how the exhibitions merge scientific knowledge with insights into the way evidence is gathered and validated. Their analyses underscore the dynamic nature of scientific understanding and display scientific research as a human endeavor.

Secondary school students' nature experiences are the focus of the chapter by Marcus Hammann, Helene Kesseler, and Justin Wonnesch (Chap. 8). The authors analyzed the nature experiences portrayed by a large sample of German secondary school students, and investigated potential links between the nature experiences and the students' ratings of their nature connectedness and the perceived personal relevance of their nature experiences. The study found that plants featured more prominently as background in recreational experiences and that animals were more often the focus of observations in exploratory activities. The study also found that students reporting experiences with plants exhibited lower nature connectedness than students reporting experiences with animals. Surprisingly, the percentage of students who reported on nature experiences without animals and plants was overwhelmingly large. This finding paves the way for new studies on what factors make nature experience involving plants and animals memorable.

The chapter written by Marian Mulcahy, Wilton Lodge and Michael Reiss (Chap. 9) ends this part broadening the focus on socioscientific issues problems in science education to include gender issues. Based on twelve semi-structured interviews with English initial teacher educators, the study provides insights into areas where university tutors feel that social justice issues may be addressed within school biology, and the role they believe school biology could play in addressing these issues. The findings align with the growing recognition that school science should contribute to social justice. The authors' main conclusion is that social justice issues could and should be addressed in biology classroom, and that biology education for a changing world requires creating spaces where students explore the historical and structural factors that have shaped our discipline.

The second group of chapters (Part 2: *Scientific Inquiry, Scientific Explanation, and Nature of Science*) focuses on various issues related to scientific inquiry, particularly the uncertainty of measurement in biology research, the nature of biological explanations, modeling as an aspect of scientific inquiry, self-produced educational videos on experimental lab techniques, and the role of science-in-the-making in curricula.

In their chapter on concepts of metrology, the science of measurement, Myriam Regent-Kloeckner, Clément Maisch, and Christophe Daussy (Chap. 10) report on findings from an exploratory study investigating the extent to which teaching goals associated with measurement are addressed and whether concepts and methods of metrology are included in the first years of university degrees in biology in France. Their analysis focuses on course materials provided to biology students as well as interviews with university teachers and program coordinators. Based on the finding that metrological concepts were rarely addressed, particularly issues of reliability and validity, the authors argue that familiarization with key concepts of metrology might enhance biology students' understanding of measurement-related issues and improve their confidence in scientific results.

The chapter by Christina Ehras and Arne Dittmer (Chap. 11) focuses on three challenging aspects of teaching biological explanations in the classroom: the complexity of biological explanations, the uncertainty of knowledge, and the absence of knowledge (non-knowledge). The authors present findings from a study in which four groups of individuals (students, student-teachers, teachers, and biology education researchers) were asked to rate three linear and three complex biological explanations. Based on the findings that the majority of participants preferred linear explanations over complex ones and tended to exclude the issue of non-knowledge and uncertainty, the authors argue that aspects of complexity, non-knowledge, and uncertain knowledge need to be integrated into biology teacher education to support students' knowledge of the nature of science and their ability to address challenging aspects of explanations in the classroom.

Modeling is an essential method for investigating and explaining biological phenomena. Kim Eleni Lobner, Tom Bielik, and Moritz Krell (Chap. 12) focus on the relationships between prospective biology teachers' knowledge of models and knowledge about the modeling process. The authors found a positive correlation between these two types of knowledge. Furthermore, they note that student teachers with a more developed understanding of the purpose of models had a better understanding of the modeling process than students who had a less developed understanding of the purpose of models. Therefore, teaching about the purpose of models and related dimensions of the modeling process is recommended for biology teacher education.

In their chapter on educational videos as visual learning tools (Chap. 13), Tetiana Krushynska and Iryna Kompanets describe a case study in which prospective biology teachers produced educational videos on experimental techniques for biology lab practicals. The authors note that it was feasible for students to create videos of lab sessions without professional equipment and additional time, and that the quality of student-produced videos was satisfactory for educational purposes. Furthermore, the students considered the creation of videos as helpful for memorizing and understanding lab procedures. The authors recommend the method of self-made educational videos because the majority of students believed that the skills they had acquired would be valuable for their future professional careers.

Scientific controversies provide an opportunity for advancing students' understanding of the nature of science. Focusing on the scientific debate about classifying chimpanzees as members of the genus *Pan* or *Homo*, Corinne Fortin (Chap. 14) analyzed how French curricula for pre-service biology students and secondary school students portray the genus *Homo*. As a main finding, the author notes that the curricula focus on fossil records to reconstruct the phylogenetic relationships between *Homo sapiens* and other modern primates, excluding other evidence, such as tool use and social organization. Furthermore, the scientific controversy surrounding the boundaries of the genus *Homo* is absent from the curricula, which is interpreted as a missed opportunity for advancing students' understanding of controversies as an essential aspect of science-in-the-making. The authors suggest conducting similar curricular analyses in other countries.

The third group of chapters (Part III: *Students' and Teachers' Conceptions*) consists of one contribution that looks at a methodology for collecting student conceptions, two contributions dealing with students' conceptions on biotechnology and on the pathway of water through the body and one contribution focusing on the beliefs of biology teachers concerning the results of biology education research.

The contribution of Bianca Reinisch, Tom Bielik, Moritz Krell, and Daniela Mahler (Chap. 15) explores whether a digital version tool of the traditional Draw-A-Scientist Test can be validly used to capture and interpret students' conceptions of biologists and their work. The alignment between verbal data and visual representations suggests that students can effectively communicate their conceptions through this digital media version, providing researchers with a valuable means of assessing conceptual understandings. However, the system categories need to be refined to better capture the emotional and personal characteristics of depicted scientists.

In her chapter, Maria Horniaková (Chap. 16) highlights the conceptions of Czech students on key concepts related to modern biotechnology. She found that students focused more on the technology aspect of biotechnology rather than understanding its biological application. They often confused genetic engineering with genealogy and artificial insemination. Furthermore, they struggled to explain how GMOs are created and often perceived cloning as a science fiction concept. The study underscores the need for targeted educational approaches that address these preconceptions and foster conceptual change.

Pernilla Granklint Enochson (Chap. 17) investigated the ways in which organs and organ systems, in particular the pathway of water through the body, are understood and perceived as challenges in teaching this subject-matter at primary school by pre-service Swedish teachers. She interprets the difficulties these students have in making links between the pathway of water through the body, the functions of the kidneys, and the importance of consuming water by the fact that they have not sufficiently problematized the functions of water in the body during their previous training. Her study confirms the need to ensure that pre-service teachers understand the relationships between different organ systems and the specific interconnections linking them.

The chapter by Alexander Büssing and Bianca Reinisch (Chap. 18) focuses on in-service biology teachers' beliefs on the nature of biology education research as well as the social certification and dissemination of biology education research results. Their qualitative interview study analyses revealed the findings that teachers' beliefs about the transfer of biology education research findings to the practice of teaching biology depends on the utility of the results. Furthermore, the teachers saw a role for themselves in the social certification of research findings. The authors stress that it may be fruitful to further theoretically elaborate on the conceptualization of beliefs and their possible differentiation from related constructs.

The fourth group of chapters (Part IV: *Teaching Strategies and Learning Environments*) comprises one article which assesses students' learning progressions and three articles which study the learning potential of game-based and cartoon-based teaching.

To challenge common assumptions about learning, François Lombard, Marie Merminod, and Laura Weiss propose in their chapter (Chap. 19) to follow the progression of pupils' learning in their explanations of biological phenomena. The authors investigated three assumptions about learning: learning paths are linear, most learners follow one learning path and teacher activities predict what is learned. Using the Reference Map Change Coding (RMCC) method to analyze individual learning paths, the authors refuted all three assumptions. Their findings rather revealed that learning progressed in a mosaic-pattern, that learning paths varied considerably across cohorts and that teaching and learning did not follow a co-linear path. In conclusion, the methods used in this study have proven valuable in guiding teachers and improving learning designs.

The chapter by Martha Georgiou (Chap. 20) explores the impact of an educational nutrition card game "Food for Good" on students' knowledge of nutrition and eating attitudes. The originality of this game created by the author is that it includes food groups that are close to those encountered by students in their daily lives, such as processed foods. By evaluating its first version through a pilot study, the author showed that the game appeared to improve students' nutritional knowledge, but this was not the case when it came to the ability to choose foods for a healthy daily diet. The author discusses ways to improve the game and the need to consider a longer implementation period of the game.

The aim of the chapter of Magali Coupaud, François Dessart, Fabienne Paulin, and Alice Delserieys (Chap. 21) is to understand the extent to which the narrative framework of the Darwinium game which was designed to address the concept of chance in the theory of evolution influences the ways in which students think about evolution. The analysis of the middle-school students' narratives who engaged in this board game showed that most of them were in line with evolutionary thinking and those who were not and showed finalist or teleological, thinking unexpectedly resorted to an argumentative discourse. The Darwinium game took students on a temporal adventure allowing them to experience a diversity of population futures, over several generations, and the role of chance in this field of possibilities. However, their interpretations often blended different concepts and combined elements of scientific and intuitive thinking.

The contribution of Marida Ergazaki and Iro Bardoutsou (Chap. 22) reports on a study concerning an edutainment series for young children, the "Dinosaur Train," which takes place in the Mesozoic era. The aim of the analysis of 30 "Dinosaur Train" episodes is to identify which biological concepts, scientific practices, and everyday life concepts are presented and how. The authors conclude that "Dinosaur Train" seems to be a valuable informal resource for engaging young children in biology learning, familiarizing them with scientific practices as tools for expanding their knowledge about nature, and having them reflect on everyday life emotions, relationships, and behaviors.

Villeurbanne, France — Olivier Morin
Villeurbanne, France — Catherine Bruguière
Münster, Germany — Marcus Hammann

Acknowledgements

We would like to kindly thank Fabien Baudier from the Congress Unit of the Université Claude Bernard Lyon 1, for helping with the logistics of the submission and review process, and for his proof reading of the chapters.

We are also extremely grateful to all the colleagues in the field of biology education who kindly contributed to the peer reviewing of the chapters in this book:

Idit Adller
Isabell Adler
Georgios Ampatzidis
Branko Andjic
Babai Reuven
Orit Ben Zvi Asaraf
Anna Beniermann
Teresa Berglund
Petra Bezeljak Cerv
Tom Bielik
Jelle Boeve-de Pauw
Bruguière Catherine
Alexander Buessing
Graça Carvalho
Nina Christenson
Magali Coupaud
Arne Dittmer
Michael Ewig
Corinne Fortin
Adiv Gal
Georgiou Martha
Niklas Gericke

Marcus Hammann
Kostas Kampourakis
Sara Klein
Konstantinos Korfiatis
Moritz Krell
Ergazaki Marida
Ralph Levinson
François Lombard
Armin Lude
Daniela Mahler
Evangelia Mavrikaki
Olivier MORIN
Marinthe Meijer
Bianca Reinisch
Michael Reiss
Carolin Retzlaff-Fuerst
Dana Sachyani
Martin Scheuch
Jelka Strgar
Tsybulsky Dina
Micha Ummels
Annette Upmeier zu Belzen
Emma Versteegh
Susanne Walan
Holger Weitzel

Finally, we would like to thank all the members of ERIDOB Academic Committee for their valuable contribution, cooperation, and support during all the stages of the preparation of the 2024 conference and of the manuscripts for the current volume.

ERIDOB Academic Committee 2024

ERIDOB Academic Committee 2026

Contents

Part I
Socioscientific Issues, Environmental Issues and Climate Change

Chapter 1
SSI/SAQ-Based Educ-ation to Develop Responsible Citizenship Towards Socio-Political Action: Ways to Deal with Controversies and Needs for Transitions

Marie-Christine P. J. Knippels and Laurence Simonneaux

1.1 Introduction

Today, society faces various challenges, such as climate change, biodiversity loss, pesticides and health, ecological and economic repercussions of agronomic practices, disappearance of bees, pandemics and vaccinations. These Socio-Scientific Issues (SSI), called Socially Acute Questions (SAQ) in the French context (Simonneaux & Pouliot, 2017), present complex, contentious, and multidisciplinary challenges deeply intertwined with scientific, social, economic, and ethical dimensions. These issues, which are often the subject of debates and media attention, highlight the intersection of scientific development, societal governance, and ethical considerations.

Dealing with SSI/SAQ sometimes requires the implementation of transitions, such as energetic transitions, ecological and societal transitions, and agroecological transitions. Transitions are difficult because socio-technical systems are locked. They are on the rails of a trajectory imposed by the socio-technical system and hardly allow alternatives to emerge. Transitions don't follow standardised models. For example, in agriculture it is possible to use almost the same chemical treatment regardless of the context. However, this is not the case in agroecology transition, where context must be taken into account, requiring customisation. Agroecological practices vary depending on the soil and climate context and previous crops. They are no longer

M.-C. P. J. Knippels (✉)
Freudenthal Institute, Utrecht University, Utrecht, Netherlands
e-mail: M.C.P.J.Knippels@uu.nl

L. Simonneaux
ENSFEA, University of Toulouse, Toulouse, France

O. Morin et al. (eds.), *Challenges of a Changing World in Biology Education*, Contributions from Biology Education Research,
https://doi.org/10.1007/978-3-032-05346-6_1

standardised, so we should talk about a plurality of transitions, rather than a single, uniform process.

Contributing to such transitions is challenging, especially in the context of a growing distrust of science and technoscience in society. It requires scientifically literate and responsible citizens.

1.2 Change of Educational Regime Towards 'Educ-action' and the Development of Action Competence

Biology education plays a crucial role in preparing young people for these issues. It should contribute to the development of scientifically literate and responsible citizens capable of making informed decisions and taking responsible actions on socio-scientific issues.

SSI/SAQ are characterised by different elements: they are problems which arise in our society and have a scientific and/or a technological component; there is no consensus on how such problems might best be solved; they are ill-structured, controversial and personally relevant issues that require scientific, evidence-based reasoning to inform decision-making; they include moral and ethical aspects, scientific uncertainties and socio-environmental risks; they concern many different stakeholders with different values and priorities; they involve opinion-forming and choice-making both at the personal and societal levels (Levinson, 2006; Ratcliffe & Grace, 2003; Sadler, 2011; Zeidler et al., 2019).

As outlined by the OECD (2018), developing 'Global Competence' is essential for addressing these types of issues. It defines global competence as the ability to critically engage with local, global and intercultural issues, to understand and appreciate different perspectives, to interact respectfully with others and to take responsible action. This requires a change in the educational regime and in the ways that science education is conceptualised and applied. Sjöström and Eilks (2018) illustrate this necessary change by elaborating on three visions of scientific literacy and science education from the concept of *Bildung*. In short, these three visions can be summarised as:

- Vision I: 'Pipe-line science' in preparation of future scientists, focuses on scientific content and scientific processes to later discover the corresponding applications.
- Vision II: 'Science for all' focuses on contextualizing scientific knowledge to make sense of its use in life and society.
- Vision III: 'Science for transformation' emphasises philosophical values, politicization and critical education for global citizenship (Sjöström & Eilks, 2018).

According to Simonneaux (2013), the educational issues of SSI/SAQ-based education can be positioned on a cold-hot continuum presented in Fig. 1.1. At one end, (the 'cold' end), the integration of SSI is used to motivate students to learn science, or even to convince them to adhere to the use of new techno-sciences (in line with Vision I). At the other end, the 'hot' end, the purpose is nurturing activist commitments, beyond aims relating to science conceptual and procedural knowledge (in line with Vision III). The necessary implementation of transitions (energy, agroecological, etc.) presupposes scientific teaching which aims for transformation and action ('hot' end and Vision III). Between these two poles are educational issues that are more or less committed, from teaching and learning underlying scientific concepts which are relatively stable at the cold end to the development of critical thinking and activism towards the 'hot' end. The 'cold' end mainly emphasises monodisciplinarity, scientific learning, and epistemic values, whereas the 'hot' end emphasises interdisciplinarity, scientific and humanities learning, political citizenship, and philosophical values (Simonneaux, 2013) (see Fig. 1.2).

From the French perspective, teaching SAQ aims to promote engaged citizenship, which implies an engaged education oriented towards the development of

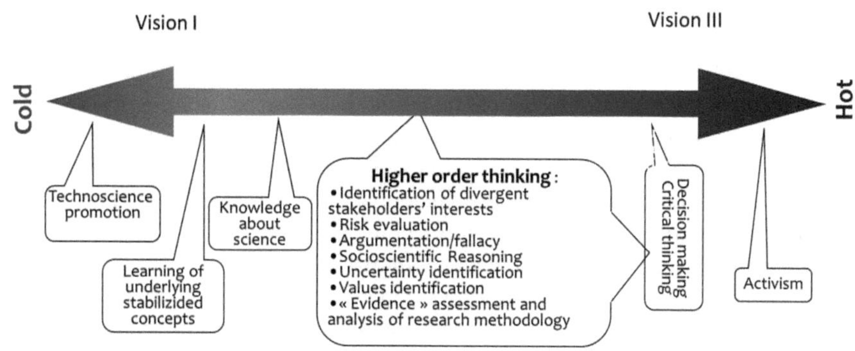

Fig. 1.1 Educational aims of SSI/SAQ based education (Simonneaux, 2013)

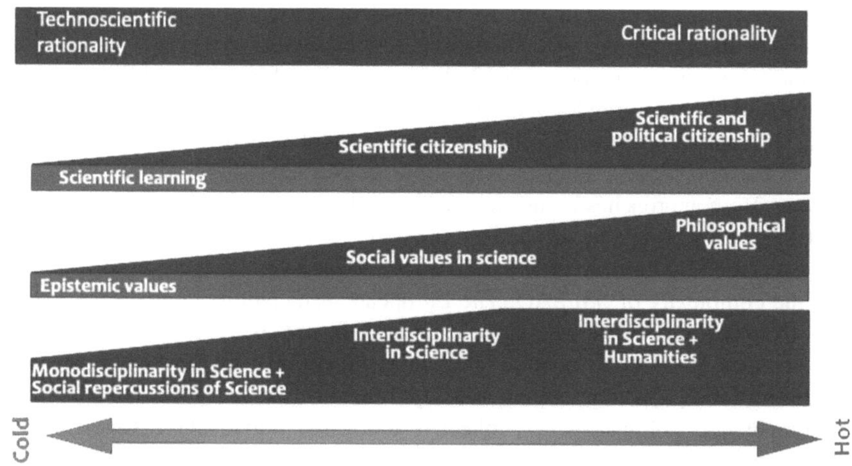

Fig. 1.2 Educational orientations of SSI/SAQ-based education (Simonneaux, 2013)

critical thinking and socio-political action: the so-called 'educ-action'. Educ-action represents pedagogical forms that aim to encourage the engagement of students and teachers in individual and collective action. It is defined as a *critical perspective* on education, since the idea is to consider education as a lever not only for acquiring knowledge, but also for making the process of technoscientific and social modernization itself an object of reflection, vigilance and problematization. It is also a *political perspective,* since the aim is to develop learners' ability to act, express themselves publicly and participate in democratic life (Simonneaux and Simonneaux, 2017). The term 'activism' in Fig. 1.1 can be confusing, because in everyday language it refers to radical and sometimes violent actions in public spaces, to militant practices whose organization calls for an erasure of the individual in a collective (Hervé, 2024). The educational works claiming to be activism are however clear: they do not promote a political education which would be linked to ideological or political movements, it is rather to work on questions of citizenship by highlighting '*the capacity and student commitment to taking appropriate, responsible and effective action on social, economic, environmental and moral-ethical issues*' (Hodson, 2003, p. 658). The term 'activism' refers to a process of collective action, which is democratic in nature and relies on research as well as deliberation to consider solutions to social or environmental problems linked to scientific and technical development. This involves students who are able to identify the causes of the problems they are working on, to consider ways to address and resolve them, individually or collectively, by exercising their rights and duties, and by taking into account the principles of social justice and environmental justice, including intergenerational justice.

We *aim for Vision III*, which is at the hot side of the continuum, to help develop students' critical and responsible citizenship and become agents of change. Science education oriented toward Vision III (critical-reflexive scientific literacy) aims to make the student capable of leading an autonomous life in his/her socio-cultural environment, participating in a democratic society and showing empathy and solidarity with others, in order to improve critical democratic participation and to shape our society and culture in a sustainable way. It relates to the concepts of subjectification and transformation (Biesta, 2012, 2020) as well as Sjöström and Eilks' (2018) definition of *Bildung*-oriented science education which is agency-oriented.

As Sjöström and Eilks (2018) indicated, this concept is also closely linked to recent educational approaches, such as Education for Sustainability and transformative learning. Both approaches focus on building skills for critical, democratic engagement and for fostering a society and culture rooted in sustainability. In the context of sustainability education, Bianchi et al. (2022) defined competences needed to deal with the complexity of SSI/SAQ: The European sustainability competence framework ('GreenComp'). They define four competence areas: (i) embodying sustainability values, (ii) embracing complexity in sustainability, (iii) envisioning sustainable futures, and (iv) acting for sustainability (Bianchi et al., 2022). The latter includes the competences political agency, collective action and individual initiatives, needed to allow students to become 'agents of change'.

This goes hand in hand with the development of *action competence,* defined by Sass et al. (2020) as a combination of knowledge, willingness, and confidence to engage in (socio-political) action. Unlike behavioural responses, action competence emphasises intentional, transformative actions that are consciously chosen by the student. An action-competent person has: (i) Knowledge of action possibilities (knowledge that helps to engage in (an) action, knowledge about the 'hows' of (an) action), (ii) Confidence in one's own influence (feeling confident that you can engage in (an) action, feeling confident that if you engage in (an) action it will produce favourable outcomes), (iii) Willingness to act (autonomously motivated to engage in (an) action, driven to engage in (an) action), (Sass et al., 2020). The goal is to allow students to become agents of change, able to act and participate in society in the private and public sphere, on a local, national and global scale, through individual and collective actions.

Education for action competence and environmental citizenship benefits from a holistic, pluralistic, and action-oriented approach (Boeve-de Pauw et al., 2015; van Harskamp, 2023). The complexity of sustainability issues (and other SSI) requires a holistic approach to education to support students' systems thinking skills (e.g., Wiek et al., 2011). It refers to the 'what' of teaching, such as adopting an interdisciplinary approach that considers temporal, ecological, and social aspects as a coherent whole when discussing the issue (e.g., Öhman, 2008; Sass et al., 2024, see Fig. 1.3). Pluralism refers to the 'how' of teaching, it concerns normative aspects, openness to different perspectives, and fostering respectful engagement with diverse viewpoints (e.g., Rudsberg & Öhman, 2010). Action-orientation refers to the opportunities given to students to discuss possible actions and contribute to addressing local real-world issues (Sinakou et al., 2019; Sass et al., 2020, 2024).

So, we see that different terminology and competence frameworks are used or emphasised in different contexts, countries, cultures, and languages. Overall, we emphasise the importance of SSI-based education to foster responsible citizenship and action taking ('educ-action') as part of scientific literacy. Fostering these competences asks for a holistic, pluralistic and action-oriented approach in education.

'*Socio-Scientific Inquiry-Based Learning*' (SSIBL) is a pedagogy that can bridge the gap between science education and societal needs by promoting critical-reflexive scientific literacy (Vision III) and the ability to critically engage with complex global issues. We will elaborate on this approach in the next section.

1.3 The Socio-Scientific Inquiry-Based Learning (SSIBL) Approach

SSIBL is a research-informed pedagogical framework, developed in the European PARRISE project, that promotes meaningful science learning and responsible citizenship. SSIBL deals with authentic questions about controversial SSI which are open-ended, involve participation and inquiry and aim at solutions which help to enact change. In other words, SSIBL promotes a pedagogy that connects the study of socio-scientific issues (SSI) with inquiry-based learning (IBL) and citizenship

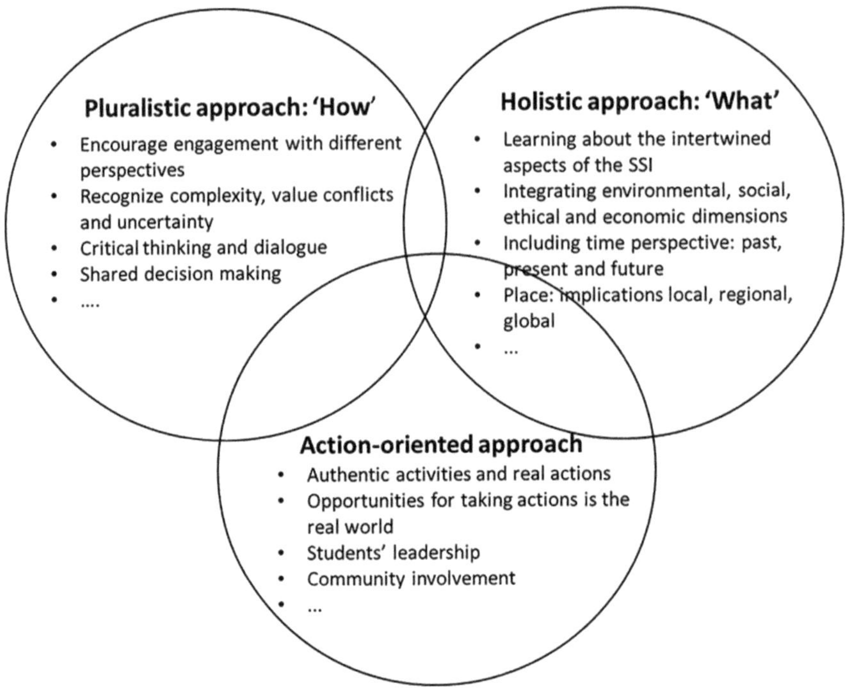

Fig. 1.3 Education for action competence and environmental citizenship (based on Sass et al., 2024; Sinakou et al., 2019)

education within the umbrella of responsible research and innovation (RRI) in science (Levinson, 2018; Levinson et al., 2017).

SSIBL consists of three interconnected stages: (i) raising authentic questions (ASK), through emergent SSI arising from the impacts of science and technology on society; (ii) integrating social, personal and scientific inquiry to explore these questions and identify solutions (FIND OUT); and, (iii) taking socially responsible action (ACT) (Knippels & van Harskamp, 2018; Levinson, 2018). The simplified representation of the SSIBL framework is shown in Fig. 1.4. It consists of three stages, with the backgrounds of social desirability, ethical acceptability and sustainability (RRI principles).

The SSIBL approach begins with an authentic and relevant scenario, i.e., a real-life issue that students are motivated to engage with and solve. This approach involves identifying an open-ended problem (SSI) that can be addressed through social, scientific and personal inquiry. It draws on multiple types of knowledge, including scientific, ethical, cultural, political, and economic perspectives, while maintaining strong connections to relevant science concepts within the curriculum. Additionally, SSIBL emphasises democratic deliberation, encouraging students to explore different viewpoints, and fosters engagement with various stakeholders both within and beyond the school community.

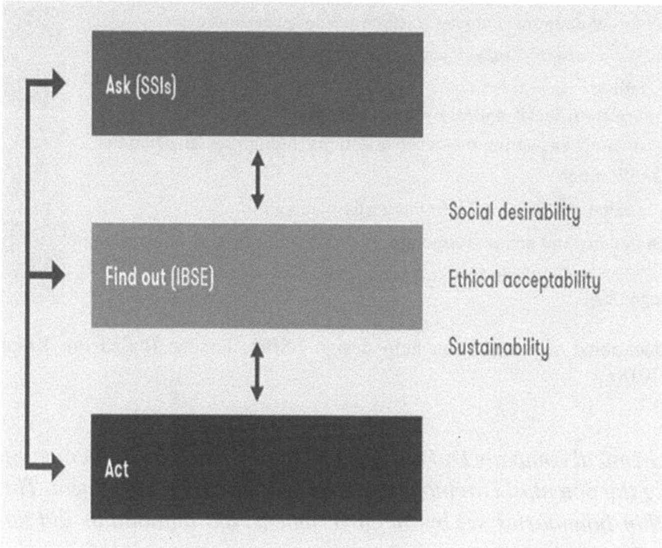

Fig. 1.4 Simplified SSIBL stages (figure from: Amos et al., 2020; Levinson et al., 2017)

Implementing SSIBL in practice requires engagement in dialogical ways of teaching and learning (Alexander, 2020), such as group discussions, dialogue, role-play, and argumentation. Strategies such as mapping controversies can scaffold students' exploration of the SSI. The use of the 'argument or controversy line' activity, in which students are asked to physically position themselves along a line of 'for' or 'against' in the classroom, can support social inquiry (e.g., what do others think about the issue?), personal inquiry (e.g., what do I feel and think about the issue?) and science inquiry (e.g., what scientific knowledge do I need to understand the issue?) (Knippels & Christodoulou, 2024; Knippels & van Harskamp, 2018). In the next section, we present educational steps that can support the design and implementation of SSIBL, illustrated with a real-life example.

1.3.1 SSIBL in Practice

Steps that might be helpful for integrating the ASK-FIND OUT-ACT stages (Ariza et al., 2021; Knippels & van Harskamp, 2018) in a lesson or module are depicted below (Fig. 1.5). It shows how each SSIBL stage in the sequence of steps might look like.

*We illustrate the steps presented in Fig. 1.5based on an example from practice (see Table 1.1), in which a Dutch secondary school science teacher chose a real life, local issue to introduce the SSI and contextualize and problematize the issue for the students (*Sarid et al., 2024, *pp. 49–50):*

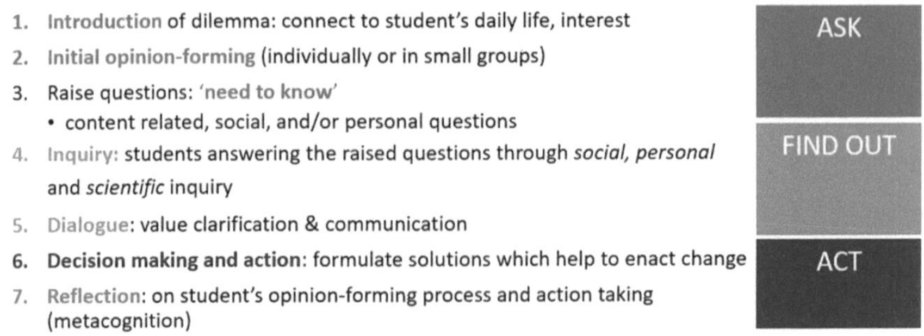

Fig. 1.5 Educational steps that can help design SSIBL lessons (based on: Knippels & van Harskamp, 2018)

Local chemical company DuPont dumped their waste into a river near the school, introducing the potential carcinogen C8 (PFAS) into the environment. The company stayed within boundaries set by the government, but inhabitants did not trust the guidelines. They feared the potential carcinogen was responsible for the inexplicable illnesses from which some of the factory workers suffered.

The teaching and learning activities implemented in this module exploring this issue arranged by the different steps (linked to the SSIBL phases) are depicted in Table 1.1.

More practical strategies for implementing SSIBL in the classroom are provided by Knippels & Christodoulou (2024)*in the context of the European COSMOS project* (www.cosmosproject.eu).

1.3.2 SSIBL in the French Context

Because the word 'inquiry' has strong connotations for French science teachers, the SSIBL approach is called 'démarche d'enquête' (translation: 'investigative approach') in France. This approach emphasises the role of schools as arenas for SAQ and socio-political education. Students are tasked with researching local issues such as water management, engaging with local stakeholders, and presenting solutions that consider scientific, social, and ethical dimensions (Simonneaux & Simonneaux, 2017). Five phases structure the inquiry of SAQ in 'démarche d'enquête' (see Fig. 1.6): collection/analysis of information, reflexivity/subjectivity of the investigators, explicitness/construction of reasoning, possible answers and actions, report on the inquiry.

Their order and recursiveness depend on the pedagogical project. The 'démarche d'enquête' proposes the broadening of scopes of inquiry, which is not necessarily to be confused with experimental approaches drawn from the sciences of matter or nature, or from the social sciences. Artistic practices and methodologies specific to

Table 1.1 Representation of the seven steps in the lesson module on C8 in river water

SSIBL phase	Educational stages	Teaching and learning activities
	(i) Introduction of the dilemma	Watch a documentary on the SSI
ASK	(ii) Initial opinion-forming	Answering questions individually during the documentary: • *Do you think the boundaries set by the government are fair?* • *Do you think residents have a reason to worry?* Students subsequently discussed their views in small groups and formulated questions they had
	(iii) Creating a need-to-know	Making students experience the different perspectives of stakeholders, by mapping the controversy
FIND OUT	(iv) Inquiry into scientific, social, and personal aspects of the dilemma	<u>Social and personal inquiry</u>: • Listing stakeholders in the issue and their motives. Discussing their views (**social and personal inquiry**) • **Controversy line** activity: they considered with which stakeholders they identified most strongly by physically positioning themselves on a line in the classroom With this activity, content-related and normative student ***questions were raised*** **<u>Scientific inquiry</u>**: • Students had to seek information on the potential toxicity of C8 • Performing a titration experiment on water samples of the polluted river, both upstream and downstream of the factory; identifying the (different) C8 concentrations in the river • Inquiring whether there are recent scientific developments that could replace C8 in the production of goods • Inquiring the trustworthiness of the set concentration boundaries by the government <u>Social Inquiry</u> • Visiting the company and interviewing factory workers and board members
	(v) Dialogue	Discussion about the dilemma, based on statements **mirroring different sides of the dilemma** e.g.: • Consequence of banning C8 from industry; the financial value of human lives; conflict of interest when companies determine the toxicity of their own processes
ACT	(vi) Decision making and action	Looking back on initial opinion and on previous activities by answering reflective questions Based on their findings in the FIND OUT phase, students **wrote an advice letter to the local government** (*indirect action*) Moreover, some students discussed this issue at home and/or decided to stop buying articles that need C8 (PFAS) in the fabrication (*direct action*)
	(vii) Reflection	Answering reflective questions about the steps taken in the learning process (self-monitoring, metacognition)

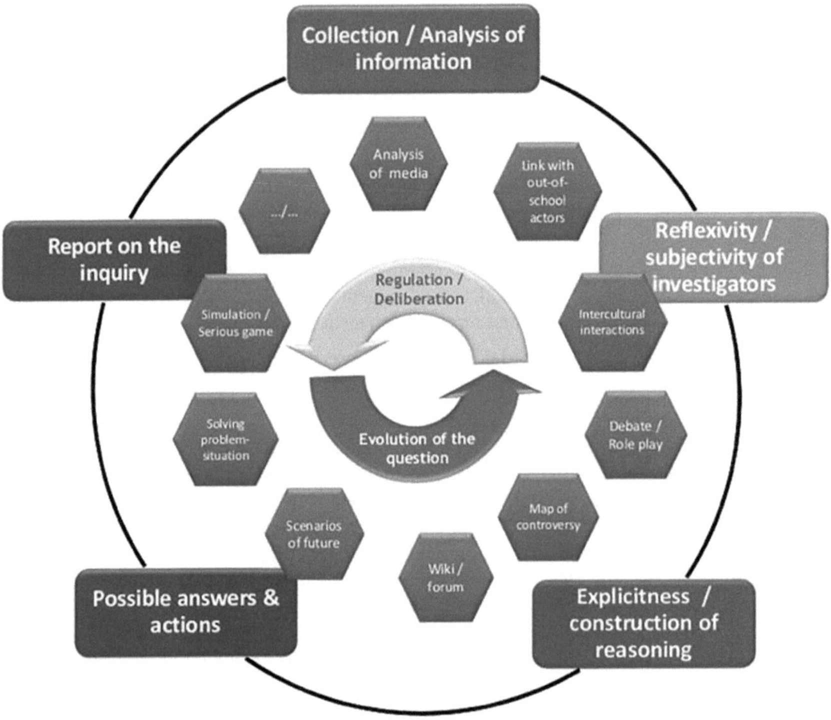

Fig. 1.6 "Démarche d'enquête" on SAQ (Simonneaux and Simonneaux, 2017, p. 149)

the humanities can also be mobilised to explore an issue and gather or analyse information. Actions undertaken are not necessarily the culmination of the investigation. They can constitute a moment in time, a stage that can lead to reformulations of the problem explored and, thus, to the relaunching of the investigation (for example, an action to communicate different options to civil stakeholders leads to their discussion, which brings to light the need to examine some of them in greater depth). In other words, the 'démarche d'enquête' approach proposes to make the inquiry itself the locus of socio-political action.

SAQ and the transitions they imply (energy, ecological and societal, agroecological transitions) are now the prescribed subject matter of teaching. In French agricultural education, the 'Teaching to produce differently 2'-plan intensifies changes to the program by placing the agroecological transition even more at the heart of the student education and training project. The 'démarche d'enquête' is developing regardless of whether within the technological baccalaureate or within the higher technician certificates in all French agricultural schools.

To respond to this new context favourable to the teaching of the agroecological transition and SAQ-based education as well as to the difficulties encountered by teachers, the *School of Controversy* emerged funded by the Ministry of Agriculture (Cancian et al., forthcoming). The School of Controversy integrates research results on SAQ pedagogy, develops trainer training resources, and runs trainer training courses. Its objective is to implement training and support for educational teams towards socio-political education supported by the 'démarche d'enquête' on SAQ.

1.4 Conclusions

The flexibility of SSIBL allows it to be adapted to various socio-scientific issues and educational contexts. It provides a pedagogical structure to help students navigate through the complexities of SSI/SAQ. By encouraging students to explore SSI through scientific, personal, and social inquiry they can develop the knowledge, skills, and dispositions necessary to take informed, responsible action in both personal and societal contexts. A distinguishing feature of SSIBL is its ability to create opportunities for action-taking as part of the learning process, fostering action competence (Sass et al., 2020) as a learning outcome. SSIBL-based education shifts the focus from the more passive knowledge acquisition (Vision I) to active participation, where students are encouraged to inquire, discuss, and act. SSIBL's emphasis on learning science within socio-scientific contexts and for responsible and active citizenship (action competence) means that SSIBL is a pedagogy that can bridge Vision II and Vision III of science education. This pedagogical shift is critical in promoting responsible citizenship, as it allows students to critically analyse societal challenges and formulate reasoned responses to them.

Biology and science education play a crucial role in empowering young citizens with competencies needed to engage with local and global issues by promoting critical-reflexive scientific literacy (Vision III). Achieving this requires a change in the educational regime and in the ways that science education is conceptualised and applied. SSIBL ('démarche d'enquête') aligns with such a demand as it represents a transformative approach to science education.

However, the successful implementation of SSIBL requires significant changes in teacher education, curriculum design, and assessment methods. Teachers need professional development opportunities that equip them to facilitate inquiry-based learning and action-taking in their classrooms. A development that is currently taking place at the School of Controversy in France.

Biology education researchers, in close collaboration with teachers, should contribute to evidence-informed educational guidelines and tools for science teachers and teacher professional development programs, to support this transition. The preliminary results of a European open schooling project called COSMOS are a promising development. In this project, SSIBL was implemented in primary and secondary science education in combination with a community-oriented approach in six countries. This was accompanied by teacher professional development in the approach. Evaluation results showed positive impact on students' attitudes towards science and their perceived action competence for sustainability.

An important aim we touched on in our keynote lecture, but do not cover in this article is the importance of education for hope. Many young people experience eco-anxiety when thinking about the future. Teaching through controversies (SSI) helps young people develop anticipation skills, such as assessing possible futures, evaluating the consequences of actions and managing risks (UNESCO, 2017). By constructing visions of the future, students can express their hopes and fears while positioning themselves in society. Critically evaluating future scenarios strengthens their belief in change. Finnegan and d'Abreu (2024) introduce the 'Hope Wheel', a model for hope-based climate education, which offers a promising framework that can enrich SSIBL pedagogy. Integrating hope-based education into SSIBL further reinforces its transformative potential, ensuring that students not only develop action competence but also sustain the optimism and perseverance needed to address complex societal challenges.

As the world continues to grapple with issues such as climate change, biodiversity loss, and technological disruption, the need for an education system that fosters informed, responsible, and action-oriented citizens has never been more pressing. SSIBL offers a promising framework for achieving this goal, helping students develop the skills, knowledge, and confidence to act as agents of change in their communities and beyond.

References

Alexander, R. (2020). *A dialogic teaching companion.* Routledge.

Amos, R., Knippels, M. C., & Levinson, R. (2020). *Socio-scientific inquiry-based learning: Possibilities and challenges for teacher education.* In J. Dillon, M. Evagorou & J. A. Nielsen (Eds.), *Science teacher education for responsible citizenship: towards a pedagogy for relevance through socio-scientific issues* (pp. 41–61). Springer.

Ariza, M. R., Christodoulou, A., Van Harskamp, M., Knippels, M. C. P. J., Kyza, E. A., Levinson, R., & Agesilaou, A. (2021). Socio-scientific inquiry-based learning as a means toward environmental citizenship. *Sustainability, 13*(20), 11059. https://doi.org/10.3390/su132011509

Bianchi, G., Pisiotis, U., & Cabrera Giraldez, M. (2022). *GreenComp the European sustainability competence framework.* In Y. Punie & M. Bacigalupo (Eds.), *EUR 30955 EN, Publications Office of the European Union, Luxembourg.* ISBN 978-92-76-46485-3, https://doi.org/10.2760/13286, JRC128040.

Biesta, G. (2012). Have lifelong learning and emancipation still something to say to each other? *Studies in the Education of Adults, 44*(1), 5–20.

Biesta, G. (2020). Risking ourselves in education: Qualification, socialization, and subjectification revisited. *Educational Theory, 70*(1), 89–104.

Boeve-de Pauw, J., Gericke, N., Olsson, D., & Berglund, T. (2015). The effectiveness of education for sustainable development. *Sustainability, 7*(11), 15693–15717. https://doi.org/10.3390/su71115693

Cancian, N., Hervé, N., Lipp, A., Vidal, M., & Simonneaux, L. (forthcoming). For a committed and critical Educ-Action towards the agro-ecological transition. In J.L. Bencze (Ed.), *The controversy school mobilizing critical and altruistic science education: Diversifying and broadening alliances for ecojustice.* Springer.

Finnegan, W., & d'Abreu, C. (2024). The hope wheel: A model to enable hope-based pedagogy in Climate Change Education. *Frontiers in Psychology, 15*, 1347392.

Hervé, N. (2024). *Penser le futur, un enjeu d'éducation pour faire face à l'Anthropocène.* Le bord de l'eau.

Hodson, D. (2003). Time for action: Science education for an alternative future. *International Journal of Science Education, 25*(6), 645–670.

Knippels, M. C. P. J., & van Harskamp, M. (2018). An educational sequence for implementing socio-scientific inquiry-based learning (SSIBL). *School Science Review, 100*(371), 46–52.

Knippels & Christodoulou. (2024).*Practical strategies for using SSIBL activities in classroom settings.* Available at: https://view.officeapps.live.com/op/view.aspx?src=https%3A%2F%2Fwww.cosmosproject.eu%2Fassets%2Ffront%2Ffiles%2Frepository%2Fpractical_strategies_for_using_ssibl_activities_in_the_classroom.docx&wdOrigin=BROWSELINK

Levinson, R. (2006). Towards a theoretical framework for teaching controversial socio-scientific issues. *International Journal of Science Education, 28*(10), 1201–1224.

Levinson, R., Knippels, M. C., van Dam, F., Kyza, E., Christodoulou, A., Chang-Rundgren, S. N., Grace, M. et al. (2017). Socio-Scientific Inquiry-Based Learning: connecting formal and informal science education. *Science and Society in Education.* Available at: http://eprints.soton.ac.uk/id/eprint/493745 or: https://elbd.sites.uu.nl/wp-content/uploads/sites/108/2017/06/parrise-en-rgb.pdf

Levinson. (2018). Introducing SSIBL. *School Science Review, 100* (371), 31–35.

OECD. (2018). *Preparing our youth for an inclusive and sustainable world. The OECD PISA global competence framework.*

Öhman, J. (2008). Environmental ethics and democratic responsibility – A pluralistic approach to ESD. In J. Öhman (Ed.), *Values and democracy in education for sustainable development: Contributions from Swedish research* (pp. 17–32). Liber.

Ratcliffe, M., & Grace, M. (2003). *Science education for citizenship.* Open University Press.

Rudsberg, K., & Öhman, J. (2010). Pluralism in practice – experiences from Swedish evaluation, school development and research. *Environmental Education Research, 16*(1), 95–111. https://doi.org/10.1080/13504620903504073

Sadler, T. D. (2011). Socio-scientific issues-based education: What we know about science education in the context of SSI. In T. D. Sadler (Ed.), *Socio-scientific issues in the classroom: Teaching, learning and research* (pp. 355–369). Springer.

Sarid, A., Knippels, M.C., Christodoulou, A., Preis, P., Boeve-de Pauw, J., Doms, M., Bastiaenssens, P., Walan, S., Gericke, N., Goldman, D., & Veldkamp, A. (2024). COSMOS open schooling framework for open schooling in science education for responsible citizenship, including generic guidelines, tools and exemplifying scenarios. D2.1 COSMOS project.

Sass, W., Boeve-de Pauw, J., Olsson, D., Gericke, N., De Maeyer, S., & Van Petegem, P. (2020). Redefining action competence: The case of sustainable development. *The Journal of Environmental Education, 51*(4), 292–305. https://doi.org/10.1080/00958964.2020.1765132

Sass, W., De Maeyer, S., Boeve-de Pauw, J., & Van Petegem, P. (2024). Effectiveness of education for sustainability: The importance of an action-oriented approach. *Environmental Education Research, 30*(4), 479–498.

Simonneaux, L. (2013). Questions socialement vives and socioscientific issues: New trends of research to meet the training needs of post-modern society. In C.Bruguière, A. Tiberghien & P. Clément (Eds.), *9th ESERA Conference Selected Contributions. Topics and trends in current science education* (pp. 37–54). Springer.

Simonneaux, L., & Simonneaux, J. (2017). STEPWISE as a vehicle for scientific and political educ-action? *Cultural studies of science education* (pp. 565–588).

Simonneaux, L., & Pouliot, C. (2017). *QSV and SAQ, Sisyphus, 5*(2), 6–9. http://revistas.rcaap.pt/sisyphus/issue/view/660

Sinakou, E., Donche, V., Boeve-de Pauw, J., & Van Van Petegem, P. (2019). Designing powerful learning environments in education for sustainable development: A conceptual framework. *Sustainability, 11*(21), 5994. https://doi.org/10.3390/su11215994

Sjöström, J., & Eilks, I. (2018). Reconsidering different visions of scientific literacy and science education based on the concept of Bildung. In Y. J. Dori, Z. R. Mevarech & D. R. Baker (Eds.), *Cognition, metacognition, and culture in STEM education. innovations in science education and technology* (Vol. 24). Springer. https://doi.org/10.1007/978-3-319-66659-4_4

UNESCO. (2017). *Education for sustainable development goals: Learning objectives.* Paris UNESCO. http://unesdoc.unesco.org/

Van Harskamp, M. (2023). *Ask, find out, and act: Fostering environmental citizenship through science education* (Doctoral dissertation, Utrecht University).

Wiek, A., Withycombe, L., & Redman, C. L. (2011). Key competencies in sustainability: A reference framework for academic program development. *Sustainability Science, 6*(2), 203–218. https://doi.org/10.1007/s11625-011-0132-6

Zeidler, D. L., Herman, B. C., & Sadler, T. D. (2019). New directions in socioscientific issues research. *Disciplinary and Interdisciplinary Science Education Research, 1*, 11. https://doi.org/10.1186/s43031-019-0008-7

Chapter 2
Towards an Educational Response to the Climate and Environment Emergency

Justin Dillon

2.1 Introduction

This talk builds on the keynote given by Marie-Christine Knippels and Laurence Simonneaux. My distinguished colleagues had four conclusions and I want to draw on two of them. The first one that I want to build on is that the important goal of our community is to foster transformation of current education towards responsible citizenship and action taking as part of scientific literacy. The second conclusion was the importance of teaching for hope.

We need a transformation of education generally, and biology education, in particular, because society faces a series of wicked problems (Achiam et al., 2021) including poverty, climate change, inequality, food security, water security, biodiversity loss, homelessness, and sustainability. Horst Rittel and Melvin Webber, from the University of California, Berkeley, coined the term 'wicked problems' in the 1970s, arguing that 'the search for scientific bases for confronting problems of social policy is bound to fail' (Rittel & Webber, 1973, p. 155). They realised that planners could deal with 'tame' problems, such as isolated incidents but not with a range of social policy problems. These problems they termed 'wicked' and while it has taken a few decades for the term to become widely accepted, it is a useful way of examining and understanding the challenges facing our society.

Rittel and Webber (ibid., pp. 161–167), identified ten distinguishing features that wicked problems displayed in the context of social planning:

1. There is no definite formulation of a wicked problem;
2. Wicked problems have no stopping rule;
3. Solutions to wicked problems are not true-or-false, but good or bad;

J. Dillon (✉)
University College London, London, UK
e-mail: justin.dillon@ucl.ac.uk

O. Morin et al. (eds.), *Challenges of a Changing World in Biology Education*, Contributions from Biology Education Research,
https://doi.org/10.1007/978-3-032-05346-6_2

4. There is no immediate and no ultimate test of a solution to a wicked problem;
5. Every solution to a wicked problem is a "one-shot operation"; because there is no opportunity to learn by trial-and-error, every attempt counts significantly;
6. Wicked problems do not have an enumerable (or an exhaustively describable) set of potential solutions, nor is there a well-described set of permissible operations that may be incorporated into the plan;
7. Every wicked problem is essentially unique;
8. Every wicked problem can be considered to be a symptom of another problem;
9. The existence of a discrepancy representing a wicked problem can be explained in numerous ways. The choice of explanation determines the nature of the problem's resolution;
10. The planner has no right to be wrong.
    (ibid., pp. 161–167).

My argument is that our current education system has not been successful in preparing young people to cope with these wicked problems which is why I agree with Knippels and Simonneaux that we need to foster transformation of education towards responsible citizenship and action taking (though I would disagree that this is limited to being part of scientific literacy. Given the severity of the problems facing us, I also agree with their conclusion that we must teach for hope—what is the alternative?

There is hope—I can look back at the changes that I have seen in my lifetime which lead me to be optimistic. When I was a child, there were electric vehicles but they were slow and only used to deliver milk to houses close to a depot. Nowadays, electric and hybrid cars are commonplace and the technology is improving all the time. Plastics have gone from being regarded as a wonder of modernity to being seen as threatening to our ocean (we only have on ocean despite giving it a number of names). So, this contribution to the conference is one of hope.

2.2 The Need for Change

2.2.1 Science Education Has Failed

COVID-19 changed the world in a number of ways. For one thing, I now see my colleagues virtually more than I see them face to face. Some of my friends and colleagues are suffering from 'long COVID'. Many people who survived the COVID pandemic have re-evaluated their lives and adopted a new work-life balance. What became abundantly clear in the months after the disease was identified in China was that many members of the general public and, even worse, their political leaders did not understand some basic scientific ideas. Even more alarming was that it became clear that significant numbers of people, including journalists and politicians, actively denied the scientific evidence presented to them. This latter phenomenon led to countless unnecessary deaths.

Science educators, myself included, have devoted many years researching children's ideas about concepts such as combustion, climate change and cells. Whether we see them as misconceptions or alternative conceptions, they allow us to see how we might teach more effectively. What COVID showed us was that despite years of what we hoped was effective teaching, many people did not have a basic, potentially life-saving, understanding of disease prevention, of the difference between vaccination and inoculation, of the characteristics of viruses and bacteria, etc. As Lucy Avraamidou and I wrote in the Journal of Activist Science and technology (JASTE):

> COVID-19; Coronavirus disease 2019; virus; SARS; pandemic; outbreak; herd immunity; the curve; flatten the curve; asymptomatic; underlying conditions; incubation period; physical distancing; social distancing; self-isolation; isolation; quarantine; lockdown; R-number; behavioural science; personal responsibility; social responsibility; political responsibility; Chinese hoax; fake news; vaccination; anti-vaccination; public engagement with science; distance education; online; virtual learning; video-conferencing; teleconferencing; personal; social; political; national; global. These terms have dominated the media over the past few months. New, technical and often unclear terms make up this glossary, which is currently used by politicians, scientists, journalists, educators, and the general public, to engage in discourses associated with the coronavirus disease 2019 (COVID-19). Some of the terms are used in an ambiguous and interchangeable manner, which serves as evidence of how unclear certain aspects of the pandemic still are. What it is clear, however, is that, as a global society, or as a group of interlinked societies, this has not been our finest hour. (Dillon & Avraamidou, 2020, p. 1).

Writing during the COVID pandemic, we went on to note that: "Science and scientists find themselves in the spotlight as both potential saviours or as untrustworthy puppets in ways that would have been unthinkable a year ago. Yet science education and science educators are invisible" (p. 1). We addressed seven questions to the science education community including: "How well has the science curriculum prepared the world's public for COVID-19?" and "Is science education research producing knowledge that protects society from catastrophic events?".

The answer to the first question is clearly "not very". Fortunately, the science education of the numerous medical researchers who fairly quickly came up with a number of vaccines was clearly very good. The same could be said of the even larger number of medical practitioners who had to cope with the worst excesses of the disease. However, far too many of the general public and their political leaders displayed a fundamental lack of understanding of basic science and, in many cases, a lack of trust in science and the scientific process (Mitchell, 2023).

We might laugh or shake our heads at those people who do not believe that humans landed on the Moon or who think that aircraft are pumping out chemicals ('Chemtrails') across the planet, but their distrust of science, of experts and of governments, helped to create a climate of uncertainty among many of their fellow public to the extent that many people refused medical treatment that would have kept them alive. Indeed, there was a case when a Wisconsin pharmacist tampered with vaccines because he was a conspiracy theorist (New York Times, 2021)—sometimes even scientific qualifications do not mean that you automatically trust the science, rather the evidence suggests that your political views may well be a good predictor of how much you deny the scientific consensus, although the issue is more complex than was originally thought (Smiley, 2016).

In summary, the public's response to COVID-19 strongly suggested that science education had failed to provide the public to cope with a pandemic. The lack of understanding is compounded by a lack of trust in science and scientists—a phenomenon that also applies to the public's attitudes towards climate change.

2.2.2 *It's the Curriculum, Stupid*

In this section, I look at climate change education and at how the National Curriculum in England is inadequate in preparing young people to understand the issue in enough detail to make sense of what is happening in the world. I identify a need for better quality climate change education for students, parents and teachers. I also look at what we know about effective climate change education.

A national curriculum was introduced into England and Wales in 1989. Science was identified as one of the Core subjects alongside English and Mathematics. Over the years, the content of the curriculum has changed, partly as a result of political influence. By 2011, the curriculum specifically mentioned climate change; Jessica Shepherd noted that "pupils demonstrate exceptional performance if they can 'describe and explain the importance of a wide range of applications and implications of science in familiar and unfamiliar contexts, such as addressing problems arising from global climate change'" (Shepherd, 2011). However, things were about to change.

The Conservative politician, Michael Gove, was appointed Education Secretary in 2010. He commissioned a review of the curriculum, which was led by Tim Oates, director of research at Cambridge Assessment. In Gove's opinion, the national curriculum was "too long [and] patronising towards teachers and stifled innovation" (Shepherd, 2011). Gove was, and probably still is, convinced that the curriculum is fundamentally a collection of knowledge to be passed on to the next generation. A number of educators, scientists and environmentalists advocated for more inclusion of climate change in the revised curriculum. Oates resisted these calls explaining why in an interview with The Guardian newspaper. Calling for the national curriculum "to get back to the science in science" he said:

> We have believed that we need to keep the national curriculum up to date with topical issues, but oxidation and gravity don't date [...] We are not taking it back 100 years; we are taking it back to the core stuff. The curriculum has become narrowly instrumentalist. (Shepherd, 2011).

In the end, attempts to remove climate change from the curriculum foundered, partly due to internal disagreements in the Conservative Party. Some of its members realised that being 'green' would be a popular move with young people so removing climate change from the curriculum was seen as being counterproductive.

2.2.2.1 Young People's Views About Climate Change Education

What has become clear over recent years is that many young people are dissatisfied with their education in terms of climate change and sustainability in the UK and elsewhere. A large-scale survey of UK school students (BSA, 2023) found that 72% of respondents "would welcome the opportunity to learn more about climate change in school and a nearly 7 in 10 (68%) felt that climate change education should be included across all subjects" (p. 9). The survey identified a real desire for more climate change education:

> Young people chose climate change as their top concern in terms of the impact it will have on their future lives, however only 1 in 4 (23%) felt 'very informed' about climate change; worryingly, this fell to 1 in 14 (7%) for 14-year-olds. (BSA, 2023, p. 10).

This sense of frustration is echoed around the world. School strikes, also known as Fridays for Future, and Youth Strike for Climate, the movement associated with Greta Thunberg, have involved more than 1 million students striking in more than 120 countries in March 2019 (Carrington, 2019). A few months later, somewhere between four and six million young people took action (Taylor et al., 2019). In the UK, a number of initiatives have focused on changing the curriculum so as to increase the amount of climate change education. In 2019, a "Teach the Future" campaign was launched at a Climate Emergency Conference organised by the National Education Union. The campaign is supported by the UKSCN and Students Organising for Sustainability (SOS-UK) and its vision is that:

"Students need to be taught about the climate emergency and ecological crisis: how they are caused, what we can do to mitigate them and what our future lives and jobs are going to look like due to them. Sustainability and these crises need to become key content in all subject areas. Educators need to be trained in how to teach about these difficult topics in a way that empowers students, and they need funding and resources to do this." (Teach the Future, n.d.)

2.2.2.2 Teachers' Views About Climate Change Education

A number of surveys have been carried out to gauge teachers' opinions and experience. Rather surprisingly, perhaps, 86% of US teachers (n = 488) surveyed thought that climate change should be taught in school (Ipsos/NPR, 2019)—given the level of denial in the US, one might have thought that figure would be lower. A survey of UK teachers (n = 352), carried out about the same time, found that 69% of participants wanted to see more teaching about climate change in schools (it is already in the curriculum in Science and Geography). Indeed, 70% of respondents wanted to see radical change in the curriculum (Oxfam/UK Student Climate Network, 2019). While 89% of the teachers agreed that UK students should always be taught about climate change, its implications for environments and societies around the world and how these implications can be addressed, only 18% said that they had received adequate training to teach about climate change and its impacts.

A more recent study by Howard-Jones et al. (2021) found that climate change was, indeed, most commonly taught in science and geography. Interestingly, despite the majority of respondents being secondary teachers (n = 626; 205 primary and 421 secondary), slightly over half (51%) the respondents 'favoured a cross curricular approach that involved six or more subjects delivering (climate change education)' (p. 1667). These results mirror other survey findings, for example, a UK-wide poll (Teach the Future, 2021) (n = 503) reported that climate change is most commonly mentioned in science (65%) and geography lessons (63%). A depressingly low figure (5%) reported that 'climate change is integral to many different aspects of the curriculum and teaching in schools' (p. 9). A slightly more recent survey (Teach the Future, 2022) found that two-thirds of the teachers (n = 6,794) agreed that "there should be more space in the curriculum to teach issues such as sustainability and global justice" (Reboot the Future, 2023, p. 13).

2.2.2.3 Effective Climate Change Education Strategies

Monroe et al. (2019) carried out a systematic review of the research on climate change education. They examined 49 sources which met their criterion of focusing on assessment of climate change education interventions. They reported that:

> Analysis of these sources examined the intervention purpose, assessment methodology, and identified strategies that might result in effective interventions. Two themes were identified that are common to most environmental education: (1) focusing on personally relevant and meaningful information and (2) using active and engaging teaching methods. Four themes specific to issues such as climate change were also generated: (1) engaging in deliberative discussions, (2) interacting with scientists, (3) addressing misconceptions, and (4) implementing school or community projects. Suggestions for addressing controversial topics like climate change are offered. (p. 791).

So, we know what seems to work—the challenge is to educate teachers in how to use them more effectively.

2.2.2.4 Citizen Science

One opportunity for students to interact with scientists is through citizen science. While the approach has been around for many years (Silvertown, 2009), the advent of Information and Communication Technology (ICT) tools, such as identification apps, offer a number of affordances for rich engagement. Of course, there are different levels of citizen science (Dillon et al., 2016) ranging from crowd-sourcing of data to much more empowering involvement. The citizen science approach is used as part of the latest sustainability and climate change initiative in England.

2.2.2.5 Recent Policy Initiatives in England

In April 2022, the UK government published *Sustainability and climate change: a strategy for the education and children's services systems* (DfE, 2022). The strategy "brought together short, medium and longer-term actions that will enable [England] to make progress towards achieving […] four strategic aims and [the] overarching vision. It is a strategy to 2030" (p. 3). The policy only referred to England, although it was noted that "[t]he UK government and the devolved governments (Northern Ireland, Scotland and Wales) are committed to climate action, and we will work together in our international action" (p. 3).

The strategy had a foreword from the then Secretary of State for Education, The Rt Hon Nadhim Zahawi, a Conservative Member of Parliament who wrote:

> The challenge of climate change is formidable. For children and young people to meet it with determination, and not with despair, we must offer them hope. Learners need to know the facts about climate change—through knowledge-rich education. They must also be given the hope that they can be agents of change, through hands-on activity and, as they progress, through guidance and programmes allowing them to pursue a green career pathway in their chosen field.

The Strategy, which is not mandatory, encourages each school to appoint a "sustainability lead", who will receive carbon literacy training with a view to developing a climate action plan. The physical education estate will be redesignated as a National Education Nature Park. The strategy involves a citizen science approach with students being involved in improving some part of the school grounds. Many teachers and environmentalists regarded this as a positive but rather unambitious initiative which failed to address the key issues in sustainability and climate change facing the world.

2.3 Is There Hope?

2.3.1 Some Examples of Effective Practice

Over the years, I have travelled to many countries and witnessed how they are addressing climate change and sustainability in and out of schools. Occasionally, I have come across schools and teachers who are carrying out outstanding work. I present three examples which I found to be inspirational.

I visited the Odyssey School in Denver, Colorado in the Spring of 2010. The school leadership were committed to organising the students' experience around the 'Learning Expedition'. Children at the lower grades took part in short excursions out of the classroom; those in higher grades took part in longer, residential visits. What was impressive about the school was the fact that evidence of these educational experiences could be found everywhere, in the classroom and on the walls of the corridors. The quality of the students' work, and their interactions with us as visitors, was exceptional—and this is not a school that is set in an affluent part of the city.

Turning closer to home, the XP Academy in Doncaster, England, which I visited in 2023, was inspired by a US school, San Diego—Hi Tech High, a similar school to the Odyssey School in Denver. The students are organised into crews of 12/13 students who meet with a member of staff for 45 min every day. The curriculum is organised around learning expeditions which start with a guiding question, then focus on up to three case studies—from human (English, history and geography) or STEAM (science, technology, engineering, arts and mathematics) perspectives. An example of a guiding question is "Are protests worth the cost?" Expeditions always have a fieldwork dimension and expert input. Each module of work takes around 12 weeks and there are three a year. As with the Odyssey School, the quality and interdisciplinary nature of the students' work was exemplary.

Finally, on a recent visit to China, I was invited to go to Haidian Experimental Number 4 Elementary School in Beijing. The school is close to a botanic garden and some of its scientists have been involved with building the school's capacity. The school takes sustainability seriously in everything that it does and its facilities are extraordinary.

The fact that these schools exist suggests that there is hope. The schools have one thing in common—committed leadership with vision and drive. If they can succeed, so can many more schools.

2.3.2 *Towards a Convergence of Science and Environmental Education*

The schools that I have seen making great strides work within the existing curriculum. Particularly in secondary (high) schools, the rigid system of subjects makes interdisciplinary working difficult. To address wicked problems effectively, schools need to work in much more interdisciplinary ways as I have argued with colleagues:

> "whole-school approaches" to sustainability and the creation of eco-schools, where different forms of learning (e.g., inquiry-based, disciplinary, and social learning) blend with the use of ICT, citizen science, and community engagement [...] involve redesigning school grounds using knowledge from [science education]—to give such spaces a more central place in teaching about health, food, and ecology—as well as using [environmental education] to strengthen community involvement and develop a sense of place. Wals et al. (2014, p. 584).

2.3.3 *Science|Environment|Health*

Another model for interdisciplinary work emerged when I worked with two Swiss colleagues on a new pedagogy for science education (Dillon, 2012; Zeyer & Dillon, 2018, 2019). We realised that the overlaps between science, the environment and health were much greater in real life than they are in the curriculum. We conceptualised a much more integrated approach which we called

Science|Environment|Health. Ideally, the line between the words would be shown as dotted to show the permeability of the division. We were careful to explain what the new notation meant:

> The label Science|Environment|Health (Dillon, 2012) is not meant to suggest that health and environmental education should be swallowed up by science education. Rather, there is a role for both beyond a reimagined science education. The label highlights a situation of mutual benefit between science education, environmental education and health education, three educational dimensions that have yet to be established in a transdisciplinary dialogue. (Zeyer & Dillon, 2014, p. 1409).

Work in this area continues and the European Science Education Research Association (ESERA) has a S|E|H Special Interest Group.

2.3.4 *Museums and Science Centres*

So far, I have focused on schools but much of my work has been with museums, science centres, botanical gardens and aquariums. Such institutions have more flexibility and more resources than schools and can educate and influence huge numbers of visitors. I was very impressed with the Natural History Museum in London which created an award-winning exhibition "Our broken planet".

Of course, such institutions, host substantial numbers of school visits and can reinforce the learning that takes place in school. Some schools take the partnership with museums exceptionally seriously. One such is the Langley Academy, near Slough in Berkshire, south east England. The school has museum quality display cases and looks more like a science centre than a school with a model dinosaur close to the school canteen. Staff are encouraged to museum pedagogies including teaching through objects and specimens.

2.4 Plants, Plants, Plants

In this final section, I want to focus on a much-neglected area of science education—the study of plants.

2.4.1 *Plant Awareness*

A number of biology educators have commented on a bias among textbook publishers in terms of animals rather than plants.

"The inattention to plants, known as 'plant blindness' or 'plant awareness disparity', is an established concern amongst biologists, but there has been no comprehensive and critical examination of the investigative literature to date." (Stagg & Dillon, 2022, p. 579).

Stagg and Dillon's review of the literature suggested that there is "no concrete evidence of innate plant awareness disparity…" (p. 579), the phenomenon is learned or acquired over time in many cases as a result of living in urban environments:

> We conclude that a decline in relevant experience with plants leads to a cyclical process of inattention that could be addressed through first-hand experiences of edible and useful plants in local environments. (Stagg & Dillon, 2022, p. 580).

Sustainability education needs to include the study of plants and their relationship with humans—students need to be more plant aware. But this is easier said than done:

> Embedding plants in sustainability education is no small challenge, since biological diversity and the ecological crisis are already sidelined in many sustainability policies and debates, with biodiversity often treated as a sub-topic to climate change. One literature analysis (Legagneux et al., 2018) found that climate change received up to eight times more media coverage than biodiversity (Stagg & Dillon, 2024, p. 11).

2.4.2 Plant Education

Strategies for improving plant awareness do exist—many have been around for a long time:

> However, we also need pedagogic approaches that foster pro-conservation behaviour […] First-hand experience of wild plants is important and can be achieved by developing ecological gardens in school and campus grounds. Informal science education and authentic enquiry can provide personally relevant learning and novel experiences with plants. Educational experience needs to extend beyond the institution and families and peers have a part to play in facilitating and contributing to behavioural change. We also need stronger links between the informal learning opportunities offered by citizen science initiatives and educational institutions. (Stagg & Dillon, 2025, p. 14).

2.5 Conclusion

I have argued that we need a transformation of education generally, and biology education, in particular, because society faces a series of wicked problems. I have tried to explain why this is the case and suggested some possible ways forward from radical reform of the education system down to individual pedagogical strategies. I live in hope.

References

Achiam, M., Dillon, J., & Glackin, M. (Eds.). (2021). *Addressing wicked problems through science education*. Springer.

British Science Association (BSA). (2023). *Future Forum. Climate change in secondary schools: Young people's views of climate change and sustainability education.* BSA.

Carrington, D. (2019, March 19). School climate strikes: 1.4 million people took part, say campaigners. *The Guardian.* https://www.theguardian.com/environment/2019/mar/19/school-climate-strikes-more-than-1-million-tookpart-say-campaigners-greta-thunberg

Department for Education (DfE). (2022). *Sustainability & climate change: A strategy for the education & children's services systems.* https://www.gov.uk/government/publications/sustainability-and-climate-change-strategy/9317e6ed-6c80-4eb9-be6d-3fcb1f232f3a.

Dillon, J. (2012). Science, environment and health education: Towards a reconceptualisation of their mutual interdependences. In A. Zeyer & R. Kyburz-Graber (Eds.), *Science|Environment|Health - towards a renewed pedagogy for science education* (pp. 87–101). Springer.

Dillon, J., & Avraamidou, L. (2020). Towards a viable response to COVID-19 from the science education community. *Journal for Activist Science and Technology Education, 11*(2), 1–5. https://doi.org/10.33137/jaste.v11i2.34531

Dillon, J., Stevenson, R. B., & Wals, A. E. (2016). Introduction to the special section: Moving from citizen to civic science to address wicked conservation problems. *Conservation Biology, 30*(3), 450–455.

Howard-Jones, P., Sands, D., Dillon, J., & Fenton-Jones, F. (2021). The views of teachers in England on an action-oriented climate change curriculum. *Environmental Education Research.* https://doi.org/10.1080/13504622.2021.1937576

IPSOS/NPR. (2019, April 22). Teachers agree that climate change is real and should be taught in schools [Press release]. https://www.ipsos.com/sites/default/files/ct/news/documents/2019-04/nprteachersclimatechangetoplineapril222019.pdf

Legagneux, P., Casajus, N., Cazelles, K., Chevallier, C., Chevrinais, M., Guéry, L., Jacquet, C., Jaffré, M., Naud, M. J., Noisette, F., Ropars, P., & Gravel, D. (2018). Our house is burning: discrepancy in climate change vs. biodiversity coverage in the media as compared to scientific literature. *Frontiers in Ecology and Evolution, 5*, 297310.

Mitchell, A. (2023, November 20). Boris Johnson was 'bamboozled' by science during the pandemic, Patrick Vallance reveals. *The Independent.* https://www.independent.co.uk/news/uk/politics/covid-inquiry-boris-johnson-patrick-valance-b2450417.html

Monroe, M. C., Plate, R. R., Oxarart, A., Bowers, A., & Chaves, W. A. (2019). Identifying effective climate change education strategies: A systematic review of the research. *Environmental Education Research, 25*(6), 791–812.

New York Times. (2021, June 9). Pharmacist accused of tampering with vaccine was conspiracy theorist. *New York Times.* https://www.nytimes.com/2021/01/04/us/pharmacist-accused-of-tampering-with-vaccine-was-conspiracy-theorist-police-say.html

Oxfam/UK Student Climate Network (UKSCN). (2019). Climate change education. https://bit.ly/3uK0mxw

Reboot the Future (2023). *The rebooting education report 2023: a transformative vision of the future from teachers.* Reboot the Future.

Rittel, H. W., & Webber, M. M. (1973). Dilemmas in a general theory of planning. *Policy Sciences. 4*(2), 155–169.

Shepherd, J. (2011, June 12). Climate change should be excluded from curriculum, says adviser. *The Guardian.* https://www.theguardian.com/education/2011/jun/12/climate-change-curriculum-government-adviser

Silvertown, J. (2009). A new dawn for citizen science. *Trends in Ecology & Evolution, 24*(9), 467–471.

Smiley, K. T. (2016). Climate change denial, political beliefs, and cities: Evidence from Copenhagen and Houston. *Environmental Sociology, 3*(1), 76–86. https://doi.org/10.1080/23251042.2016.1236884

Stagg, B. C., & Dillon, J. (2022). Plant relevance and plant blindness: A review of ethnobiological and educational research literature (1998–2020). *People, Plants, Planet.* https://doi.org/10.1002/ppp3.10323

Stagg, B., & Dillon, J. (2024). Teaching the power of plants. *The Biologist, 71*(1), 10–11. https://www.rsb.org.uk/biologist-opinion/teaching-the-power-of-plants?fbclid=IwAR13tfamLu2G0DbBHYSRI3tkd4QO39sJv6m8GHksZYkDnl9K_ITId7DRu_g

Stagg, B., & Dillon, J. (2025). Plants and the Kunming-Montreal Global Biodiversity Framework: Educational approaches to support pro-conservation behaviours. *Journal of Biological Education.* https://doi.org/10.1080/00219266.2024.2332728

Stagg, B. C., Hetherington, L., & Dillon, J. (2024). Towards a model of plant awareness in education: A literature review and framework proposal. *International Journal of Science Education.* https://doi.org/10.1080/09500693.2024.2342575

Taylor, M., Laville, S., Walker, A., Noor, P., & Henley, J. (2019, February 15). School pupils call for radical climate action in UK-wide strike. *The Guardian.* https://www.theguardian.com/environment/2019/feb/15/uk-climate-changestrike-school-pupils-children-environment-protest

Teach the Future. (n.d.). *Teach students about climate change.* https://www.teachthefuture.uk/vision

Wals, A. E. J., Brody, M., Dillon, J., & Stevenson, R. B. (2014). Convergence between science and environmental education. *Science, 344*, 583–584.

Zeyer, A., & Dillon, J. (2014). Special issue: Science|Environment|Health—Towards a reconceptualization of three critical and inter-linked areas of education. *International Journal of Science Education, 36*(9), 1409–1411.

Zeyer, A., & Dillon, J. (2018). Can – and should - empathy play a role in teaching complex science|environment|health contexts? *International Journal of Science Education.* https://doi.org/10.1080/09500693.2018.1549371

Zeyer, A., & Dillon, J. (2019). Science|Environment|Health – the emergence of a new pedagogy of complex living systems. *Disciplinary and Interdisciplinary Science Education Research, 1*(9). https://doi.org/10.1186/s43031-019-0014-9

Chapter 3
Norwegian Primary School Students' Decision-Making Related to a Local and Authentic Sustainability Dilemma

Maren Skjelstad Fredagsvik, Marthe Lønnum, and Ragnhild Lyngved Staberg

3.1 Introduction

A major part of the information that young people encounter in school and in their daily lives is connected to complex and often controversial issues, often in the form of socio-scientific issues (SSIs). SSIs are defined as complex, multidimensional issues that often involve conflicting perspectives and lack clear-cut solution (Aziz & Johari, 2023). These are issues that, in addition to scientific knowledge, are often affected by social, ethical, political, and economic perspectives. Sustainability dilemmas can be examples of such issues, where something, in an ideal world, is perceived as sustainable if the environmental, social, and economic perspectives are considered (Wals, 2009). To navigate such complex dilemmas, and in turn be able to make informed decisions, young people need to identify and explore different perspectives and evaluate these perspectives through critical argumentation.

Critical thinking has been proposed as a fundamental competence in sustainability education, essential for developing students into action-competent individuals capable of addressing in complex environmental issues (Rieckmann, 2018).

M. S. Fredagsvik (✉) · R. L. Staberg
Department of Teacher Education, Faculty of Social and Educational Sciences, Norwegian University of Science and Technology (NTNU), Trondheim, Norway
e-mail: maren.s.fredagsvik@ntnu.no

R. L. Staberg
e-mail: ragnhild.l.staberg@ntnu.no

M. Lønnum
The Norwegian Centre for Writing Education and Research, Norwegian University of Science and Technology (NTNU), Trondheim, Norway
e-mail: marthe.lonnum@ntnu.no

O. Morin et al. (eds.), *Challenges of a Changing World in Biology Education*, Contributions from Biology Education Research,
https://doi.org/10.1007/978-3-032-05346-6_3

Internationally, critical thinking is recognized as a core competence for the twenty-first century, which means that people should be able to navigate different types of information and make informed decisions in increasingly complex issues (OECD, 2016). This is also reflected in the Norwegian curriculum (Ministry of Education and Research, 2017). Critical thinking is an essential prerequisite for any action related to sustainability and should therefore be taught and trained throughout all school years (Pache & Rouiller, 2022). Several studies have investigated conditions that deal with critical thinking in teaching contexts for older students (e.g., Davies & Barnett, 2015), but there is a lack studies that shed light on teaching aimed at younger students (Pollarolo et al., 2023).

Globally, there is a growing awareness of the importance of integrating sustainability education into curricula (Monika, 2024). In Norway, a new curriculum was introduced in 2020, in which sustainability was incorporated as an interdisciplinary topic for all subjects in primary and secondary school, including biology. In grades 1–11, the subject of natural science merges biology, chemistry, and physics, and the same teacher is responsible for teaching all topics. Despite the increased interest in incorporating sustainability in school education (Monika, 2024), many teachers face challenges when teaching sustainability (Parry & Metzger, 2023). Student teachers feel less prepared to teach about sustainability issues (Dahl, 2019), and the complexity of these issues is thus often lost in teaching (Sund & Gericke, 2020). At the same time, students need scaffolding to be able to deal with such complex issues. To make informed decisions, they need to see different perspectives and to learn how to argue critically. Argumentation is a central practice in the natural sciences but is almost absent in Norwegian science classrooms (cf. Mork & Erlien, 2017). According to PISA 2022, the performance of Norwegian students is also below the OECD average in terms of scientific literacy, including the ability to draw evidence-based conclusions about science-related issues (Jensen et al., 2023). At an international level, there is limited research on the assessment of students' argumentation skills (Osborne et al., 2016). Thus, there is a need to further investigate what characterizes students' argumentation, both internationally and in Norwegian science classrooms, to be able to facilitate and improve students' argumentation abilities and critical thinking.

There is a common perception in critical thinking that emotions have no place in the critical assessment of information and knowledge (Hufnagel, 2022). However, it is important to take into account students' attitudes, deep-rooted ethics and values when making decisions on SSIs (Aikenhead, 2003), including their feelings and emotions (e.g., Chang & Lee, 2010). Emotions are seen as perceptions of value (Mulligan, 2009). Emotions are, hence, important evaluative mechanisms that we use to understand the world (Hufnagel, 2022) and should be emphasized when teaching critical thinking. Emotions are strongly connected to our identity and how we see the world and are thus connected to how we perceive knowledge and truth. The assessment of knowledge will be largely influenced by emotions such as optimism, fear, and anger (cf. Hufnagel, 2022). On this basis, it would be impossible to think that critical thinking is an emotionless act. Previous studies on students' reasoning have mainly focused on college and secondary school students (Aziz & Johari, 2023;

Dawson & Venville, 2009; Öhman & Östman, 2008; Sadler & Zeidler, 2005), but few studies have addressed the reasoning of primary school students.

Our research question is: *How does engaging with different perspectives influence students' decisions in a local and authentic sustainability dilemma?*

Engaging refers, in this context, to students actively interacting with and considering different perspectives to make informed decisions in a sustainability dilemma.

3.1.1 Argumentation as a Key Competence for Practicing Critical Thinking

Critical thinking can be defined as reflecting on and reasoning about what to do or believe when confronted with complex problems (Ennis, 1987). A central part of critical thinking is hence to identify and evaluate the different factors that are involved in a complex issue, to effectively be able to make a decision or find a solution (Iordanou, 2022). A key component in critical thinking is, therefore, to engage in argumentation (Checchi et al., 2022). Argumentation plays a major role in decision-making, and especially when working with SSIs (Bayat et al., 2022; Toulmin, 2003). Argumentation is about evaluating claims and opinions that are both scientific and beyond the scientific (Jiménez-Aleixandre & Puig, 2012), i.e., identifying different perspectives and assess these based on reason and scientific evidence (Kuhn & Cowell, 2011). Further, argumentation includes understanding, assessing, and evaluating different opinions through critical listening and reading (Ferguson & Krange, 2020). According to Toulmin (2003), an argument consists of the following components: (a) claim—the statement the person wants to prove, (b) warrant—a logical and persuasive connection between a claim and the evidence, and (c) data/grounds—the evidence supporting the claim. In addition, more advanced arguments can also contain (d) rebuttals—an exception to the claim, (e) qualifiers—that place limits on claims, and (f) backing/theoretical assumptions—reasons that support the warrant (Toulmin, 2003). In the process of argumentation, a good argument is therefore expected to take into account competing or conflicting perspectives (counterarguments) and evidence that supports the claim.

Sadler and Zeidler (2005) investigated explicit reasoning in the context of SSIs among 30 college students in the USA and found that the students justified their claims with information based on rationality (logic and scientific facts), emotions (empathy, sympathy, and caring for others' well-being), and intuition (gut feelings that cannot be explained rationally). They also used integrated justifications (combinations of the three primary forms of reasoning given above). The same forms of reasoning have been found in studies of lower secondary students (aged 14–16) in Brunei (Aziz & Johari, 2023). Australian high school students (aged 12–17) have been found to often base their reasoning on intuition or emotions when creating arguments in different SSIs, even after participating in interventions (Dawson & Venville, 2009). Öhman and Östman (2008) found that, with regard to sustainability issues, secondary school students had a tendency to reflect on the basis of ethics, morality, and "correct" behavior rather than knowledge. These examples suggest that students frequently justify their claims with moral views, personal experience, and emotional factors.

3.1.2 Authenticity as a Teaching Context for Critical Thinking

In this study, by authenticity we mean educational practices that use activities or issues from the real world (Fought et al., 2019). From a European point of view, it is particularly recommended to introduce realistic challenges and projects into schools, and preferably to involve actors from the local community (Hazelkorn, 2015). This is in line with a general understanding that realistic situations and activities can contribute to giving pupils a more holistic understanding of sustainable development (Sinnes, 2020). Authentic problems or activities provide natural problem-solving contexts with a high degree of complexity (Lee & Butler, 2003), and critical thinking thus becomes necessary in order to understand, assess, and solve such problems (Volman & Ten Dam, 2015). Moreover, previous studies have concluded that authenticity is an important prerequisite for teaching critical thinking effectively (Abrami et al., 2015). In authentic problems, different actors and stakeholders are involved, which makes it possible to explore different perspectives (Newton & Zeidler, 2020). Working with such problems has therefore been shown to promote students' ability to take a perspective (Newton & Zeidler, 2020), which is an important component of critical thinking. Interaction with authentic actors has also been shown to increase the quality of students' argumentation (Åkerblom & Lindahl, 2017). In this study, we therefore start from a local sustainability challenge that can be perceived as authentic for the students: the question of whether a roller-skiing track should be built in the forest close to the school.

3.2 Research Design and Method

This study is part of an intervention within the CriThiSE project (Critical Thinking in Sustainability Education). In this study, we followed a teaching sequence where a team of five 7th grade teachers from the subjects of science, social science, and Norwegian, and their students, worked with an interdisciplinary sustainability dilemma for three weeks. The teachers planned this project using a planning tool for teaching critical thinking in the context of sustainability education that was developed in the CriThiSE project. The starting point for the planning was to investigate a complex, value-laden issue (dilemma), preferably authentic and locally rooted (cf.

Abrami et al., 2015)—in this case: Should parts of the city's forest area be asphalted to make roller-skiing tracks?

3.2.1 The Teaching Design

In the project, the students were first introduced to the dilemma and explored it through various activities. The students started to identify actors who might have an interest in the locally rooted dilemma and their justifications for and against the suggested action. They visited and observed the forest, and they read four authentic articles that highlighted the perspectives of several actors. Two articles highlighted one-sided perspectives presented by the actors, such as the skiers/skiing team and neighbors, and the other two articles presented a more balanced view from, e.g., the actors skiers/skiing team, the chair of the municipality, the tourist association, the neighbors, and environmental organizations, even though these actors had taken a clear stance in the case. The articles focused mostly on the social dimension of sustainability (27 justifications) and less on the environmental dimension (10) and the economic dimension (3). The students then wrote balanced news articles with the support of a writing frame and teachers, and peer-assessment based on a set of criteria concerning content, genre, structure, language, and grammar. They ultimately wrote texts in which they made and justified their own decisions.

3.2.2 Data Collection and Analysis

The data sources in this study were the students' news articles (n = 15) and their decision-making texts (n = 16). The data were deductively analyzed using qualitative content analysis (Patton, 2002). When analyzing the students' news articles, we first identified which actors were represented in each article, and the actors' stance in the case. After that, we identified what dimension(s) of sustainability the actors' stance represented, based on the three dimensions of sustainability (social, economic, and environmental) (cf. Wals, 2009). For example, in one of the news articles, the student highlighted the perspectives of the local skiing team, the neighbors, the municipality and the tourist association. When identifying which sustainability dimension the actors represented, we analyzed whether their reasoning was based on economic, environmental or social interests. In the following excerpt, the actor "neighbors" from one of the students' news articles represented environmental interests when expressing: "In the area, there is a lot of untouched nature and rich wildlife. The construction is also not good for the environment." In addition, they represented a social dimension when expressing: "There will also be a lot of noise and disturbance in the area."

When analyzing the decision-making texts, we used the patterns of informal reasoning that were described in Sadler and Zeidler (2005), which characterize students' justifications in arguments in an SSI context as rational, emotional, intuitive, or integrated. Justifications that are based on factual information and logic were categorized as rationality, as shown in the extracts from two of the decision-making texts: "[I'm against roller-skiing tracks, because] they cost a lot of money" and "it's the habitat to various animals." Justifications that are based on feelings, empathy, sympathy, caring for others' well-being, and values were categorized as emotional justifications, as shown in this extract: "[I'm against] because if you are walking in the forest, it can be scary and unpleasant for many to encounter roller-skiers at great speed". Further, justifications that are based on a gut feeling that cannot be explained rationally were categorized as intuitive justifications, as shown in this extract: "[I'm against] because we already have enough tracks. We don't need to use [the local forest area] for this." Rationality and intuition can be difficult to separate at this level, especially because the students' justifications were not especially detailed. The students treated the information they received through their source material as facts or truths, which made this information appear as rational justifications in their decision texts—and we chose to code these justifications as such (rationality). Consequently, rationality or facts form the basis of more justifications than intuition (14 vs 12, respectively) and are more frequently represented in the integrated justifications.

Integrated justifications are those that include a combination of types of reasoning, for example a combination of rational and emotional justification, as shown in this extract: "[I don't want roller-skiing tracks] because wherever you go you walk on asphalt. Walking on asphalt is very tiresome for the feet." This extract shows that the justification is a combination of facts regarding the use of asphalt and the implications of the amount of asphalt, and a sense of sympathy for the users of the forest.

In the same way as for the news articles, we also analyzed which sustainability dimensions the students highlighted in their justifications in the decision-making texts, based on the dimensions described by Wals (2009).

To ensure the quality of the study and data accuracy, all three authors collaborated in conducting the analysis. Any divergence of opinion related to the coding was fully discussed until consensus was achieved. Ethical approval for the study was obtained from SiKT (Norwegian Agency for Shared Services in Education and Research).

3.3 Findings

The findings of this study are based on the analysis of students' news articles and decision-making texts. We will first present the results from the news articles.

3.3.1 Actors and Perspectives Found in the Students' News Articles

The analysis of students' news articles resulted in the identification of the different actors represented in the texts and of the different arguments presented by the actors, as well as the stance and sustainability dimension represented by each argument. Table 3.1 presents an overview of the results from analyzing all 15 news articles.

Each text presented perspectives that were for and against the proposition, in addition to showcasing 3–4 actors' opinions, in a balanced manner. The actors with the greatest number of arguments in the texts were the first three actors shown in Table 3.1 (skiing teams/skiers, the chair of the municipality, and the neighbors), and the first two actors were mentioned in all texts. Only three students did not include the chair of the municipality in their texts.

With regard to the analysis of the texts considering the sustainability dimensions, we found that the dimension that was highlighted the most was the social dimension (71 justifications), followed by the environmental dimension (26 justifications) and economic dimension (9 justifications). In addition, we found that all but one of the students included more than one dimension in their texts, with the remaining student only including the social dimension. Four of the students included all three dimensions in their texts. However, this is not represented in the table above.

3.3.2 What characterizes the Students' decision-Making texts?

In the analysis of the students' decision-making texts (in which they were supposed to take their own stance in the case), we found that the students also considered different actors or interest groups when presenting their arguments both for and against the proposition. Only three students considered their own interest in the proposition in their decisions but made use of arguments made by other actors. Table 3.2 presents an overview of the results of the analysis of 16 decision-making texts.

The analysis showed that, by considering different interests (actors) in their justifications, the students presented 68 arguments that highlighted two or all three sustainability dimensions in their decision-making texts. Eight students highlighted both the social and the environmental dimensions, three students highlighted both the social and the economic dimensions, and two students highlighted all three sustainability dimensions in their decisions. Only three students highlighted just the social dimension in their decisions. The social dimension of sustainability was highlighted the most throughout the decision-making texts ($n = 43$), in both the arguments for and against the proposition, followed by the environmental dimension ($n = 18$) and the economic dimension ($n = 7$). The economic dimension was represented more in the for-arguments (5 of 7), and the environmental dimension was represented far more in the against-arguments (17 of 18).

Table 3.1 Overview of actors, their stance (decision), number of arguments, and sustainability dimensions in students' news articles

Actor	Student ID	Decision	No. arguments	Sustainability dimension in the justifications and frequency
Skiing teams/ skiers	1, 2, 3, 4, 5, 6, 7, 8, 9, 10, 11, 12, 13, 14, 15	For	31	Social (28), Environmental (2), no justification (1)
		Against	0	–
Neighbors (incl. hikers etc.)	1, 2, 3, 4, 5, 6, 7, 8, 9, 10, 11, 12, 13, 14, 15	For	8	Social (7), no justification (1)
		Against	37	Social (23), Environmental (13), no justification (1)
The chair of the municipality	1, 3, 5, 6, 8, 10, 11, 12, 13, 14, 15	For	3	Social (3)
		Against	12	Social (1), Environmental (4), Economic (7)
The tourist association	1, 5, 11, 12	For	6	Social (6)
		Against	2	Social (1), Environmental (1)
Environmental organizations	4, 9, 10, 15	For	0	–
		Against	4	Environmental (4)
Landowner	7	For	0	–
		Against	2	Social (1), Environmental (1)
Hiking cabin owner	8	For	1	Economic (1)
		Against	0	–
Unknown actors	8, 14	For	1	Social (1)

(continued)

Table 3.1 (continued)

Actor	Student ID	Decision	No. arguments	Sustainability dimension in the justifications and frequency
		Against	2	Environmental (1), Economic (1)
Sum (all the texts)		For Against	50 59	Social (71) Environmental (26) Economic (9) No justification (2)

Looking more closely at what the students based their justifications on, we could see that most justifications were based on emotions—either alone (n = 24) or in combination with rationality or intuition (n = 19)—followed by rationality (n = 14) and intuition (n = 12). Emotions were included in all integrated justifications, and rationality was more frequently represented than intuition in the students' decision-making texts. We also found that all but four students based their reasons on more than one reasoning pattern. Five students included all patterns in their reasoning, while four students included only two (rationality and emotions).

The analysis also showed that the students' arguments highlighting the social dimension of sustainability were mostly based on emotions (n = 29), followed by intuition (n = 12) and rationality (n = 10). The arguments that highlighted the environmental dimension of sustainability were mostly based on rationality (n = 13), followed by emotions (n = 11) and intuition (n = 3), where the emotions were more frequently represented in the arguments concerning the well-being of animals and plants. Finally, we found that the arguments that highlighted the economic dimension of sustainability were mostly based on rationality (n = 6), followed by emotions (n = 3). However, emotions were always represented together with rationality.

3.4 Discussion and Conclusion

Our findings show that the use of authentic dilemmas was effective in enabling students to identify different perspectives. Working with different actors and news articles made it possible to highlight a range of arguments. By studying the students' texts, we found that they highlighted several actors in their news articles, as well as in the decision-making texts. Looking more closely at which actors were highlighted, however, revealed that these were mostly the same actors as in the text material they had read prior to writing their own articles—even though they did

Table 3.2 Overview of actors, their stance (decision), number of arguments, sustainability dimensions, and patterns of informal reasoning (justification) of arguments in students' decision-making texts

Actor	Student ID	Decision	No. argu-ments	Sustainability dimension in the justifications and frequency	What the justifications are based on (Rationality (R), Emotions (E), Intuition (I)) and frequency
Skiers	1, 6, 7, 8, 10, 13, 15	For	9	Social (7), Economic (2)	R (3), E (4), R&E (2)
		Against	0	–	–
Hikers and other users of the area	2, 3, 4, 5, 6, 7, 8, 11, 12, 13, 14, 15, 16	For	9	Social (9)	E (4), I (1), R&E (2), E&I (2)
		Against	21	Social (21)	R (1), E (9), I (5), R&E (5), E&I (1)
Landowner	6	For	1	Economic (1)	I (1)
		Against	0	–	–
Hiking cabin owner	11	For	1	Economic (1)	R (1)
		Against	0	–	–
Schools and kindergartens	7	For	1	Economic (1)	R&E (1)
		Against	0	–	–
Society as a whole	5, 7, 11	For	0	–	–
		Against	4	Social (2), Economic (2)	I (2), R (2)
Environment	1, 2, 3, 5, 7, 8, 9, 12, 15, 16	For	1	Environmental (1)	R (1)
		Against	17	Environmental (17)	R (6), E (4), I (1), R&E (4), E&I (1), R&E&I (1)
Own interests	7, 9, 16	For	0	–	–
		Against	4	Social (4)	E (3), I (1)
Sum (all the texts)		For Against	22 46	Social (43) Environmental (18) Economic (7)	Rationality (14) Emotions (24) Intuition (19) R&E (14) E&I (5) R&E&I (1)

not make explicit references to these sources, the influence was evident. This indicates that young students were highly influenced by the source material provided by their teachers. The same pattern was also found in the types of sustainability dimensions that were highlighted in the students' texts and in their source material, as the students highlighted the same types of dimensions as were used in their sources. We also found the same pattern in the decision-making texts, which meant that the social dimension dominated the students' arguments. In the case of our specific dilemma, it was natural to highlight environmental challenges, but this was less evident in the students' source material, and later in their own texts. This illustrates the importance of teachers selecting and providing sources that reflect different perspectives, including in primary school, as this allows the students to evaluate the sources in a critical manner. If teachers are not prepared to make this choice, this could lead to the complexity of sustainability issues being lost in their teaching, as has been warned by Sund and Gericke (2020). Teachers need to be more critical of the sources that they present to the students (cf. Iordanou, 2022). It is important to include sources that highlight different perspectives of the dilemma, and if these do not exist, it should be addressed in the classroom.

This study shows that the young students based their justifications on emotions, rather than knowledge and logic. Altogether 43 of the students' justifications (of a total of 68) were either fully or partially based on emotions. This result corresponds with previous research on older students' patterns of reasoning in SSIs (Aziz & Johari, 2023; Dawson & Venville, 2009; Sadler & Zeidler, 2005) and sustainability dilemmas (Öhman & Östman, 2008). We also see that social justifications were mostly based on emotions, while economic and environmental perspectives were more based on rationality. This is probably not surprising, as the students knew and could use factual justifications related to economics and the environment. In addition, feelings were included a lot in the environmental reasoning, in the form of concern for animals and plants. The difficulty of justifying their claims with factual information and scientific knowledge can also be explained by the context of this project. Leung (2020) found that the context in which college students in Hong Kong worked affected how they used scientific knowledge to evaluate scientific claims. Leung's study points out that it is unrealistic to assume that students are able to put their epistemic understanding into practice without being taught how to do so. The difficulty experienced by older students explains why this is also problematic for much younger students. It also points to a need for teachers to explicitly support this in their teaching.

Although we see a lack of knowledge-based reasoning in the students' arguments, we also need to acknowledge the importance of valuing emotions as an important dimension that affects the decisions concerning sustainability dilemmas and social-scientific issues, and hence functions as an important evaluative mechanism in understanding the issue (Hufnagel, 2022). How the actors' emotional responses to an issue and dilemma affect their points of view should be addressed and highlighted in the teaching.

This study shows that working with perspective-taking within the context of a sustainability dilemma seems to make it possible for students to be aware of which perspectives and interests are involved in an issue, and how they must assess the different perspectives against each other in order to make a decision. This study also provides greater insights into what students base their reasoning on—whether it is based on facts, feelings, intuition, or a combination of these. The study shows how emotions affect primary school students' opinions, so we can therefore suggest that emotions should be an important part of the teaching of critical thinking, in addition to promoting the need for justifications based on logic and evidence.

Acknowledgements This paper is based on work conducted within the CriThiSE project (Critical Thinking in Sustainability Education), which received funding from The Research Council of Norway under grant agreement no. 302774.

References

Abrami, P. C., Bernard, R. M., Borokhovski, E., Waddington, D. I., Wade, C. A., & Persson, T. (2015). Strategies for teaching students to think critically: A meta-analysis. *Review of Educational Research, 85*(2), 275–314. https://www.jstor.org/stable/24434258

Aikenhead, G. S. (2003, August). Review of research on humanistic perspectives in science curricula. Paper presented at the meeting of the European Science Education Research Association, Noordwijkerhout, The Netherlands.

Åkerblom, D., & Lindahl, M. (2017). Authenticity and the relevance of discourse and figured worlds in secondary students' discussions of socio scientific issues. *Teaching and Teacher Education, 65*, 205–214. https://doi.org/10.1016/j.tate.2017.03.025

Aziz, A. A., & Johari, M. (2023). The effect of argumentation about socio-scientific issues on secondary students' reasoning pattern and quality. *Research in Science Education, 53*, 771–789. https://doi.org/10.1007/s11165-023-10099-5

Bayat, M., Banihashem, S. K., & Noroozi, O. (2022). The effects of collaborative reasoning strategies on improving primary school students' argumentative decision-making skills. *The Journal of Educational Research, 115*(6), 349–358. https://doi.org/10.1080/00220671.2022.2155602

Chang, H. S., & Lee, H. J. (2010). College students' decision-making tendencies in the context of socioscientific issues (SSI). *Journal of the Korean Association for Science Education, 30*(7), 887–900. https://doi.org/10.14697/jkase.2010.30.7.887

Checchi, K. D., Pallarés, G., Taras, V., & Bächtold, M. (2022). Epistemic belief as a means of understanding critical thinking in the context of education for sustainable development. In B. Puig & M. P. Jiménez-Aleixandre (Eds.), *Critical thinking in biology and environmental education: Facing challenges in a post-truth world* (pp. 229–248). Springer.

Dahl, T. (2019). Prepared to teach for sustainable development? Student teachers' beliefs in their ability to teach for sustainable development. *Sustainability, 11*(7), 1993.

Davies, M., & Barnett, R. (Eds.). (2015). *The Palgrave handbook of critical thinking in higher education*. Palgrave MacMillian.

Dawson, V., & Venville, G. J. (2009). High-school students' informal reasoning and argumentation about biotechnology: An indicator of scientific literacy? *International Journal of Science Education, 31*(11), 1421–1445. https://doi.org/10.1080/09500690801992870

Ennis, R. (1987). A taxonomy of critical thinking dispositions and abilities. In J. Baron & R. Sternberg (Eds.), *Teaching thinking skills: Theory and practice* (pp. 9–26). Freeman.

Ferguson, L. E., & Krange, I. (2020). Hvordan fremme kritisk tenkning i grunnskolen? Forskningsbaserte forslag. *Norsk pedagogisk tidsskrift, 104*(2), 194–205.

Fought, S. S., Misfeldt, M., & Shaffer, D. W. (2019). Realistic authenticity. *Journal of Interactive Learning Research, 30*(4), 477–504. https://www.learntechlib.org/primary/p/184664/.

Hazelkorn, E. (2015). *Science education for responsible citizenship: Report to the European Commission of the Expert Group on Science Education*. Publications Office of the European Union.

Hufnagel, E. (2022). Emotional sense-making and critical thinking in the era of post-truth: The case of climate change. In B. Puig & M. P. Jiménez-Aleixandre (Eds.), *Critical thinking in biology and environmental education: Facing challenges in a post-truth world* (pp. 41–54). Springer. https://doi.org/10.1007/978-3-030-92006-7_3

Iordanou, K. (2022). Supporting critical thinking through engagement in dialogic argumentation: Taking multiple considerations into account when reasoning about genetically modified food. In B. Puig & M. P. Jiménez-Aleixandre (Eds.), *Critical thinking in biology and environmental education: Facing challenges in a post-truth world* (pp. 93–112). Springer.

Jensen, F., Pettersen, A., Frønes, T. S., Eriksen, A., Løvgren, M., & Narvhus, E. K. (2023). *PISA 2022: Norske elevers kompetanse i matematikk, naturfag og lesing* (p. 40). Cappelen Damm Akademisk/NOASP (Nordic Open Access Scholarly Publishing).

Jiménez-Aleixandre, M. P., & Puig, B. (2012). Argumentation, evidence evaluation and critical thinking. In B. Fraser, K. Tobin & C. McRobbie (Eds.), *Second international handbook of science education*(pp. Vol. 24, 1001–1015). https://doi.org/10.1007/978-1-4020-9041-7_66

Kuhn, D., & Crowell, A. (2011). Dialogic argumentation as a vehicle for developing young adolescents' thinking. *Psychological Science, 22*, 545–552. https://doi.org/10.1177/095679761140 2512

Leung, J. S. C. (2020). Students' adherences to epistemic understanding in evaluating scientific claims. *Science Education, 104*(2), 164–192.

Lee, H. S., & Butler, N. (2003). Making authentic science accessible to students. *International Journal of Science Education, 25*(8), 923–948. https://doi.org/10.1080/09500690305023

Ministry of Education and Research. (2017). Core curriculum – Values and principles for primary and secondary education. Laid down by Royal decree. The National Curriculum for the Knowledge Promotion 2020. https://www.regjeringen.no/contentassets/53d21ea2bc3a4202b86b83cf e82da93e/core-curriculum.pdf

Monika, J. (2024). Sustainability education in the 21st century: Incorporating environmental awareness in K-12 curriculum. *International Journal of Post Axial: Futuristic Teaching and Learning, 2*(3), 120–133.

Mork, S., & Erlien, W. (2017). *Språk, tekst og kommunikasjon i naturfag*. Universitetsforlaget.

Mulligan, K. (2009). Emotions and values. In P. Goldie (Ed.), *The Oxford handbook of philosophy of emotion*. Oxford Academic. https://doi.org/10.1093/oxfordhb/9780199235018.003.0022

Newton, M. H., & Zeidler, D. L. (2020). Developing socioscientific perspective taking. *International Journal of Science Education, 42*(8), 1302–1319. https://doi.org/10.1080/09500693.2020.175 6515

OECD. (2016). *Education at a glance: OECD indicators*. http://www.oecd.org/education/skillsbey ond-school/education-at-a-glance-2016-indicators.htm

Öhman, J., & Östman, L. (2008). Clarifying the ethical tendency in education for sustainable development practice: A Wittgenstein-inspired approach. *Canadian Journal of Environmental Education, 13*(1), 57–72. https://openjournal.lakeheadu.ca/index.php/cjee/article/view/444

Osborne, J. F., Henderson, J. B., MacPherson, A., Szu, E., Wild, A., & Yao, S. Y. (2016). The development and validation of a learning progression for argumentation in science. *Journal of Research in Science Teaching, 53*(6), 821–846. https://doi.org/10.1002/tea.21316

Pache, A., & Rouiller, S. (2022). Complexity and criticality in relation to ESD competences. In P. Vare, N. Lausselet, & M. Riekmann (Eds.), *Competences in education for sustainable development: Critical perspectives* (pp. 53–60). Springer International Publishing.

Parry, S., & Metzger, E. (2023). Barriers to learning for sustainability: A teacher perspective. *Sustainable Earth Reviews, 6*(1), 2.

Patton, M. Q. (2002). *Qualitative research & evaluation methods* (3rd ed.). Sage.

Pollarolo, E., Størksen, I., Skarstein, T. H., & Kucirkova, N. (2023). Children's critical thinking skills: Perceptions of Norwegian early childhood educators. *European Early Childhood Education Research Journal, 31*(2), 259–271.

Rieckmann, M. (2018). Learning to transform the world: Key competencies in education for sustainable development. *Issues and Trends in Education for Sustainable Development, 39*(1), 39–59.

Sadler, T. D., & Zeidler, D. L. (2005). Patterns of informal reasoning in the context of socio-scientific decision making. *Journal of Research in Science Teaching, 42*(1), 112–138. https://doi.org/10.1002/tea.20042

Sinnes, A. T. (2020). *Action takk! Hva kan skolen lære av unge menneskers handlinger for bærekraftig utvikling?* Gyldendal.

Sund, P., & Gericke, N. (2020). Teaching contributions from secondary school subject areas to education for sustainable development – A comparative study of science, social science and language teachers. *Environmental Education Research, 26*(6), 772–794.

Toulmin, S. E. (2003). *The uses of argument* (2nd ed.). Cambridge University Press. https://doi.org/10.1017/CBO9780511840005

Volman, M., & Ten Dam, G. (2015). Critical thinking for educated citizenship. In *The Palgrave handbook of critical thinking in higher education* (pp. 593–603). Palgrave Macmillan.

Wals, A. (2009). *Review of contexts and structures for education for sustainable development, 2009.* UNESCO: United Nations Educational, Scientific and Cultural Organisation. France. Retrieved from https://coilink.org/20.500.12592/0rwmg3 on 05 Oct 2024. https://coilink.org/20.500.12592/0rwmg3.

Chapter 4
Students' Knowledge of Ocean Acidification and Its Impact on Marine Organisms

Theodora Boubonari and Athanasios Mogias

4.1 Introduction

The increased atmospheric carbon dioxide (CO_2) levels due to human fossil fuel combustion and deforestation, which is mitigated by increased oceanic uptake (Moran, 2011), brings about pH reductions and alterations in fundamental chemical balances commonly called ocean acidification (Moran, 2011). According to the global framework for ocean sustainability (UNESCO, 2017), one of the future targets is to minimize and address the impacts of ocean acidification. Furthermore, considering that citizens through their everyday lives contribute to the increased CO_2 emissions, they have a responsibility to make informed lifestyle choices to minimize this impact. To do so, citizens need to have a solid understanding of this complex environmental issue and hold certain attitudes that shape their behaviour, to contribute to its amelioration. However, the dearth of research concerning either the public's or students' knowledge of ocean acidification or the carbon cycle has articulated participants' low awareness of these subjects (e.g. Danielson & Tanner, 2015; Düsing et al., 2019; Hartley et al., 2011; Spence et al., 2018; Zangori et al., 2017), while, to our knowledge, there are no relevant studies concerning primary students' knowledge of these issues.

School is one of the leading institutions responsible for communicating environmental knowledge and shaping attitudes to students, who are regarded as future decision-makers. The development of their knowledge, attitudes, and pro-environmental behavior would, therefore, have a significant effect on sustainable development. To the best of our knowledge, there are no relevant studies concerning the teaching of ocean acidification to primary students. According to Ocean Literacy

T. Boubonari (✉) · A. Mogias
Department of Primary Education, Laboratory of Environmental Research & Education, Democritus University of Thrace, Alexandroupolis, Greece
e-mail: tmpoumpo@eled.duth.gr

O. Morin et al. (eds.), *Challenges of a Changing World in Biology Education*, Contributions from Biology Education Research,
https://doi.org/10.1007/978-3-032-05346-6_4

Framework (National Oceanic & Atmospheric Administration, 2013) and Scope and Sequence for Ocean Literacy (National Marine Educators Association, 2010), ocean acidification could be taught to 11-12 year-old primary students.

A preliminary survey of all the Greek formal educational material for the primary level revealed that there is no material or program specifically concerning the carbon cycle and/or the ocean acidification. In the textbooks, there are fragmented concepts concerning the carbon cycle namely photosynthesis, respiration, fossil fuels, acids and bases, the flow of energy, the cycling of matter, and food chains (Hellenic Pedagogical Institute, 2003), and, thus, a sixth grader is expected to know about them.

Concerning these concepts, previous studies noted that primary students at least partially know what is made during photosynthesis, but they believe that the main purpose of photosynthesis is the formation of oxygen, followed by the formation of glucose (e.g., Dimec & Strgar, 2017; Lin & Hu, 2003). Respiration is often confused with breathing, and one of its major products, energy, is quite vague for younger pupils, probably due to their poor understanding of the cycling of matter and flow of energy (e.g., Lin & Hu, 2003). Decomposition and the role of micro-organisms in the cycling of matter are the most commonly misunderstood concepts (Byrne et al., 2009). Decomposition is seen as a process during which matter disappears over time and dead organisms are converted into minerals or soil without involving other organisms and without acknowledging physiological processes (Hellden, 1992; Özkan et al., 2004; Sequeira & Freitas, 1986; Smith & Anderson, 1986). Additionally, students do not recognise the trophic level of decomposers (Leach et al., 1996). On top of all this, it should be mentioned that the carbon cycle is highly complex because these carbon-compound transforming processes need to be integrated, which is problematic, since primary school students lack knowledge about the chemical aspects of them. Thus, it is unlikely to teach subject matter about each process and integrate it at the very beginning of learning about complex environmental problems, such as ocean acidification.

Given the lack of research on primary students' knowledge of the carbon cycle and of ocean acidification, the purpose of this study is to investigate the impact of a teaching intervention, based on the carbon cycle, on elementary school students' construction of knowledge on ocean acidification. The main focus of the study is students to achieve a basic understanding of the following concepts of the carbon cycle: (a) carbon is a chemical element found in the atmosphere as carbon dioxide (CO_2), as well as in all living things and fossil fuels, (b) the carbon cycle is a natural process where carbon moves through the environment via photosynthesis, respiration and decomposition; in photosynthesis, plants take in carbon dioxide from the air to make food and release oxygen, in respiration animals breathe in oxygen and exhale carbon dioxide back into the atmosphere, and during decomposition decomposers break down dead plants and animals, returning carbon to the soil and atmosphere, (c) the sources and sinks of carbon, (d) human activities like burning fossil fuels and deforestation can increase carbon levels in the atmosphere, contributing to ocean acidification, (e) ocean acidification is the process where the ocean becomes more acidic due to increased levels of carbon dioxide, harming sea creatures, especially those with shells or skeletons made of calcium carbonate, and (f) the importance of reducing emissions of carbon dioxide for a healthy ocean. These concepts were the content of the teaching intervention, without focusing on the chemical aspects of the processes.

For the needs of this study, we used a mixed-method approach with qualitative and quantitative research tools. In the qualitative approach, students were asked to draw the carbon cycle including ocean acidification, before and after the intervention (Boubonari et al., 2023). In their post-drawings, students included in their post-drawings all four earth systems in the carbon cycle, increased the number and variety of the processes and improved their ability to identify dynamic relationships among the components and the processes. However, although they understood the input of carbon, they still had difficulties to understand and include in their post-pictures the output mechanisms of the carbon cycle, mostly respiration and less decomposition. Also, students achieved to understand and integrate ocean acidification into the carbon cycle. They highlighted human's impact on the carbon cycle, which indicates that they realized that humans' everyday actions concerning CO_2 emissions may affect ocean acidification and, consequently, the ocean's health.

In the present study, we focus on the quantitative research tools, addressing the following research questions:

1 To what extent did the intervention lead to an increase in students' knowledge of the carbon cycle including ocean acidification from pre-test to post-test?
2 To what extent did students' self-assessments of knowledge about the terms and concepts of the carbon cycle change from pre-test to post-test?

4.2 Methodology

4.2.1 Sample

The study was conducted during the school year 2018 with a convenient sample of 85 sixth-grade students (11–12 years-old) of 5 classes from 2 public primary schools located in a coastal provincial town, in Greece. Both schools are considered regular ones representing the vast majority of Greek primary schools in relation to both students and teachers.

4.2.2 Research Design

The authors developed a teaching intervention for the age group of the present study taking into account (a) the Ocean Literacy Guide (National Oceanic & Atmospheric Administration, 2013), especially the fundamental concepts which concern the domination of the ocean on the carbon cycle and the balance of pH (Principle 1, concept

E; Principle 2, concept D; Principle 3, concepts A, E, F); (b) the Ocean Literacy Scope and Sequence for the corresponding age (grade band 6–8) (National Marine Educators Association, 2010), according to which sixth graders should accomplish to understand the role of the ocean in the carbon cycle, as well as the effects of the increased CO_2 on the pH of the seawater, and (c) educational sources concerning both the carbon cycle and the ocean acidification (e.g., Matracia & Zillmer, 2012).

The intervention included 8 h of inquiry-based and knowledge-integration activities, particularly experiments, concept maps, virtual laboratories, and interactive online activities, concerning photosynthesis, respiration, web chain, the carbon cycle, pH and ocean acidification. In these activities, students were asked to present their knowledge concerning the carbon cycle, emphasizing the effects of CO_2 increase on ocean acidification. The intervention was implemented by the first author (for detailed information on the teaching intervention see Boubonari et al., 2023).

4.2.3 Research Tools

The tools were a structured knowledge questionnaire and a self-assessment tool, concerning the carbon cycle and ocean acidification. Both tools were administered at the beginning and immediately after the last session of the intervention.

Questionnaire: Since no published questionnaires were found, the authors developed a questionnaire consisting of 18 items to examine students' knowledge of the components and processes of the carbon cycle and ocean acidification. Field experts checked the questionnaire, while a pilot study using a sample of 10 respondents helped to time the questionnaire, identify possible problems in terms of clarity and accuracy, and improve the final version of it. Participants were to answer "agree", "disagree" and "do not know", with the last option given in order to exclude the choice of random response. Each correct answer was coded as "1", and each incorrect answer, as well as the "I do not know" option, as "0". As for the incorrect items, it was expected that a greater number of students would disagree with them at the post-test (e.g. item 12).

Self-assessment tool: At the beginning and the end of the intervention, students were given a list of 16 terms and concepts concerning the carbon cycle and ocean acidification. They were asked how well they know them, ranking their level of knowledge using a 5-point Likert scale ranging from "I have never heard of this" to "I know this so well, I could teach someone else".

4.2.4 Data Analysis

The changes in students' content knowledge, as well as in their answers in the self-assessment tool were estimated using both descriptive and inferential statistics. Normality tests were performed prior to inferential statistics revealing normal distributions, and therefore parametric tests were applied. Cronbach's alpha was applied to assess the internal reliability of the knowledge questionnaire and the self-assessment tool. The "I don't know" responses of the questionnaire were treated as incorrect answers when calculating the Cronbach's alpha, so as to maintain a complete data matrix while acknowledging that a lack of knowledge represented an incorrect response in our assessment framework. Data analysis was performed with SPSS v23.0.

4.3 Results

In the following sections, the main results are presented separately for the two research questions.

4.3.1 To What Extent Did the Intervention Lead to an Increase in Students' Knowledge of the Carbon Cycle Including Ocean Acidification from Pre-test to Post-test?

The value of Cronbach's alpha showed a low internal consistency of the content knowledge questionnaire before the intervention ($\alpha = 0.509$, $n = 85$) and an acceptable internal consistency after the intervention ($\alpha = 0.619$, $n = 83$). The results of the independent-samples T-test showed that the pre- and post-intervention measurement was statistically significant for the knowledge test as the mean scores increased (Fig. 4.1). During the initial administration the students revealed a low mean percentage of correct answers (38.4%), while after the intervention this value significantly increased (62.4%) (Table 4.1).

Before the intervention, items 16, 7, 3 and 2 concerning the "increased seawater pH because of the increased CO_2, carbon release during decomposition, and the role of carbon in respiration and photosynthesis", were particularly difficult for the students (6.1%, 12.3%, 15.3%, and 16.5%, respectively) (Table 4.1). In addition, the students were not aware of the "role of decomposers in the carbon cycle" (item 9: 22.0%), the "dissolution of CO_2 in seawater" (item 14: 23.5%), and the "release of carbon into the deep during decomposition of dead organisms" (item 8: 27.4%). Students' performance was also low on items 5 and 12 (32.9% for both items), which concerned the "continuous recycle of the carbon on Earth" and that "the ocean is the biggest carbon sink" (Table 4.1) Finally, in 7 out of 18 items, the relative frequencies of the "I don't know" answers were high (42.9 to 75.0%).

The easiest items before the intervention concerned the "presence of carbon on Earth in abiotic and biotic components" (item: 82.1%), "oxygen inhalation and CO_2 exhalation of animals" (item 4: 76.5%), the "increase of CO_2 on Earth" (item

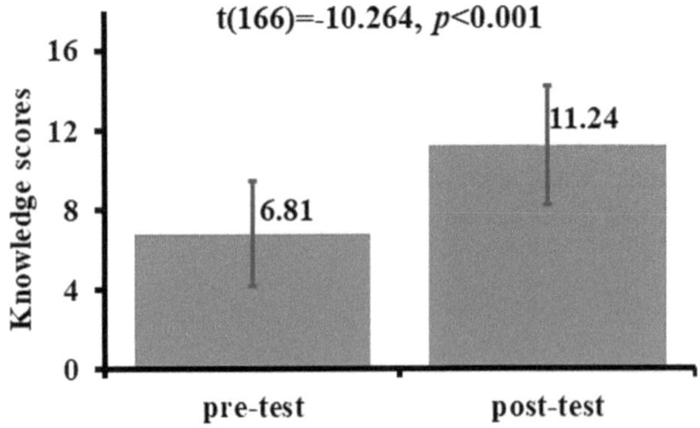

Fig. 4.1 Students' mean knowledge scores in pre- and post-questionnaire

13: 66.3%), and "carbon release by combustion of fossil fuels" (item 10: 53.7%) (Table 4.1).

After the intervention, students' performance increased in all items, exhibiting statistically significant difference in 13 out of 18 statements (Table 4.1). Moreover, the frequencies of the "I don't know" answers were very low in all items (2.4–8.9%). However, their scores presented only a slight increase on items 2, 16, and 9, which concerned the role of "CO_2 in plants' respiration, the increased seawater pH due to increased CO_2 levels and the role of decomposers in the carbon cycle" (Table 4.1).

4.3.2 *To What Extent Did the Students' Self-assessments of Knowledge About the Terms and Concepts of the Carbon Cycle Change from Pre-test to Post-test?*

The value of Cronbach's alpha showed a high internal consistency of the self-assessment tool before and after the intervention ($\alpha = 0.873$ and 0.903, respectively). Independent-samples T-test showed a statistically significant difference in students' perceived knowledge between pre- and post-self-assessment tool surveys (Fig. 4.2). Before the intervention, the mean value of all items was relatively low (M 2.87; SD 0.65) the less known concepts were alkalic, decomposers, pH, acidic, and ocean acidification (ranging from 1.57 to 2.14) (Table 4.2).

After the intervention, with an evident increase in the mean value (M 3.80; SD 0.66), students reported that they knew most of the concepts quite well, while the concepts "decomposers", "alkalic" and "fossil fuels" remained rather difficult (scores ranged between 3.04 and 3.39) (Table 4.2).

Table 4.1 Relative frequencies of correct and "I don't know" answers for the pre- and post-questionnaire. The italicized items are incorrect

Knowledge questions	Pre-test (n = 85)		Post-test (n = 83)	
	Correct answers	I don't know	Correct answers	I don't know
1. There is carbon in the atmosphere, in plants and animals, in soil, in subsoil and in the ocean	82.1	14.3	97.6	2.4
2. *Plants use carbon dioxide from the air for their respiration*	16.5	14.1	24.1	1.2
3. Phytoplankton use carbon dioxide from water in photosynthesis	15.3	76.5	60.2	28.9
4. Humans and animals breathe in oxygen and breathe out carbon dioxide	76.5	14.1	86.7	1.2
5. Carbon is continually recycled in the earth	32.9	34.1	55.4	16.9
6. Photosynthesis in land and marine plants is part of the carbon cycle	40.0	47.1	68.7	13.3
7. Carbon is released during decomposition	12.3	44.4	69.9	6.0
8. Some carbon gets into the deep ocean when living things in the ocean die	27.4	42.9	47.0	24.1
9. *Decomposers (such as bacteria) break down dead organisms and release oxygen*	22.0	58.5	36.1	24.1
10. Fossil fuels are sinks for carbon until they are burned and then they become sources	53.7	35.4	71.1	18.1
11. Ocean is the biggest carbon sink	43.2	21.0	75.9	4.8
12. *Land is the biggest carbon sink*	32.9	25.6	72.3	4.8
13. Carbon dioxide in the atmosphere has increased in the last years	66.3	25.3	78.3	14.5
14. *Carbon dioxide is hardly dissolved in ocean water*	26.2	38.1	51.8	16.9
15.*Increased carbon dioxide emissions have no effect on the ocean environment*	51.9	32.1	61.4	25.3
16. *Increased carbon dioxide levels increase seawater pH*	6.1	72.0	24.1	15.7
17. *Increased amounts of dissolved carbon dioxide in the ocean do not affect organisms that have shells*	50.0	37.8	74.7	7.2
18. Increased amounts of dissolved carbon dioxide in the ocean are harmful to the corals	36.1	55.4	68.7	20.5
Mean percentage	38.4	38.3	62.4	13.7

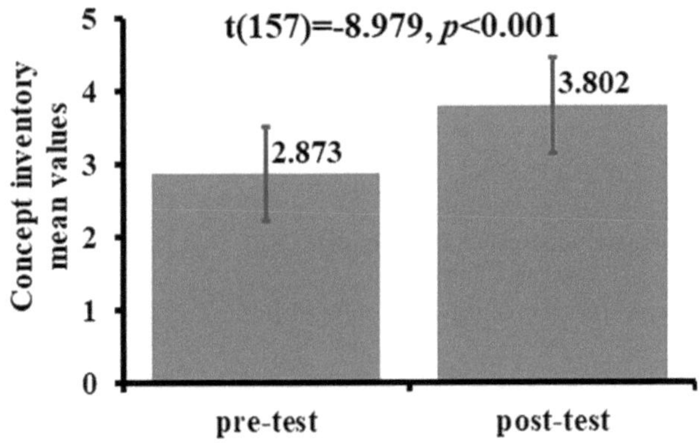

Fig. 4.2 Students' mean scores in pre- and post- self-assessment tool

Table 4.2 Mean values of the pre- and post-intervention self-assessment tool

Terms and concepts	Pre-test (n = 78)		Post-test (n = 81)	
	M	SD	M	SD
Photosynthesis	3.08	1.125	4.00	0.880
Carbon	3.44	0.975	4.23	0.795
Oxygen	4.40	0.631	4.44	0.707
Respiration	4.44	0.713	4.49	0.656
Carbon cycle	2.44	1.241	4.00	0.894
Food chain	3.63	1.353	4.19	0.956
pH	1.96	1.057	3.53	1.073
Acidic	2.00	1.294	3.44	1.123
Alkalic	1.57	0.895	3.14	1.159
Marine organisms having a shell	3.07	1.436	3.95	1.117
Phytoplankton	2.62	1.328	3.86	0.978
Fossil fuels	2.66	1.304	3.39	1.108
Ocean acidification	2.14	1.085	3.46	1.102
Decomposers	1.67	0.892	3.04	1.214
Producers	3.32	1.400	3.72	1.143
Consumers	3.36	1.309	3.86	1.118
Mean value	2.86	0.886	3.80	0.435

4.4 Discussion

The statistically significant increase in the mean correct answers, the dramatically decreased mean score in the "I don't know" responses and the statistically significant improvements in 13 out of 18 items in the post-questionnaire demonstrated students' overall knowledge improvement. Additionally, the analysis in the self-assessment tool revealed a statistically significant difference in their perceived knowledge.

Specifically, the results of the post- questionnaire serve as evidence that the intervention had a positive effect on students' understanding of the subject matter and that their answers weren't based on conjecture, thus, the instructional methods used were successful. Additionally, the noteworthy reduction in the "I don't know" responses after the intervention suggests that students became more confident in their knowledge following the instructional changes, indicating increased engagement with the material. However, it must also be said that the close-ended questions of the questionnaire in combination with the treatment helped them give the correct answers, which concerned factual knowledge and not differentiated understanding. Thus, to have more explicit results, we also used qualitative tools (Boubonari et al., 2023).

Some of the most challenging concepts which significantly improved after the intervention were the release of carbon during decomposition, the understanding of carbon's role in respiration and photosynthesis, the continuous recycling of carbon on Earth and the grasp of the ocean as a carbon sink. Students' knowledge gaps about photosynthesis and respiration have also been identified in previous studies (e.g., Dimec & Strgar, 2017; Lin & Hu, 2003; Marmaroti & Galanopoulou, 2006), and have been attributed to pupils' poor understanding of the cycling of matter and the flow of energy (e.g., Lin & Hu, 2003). Additionally, the fact that there was a low score in the item concerning photosynthesis by phytoplankton probably could be ascribed to the fact that in Greek primary textbooks the process of photosynthesis is mainly discussed for terrestrial plants, and explicit references or connections to the oceans are rare in any thematic sections of the Greek science education curricula (Mogias et al., 2021). The experiments, concept maps, virtual laboratories, and interactive online activities, concerning photosynthesis, respiration, food chain and decomposition, using paradigms from land, as well as ocean life during the intervention helped students improve their factual knowledge about these concepts. Students' knowledge improvement in these concepts is important because it underlines the feasibility of teaching such complex subject matter at the primary school level.

The best-understood concepts concerned carbon's presence in biotic and abiotic components, animal respiration (oxygen intake/CO_2 output), increasing CO_2 levels on Earth and carbon release from fossil fuel combustion, indicating that students had grasped more visible or commonly discussed aspects of carbon, thus leaving little room for improvement after the intervention.

Although there were overall gains, it's important to note that certain items related to more complex concepts, i.e. CO_2's role in plant respiration and the role of decomposers in the carbon cycle, showed only slight improvements. These processes are not evident to the unassisted eye, and children's knowledge seems firmly anchored in things they can directly see (Ero-Tolliver et al., 2013). Furthermore, previous studies also highlighted students' difficulty with decomposition and concluded that children have little to no understanding of what happens during this process (e.g., Leach et al., 1996). They believe either that matter from dead animals and plants simply disappears (Hellden, 1992; Sequeira & Freitas, 1986), or decay is a state that happens to materials, and no explanation is needed (Smith & Anderson, 1986), or they think about decomposers as "sweepers of nature" (Özkan et al., 2004). Students' gaps in plant respiration and decomposition underline the need for educators to focus on these processes. Respiration could be approached in a carbon cycle context, focusing on the sub-system photosynthesis-respiration, instead of teaching them in different grade levels or lessons, so as students can interrelate the two topics. In addition, the process of decomposition requires specialized teaching techniques, such as videos and animation, to help children understand this process.

It was expected that students wouldn't know about ocean acidification before the intervention, since they had never been taught about it. Their post-questionnaire answers indicate that they achieved to understand that the increased carbon dioxide in the atmosphere is transferred to the sea and brings the carbon system out of balance, affecting the marine organisms, thus realizing the human's impact on ocean acidification and its effects on the marine organisms. However, although they understood that the seawater becomes more acidic, they interpreted acidification as an increase in pH values and not a decrease; this misunderstanding implies that more activities concerning the pH scale should be added, so as to help students realize that more acidity means lower pH values.

Overall, while our current intervention may have effectively covered numerous aspects of factual knowledge concerning the carbon cycle and ocean acidification, it may not provide sufficient time for students to internalize and reconstruct their understanding of these complex concepts. This reflects a need for a shift in instructional priorities for better educational outcomes. Future evaluations should prioritize depth of teaching over breadth. This means narrowing the focus on fewer concepts but exploring them thoroughly to ensure that students not only know the facts but can apply their understanding critically. A curriculum design that incorporates lessons dedicated to deeper explorations of the most difficult concepts of the carbon cycle might lead to better long-term retention and understanding.

The relatively low Cronbach's alpha values observed in the questionnaire warrant careful consideration when interpreting our findings (Taber, 2018). Several factors may have contributed to these reliability coefficients. Firstly, treating "I don't know" responses as incorrect answers, while methodologically necessary for maintaining a complete data matrix, could have introduced measurement noise (Denman et al., 2018). Alternative approaches to handling "I don't know" responses, such as incorporating them as a separate response category or using item response theory models that can accommodate missing data, might provide more nuanced insights into student knowledge development. Moreover, the questionnaire itself contains items that measure different facets of knowledge about the carbon cycle, resulting in lower internal consistency (Taber, 2018). This heterogeneity could indicate that our assessment captured a broader construct than initially anticipated, suggesting that future research might benefit from developing more homogeneous subscales. Additionally, the sample size for the reliability analysis may have affected the stability of the Cronbach's alpha estimates. Typically, larger sample sizes provide more precise reliability coefficients. Despite these limitations, the pre-post improvements observed in our study remain meaningful, though the magnitude of these effects should be interpreted with appropriate caution given the reliability considerations discussed above.

Furthermore, the statistically significant difference in perceived knowledge in the self-assessment tool indicates that the intervention had a meaningful impact and students felt more confident about their understanding of the concepts after it, which is important when dealing with environmental problems, since previous studies have demonstrated a relationship between self-assessment of knowledge and motivation (e.g. Sitzmann et al., 2010). However, although there was an overall improvement, certain concepts like "decomposers," "alkalic," and "fossil fuels" still posed challenges, indicating that while students improved, there are specific areas that require additional focus in future teaching efforts. Students' replies to the pre- and post-questionnaire and self-assessment tool typically matched, indicating that they probably had a fairly accurate understanding of their performance and learning requirements, providing guidance for future learning (Baas et al., 2015). On the other hand, studies have shown that summative self-assessment tends to be inconsistent with external judgements the younger the learners are (Andrade, 2019), and, thus, students of the present study might have slightly overestimated their knowledge.

Overall, students' significant progress which was measured both with qualitative (Boubonari et al., 2023) and quantitative tools supports the suggestion of Scope and Sequence for Ocean Literacy (NMEA, 2010) that the carbon cycle and ocean acidification can be taught in primary education. Moreover, our findings provide insight into developing primary students' understanding of ocean acidification and the carbon cycle, which is a crucial first step towards taking responsibility to make informed lifestyle choices to minimize this impact. However, the small, localized sample limits the applicability of the findings to broader contexts and, thus, more studies are needed to support our conclusions.

References

Andrade, H. L. (2019). A critical review of research on student self-assessment. In *Frontiers in education* (Vol. 4, p. 87). Frontiers Media SA.

Baas, D., Castelijns, J., Vermeulen, M., Martens, R., & Segers, M. (2015). The relation between assessment for learning and elementary students' cognitive and metacognitive strategy use. *British Journal of Educational Psychology, 85*(1), 33–46.

Bielik, T., Damelin, D., & Krajcik, J. S. (2019). Shifting the balance: Engaging students in using a modeling tool to learn about ocean acidification. *EURASIA Journal of Mathematics, Science and Technology Education, 15*(1), em1652.

Boubonari, T., Papazoglou, D. N., Mogias, A., & Kevrekidis, T. (2023). Challenging Greek primary students' knowledge of ocean acidification using the carbon cycle context. *International Journal of Science and Mathematics Education*, 1–24.

Byrne, J., Grace, M., & Hanley, P. (2009). Children's anthropomorphic and anthropocentric ideas about micro-organisms: Educational research. *Journal of Biological Education, 44*(1), 37–43.
Ero-Tolliver, I., Lucas, D., & Schauble, L. (2013). Young children's thinking about decomposition: Early modeling entrees to complex ideas in science. *Research in Science Education, 43*(5), 2137–2152.
Danielson, K. I., & Tanner, K. D. (2015). Investigating undergraduate science students' conceptions and misconceptions of ocean acidification. *CBE-Life Sciences Education, 14*(3), 1–11.
Denman, D. C., Baldwin, A. S., Betts, A. C., McQueen, A., & Tiro, J. A. (2018). Reducing "I don't know" responses and missing survey data: Implications for measurement. *Medical Decision Making, 38*(6), 673–682.
Dimec, D. S., & Strgar, J. (2017). Scientific conceptions of photosynthesis among primary school pupils and student teachers of biology. *Center for Educational Policy Studies Journal, 7*(1), 49–68.
Düsing, K., Asshoff, R., & Hammann, M. (2019). Students' conceptions of the carbon cycle: Identifying and interrelating components of the carbon cycle and tracing carbon atoms across the levels of biological organisation. *Journal of Biological Education, 53*(1), 110–125.
Hartley, L. M., Wilke, B. J., Schramm, J. W., D'Avanzo, C., & Anderson, C. W. (2011). College students' understanding of the carbon cycle: Contrasting principle-based and informal reasoning. *BioScience, 61*(1), 65–75.
Hellenic Pedagogical Institute. (2003). Διαθεματικό Ενιαίο Πλαίσιο Προγραμμάτων Σπουδών και Προγράμματα Σπουδών της Υποχρεωτικής Εκπαίδευσης (Cross-Thematic Curriculum Framework and Curricula of Compulsory Education). http://www.pi-schools.gr/programs/depps/
Hellden, G. (1992). *Pupils' understanding of ecological process* (The LISMA group learning in science and mathematics Report No. 2). Kristianstad University College.
Leach, J., Driver, R., Scott, P., & Wood-Robinson, C. (1996). Children's ideas about ecology 3: Ideas found in children aged 5–16 about the interdependency or organisms. *International Journal of Science Education, 18*(2), 129–141.
Lin, C. Y., & Hu, R. (2003). Students' understanding of energy flow and matter cycling in the context of the food chain, photosynthesis, and respiration. *International Journal of Science Education, 25*(12), 1529–1544.
Marmaroti, P., & Galanopoulou, D. (2006). Pupils' understanding of photosynthesis: A questionnaire for the simultaneous assessment of all aspects. *International Journal of Science Education, 28*(4), 383–403.
Matracia, K., & Zillmer, C. M. (Eds.). (2012). *Changing climate. A guide for teaching climate change in grades 3 to 8*. Environmental literacy teacher guide series. National Geographic Society.
Mogias, A., Boubonari, T., & Kevrekidis, T. (2021). Examining the presence of ocean literacy principles in Greek primary school textbooks. *International Research in Geographical and Environmental Education, 30*(4), 314–331.
Moran, J. M. (2011). *Ocean studies. Introduction to Oceanography* (3rd ed.). The American Meteorological Society.
National Marine Educators Association. (2010). *Ocean literacy scope and sequence for grades K-12, published in the National Marine Educators Association, U.S.A.* (Special Report #3 on The Ocean Literacy Campaign Featuring the Ocean Literacy Scope & Sequence for Grades K-12). Retrieved from Ocean literacy website: http://www.oceanliteracy.net
National Oceanic and Atmospheric Administration. (2013). Ocean literacy: The essential principles and fundamental concepts of ocean sciences for learners of all ages. Version 2, U.S.A.
Özkan, O., Tekkaya, C., & Geban, Ö. (2004). Facilitating conceptual change in students' understanding of ecological concepts. *Journal of Science Education and Technology, 13*(1), 95–105.

Sequeira, M., & Freitas, M. (1986). *Death and decomposition of living organisms; children's alternative frameworks*. Paper presented at the annual meeting of the Association for Teacher Education in Europe. France: Toulouse.

Sitzmann, T., Ely, K., Brown, K. G., & Bauer, K. N. (2010). Self-assessment of knowledge: A cognitive learning or affective measure? *Academy of Management Learning & Education, 9*(2), 169–191.

Smith, E. L., & Anderson, C. W. (1986). *Alternative student conceptions of matter cycling in ecosystems*. Paper presented to the *Annual Meeting of the National Association for Research in Science Teaching*. San Francisco: California.

Spence, E., Pidgeon, N., & Pearson, P. (2018). UK public perceptions of Ocean Acidification—The importance of place and environmental identity. *Marine Policy, 97*, 287–293.

Taber, K. S. (2018). The use of Cronbach's alpha when developing and reporting research instruments in science education. *Research in Science Education, 48*, 1273–1296.

UNESCO. (2017). Education for sustainable development goals. Learning objectives. Paris.

Zangori, L., Peel, A., Kinslow, A., Friedrichsen, P., & Sadler, T. D. (2017). Student development of model-based reasoning about carbon cycling and climate change in a socio-scientific issues unit. *Journal of Research in Science Teaching, 54*(10), 1249–1273.

Chapter 5
Environmental Issues, Causes and Solutions in Greek Secondary Education Biology Textbooks: How Do Textbooks Contribute to Students' Environmental Literacy?

Georgios Ampatzidis and Konstantinos Korfiatis

5.1 Introduction

The term "environmental literacy" was introduced by Roth (1968) in an issue of Massachusetts Audubon, where he posed the question, "How shall we know the environmentally literate citizen?" While it was coined over five decades ago and has been a subject of numerous studies, there is no universally accepted definition for environmental literacy. It is an abstract concept which has been applied in a broad and inconsistent way by different authors around the world (Fang et al., 2023; McBride et al., 2013). Nonetheless, some general principles are common to most definitions, such as familiarity with the environment and ecology, ability to analyse environmental challenges, and behaviours aimed at reducing individual environmental footprint or supporting collective actions to protect the environment (Ardoin et al., 2023; Negev et al., 2008).

Hollweg et al. (2011) argued that environmental literacy encompasses five distinct categories of knowledge that individuals must use to effectively address environmental issues or concerns. These concern an understanding of: (a) political, societal and cultural systems, (b) natural and ecological systems, (c) environmental challenges, (d) a range of approaches to environmental problems and (e) strategies for civic engagement and action. Moreover, Erdogan and Ok (2011) argued that

G. Ampatzidis (✉)
University of Thessaly, Volos, Greece
e-mail: gampatzidis@uth.gr

K. Korfiatis
University of Cyprus, Nicosia, Cyprus
e-mail: korfiatis.konstantinos@ucy.ac.cy

O. Morin et al. (eds.), *Challenges of a Changing World in Biology Education*, Contributions from Biology Education Research,
https://doi.org/10.1007/978-3-032-05346-6_5

most researchers agree on four dimensions of environmental literacy, namely: (a) behaviour, (b) affect, (c) skill, and (d) knowledge. Knowledge, in particular, concerns (i) knowledge of environmental issues, (ii) socio-political knowledge and (iii) ecological knowledge. Focusing on (i), Erdogan and Ok (2011) noted that it entails a comprehension of environmental problems and issues stemming from human interactions with the environment, along with knowledge associated with alternative solutions for these issues.

As new views of the ecosystems emerge—e.g. the Balance of Nature concept (Ampatzidis & Ergazaki, 2018) has been replaced by modern views of nature which underline the fact that ecosystems may experience unpredictable changes when certain tipping point are reached (Pang et al., 2025)—and environmental challenges increasingly dominate global discussions, there's growing scrutiny of public understanding of the natural world. The recognition that fostering a deeper ecological awareness and understanding is crucial for addressing environmental issues has highlighted the critical role of developing environmental literacy through education within the broader framework of sustainability. Sustainability is about ensuring a good quality of life for everyone today without jeopardizing the well-being of future generations, i.e. finding a balance between environmental protection, social progress, and economic growth (Thiele, 2024). The importance of integrating environmental literacy into educational initiatives and sustainability efforts has been emphasized and, consequently, there has been a considerable effort to incorporate these principles into education and training programs, as well as the creation of textbooks that align with these objectives (Şeker, 2024).

Textbooks structure and content are crucial in advancing a particular curriculum vision (Okeeffe, 2013; Yalçınkaya-Önder et al., 2022). The "delivered" curriculum is made up of textbooks, which specify what teachers can teach and what students can learn (Chiappetta et al., 2002; Huang et al., 2022). This highlights the significant role textbooks play in shaping the actual educational experience in classrooms, as they often serve as the primary resource for both teachers and students. The content selection and organization within textbooks thus have a direct impact on the implementation of curriculum objectives and the scope of knowledge presented to learners. Overall, science textbooks play a pivotal role in developing scientific literacy, providing students with a structured and reliable foundation of knowledge (Isaksen et al., 2024).

Considering that (a) an understanding of environmental problems and relevant solutions is an important part of environmental literacy, (b) environmental literacy has a critical role in enhancing environmental awareness and promoting sustainability and (c) textbooks play a significant role in shaping the actual educational experience in classrooms, we decided to explore how environmental problems along with their causes and suggested solutions are discussed in Greek secondary education biology textbooks. Our research covers a period from 1970s to present in order to detect possible differences across the years.

5.2 Methods

Access to textbooks was made possible through the historic collection of the Greek Institute of Educational Policy. The institute's website hosts a database of digitized old Greek textbooks, which can be searched based on criteria such as subject, language, education level, and date. We conducted a search for biology textbooks written in Greek and used in secondary education. Our search, conducted in September 2023, yielded 45 ebooks, which we reviewed alongside six biology textbooks currently in use, to identify parts discussing environmental issues. Some of the files we reviewed were different editions of the same book; in such cases we used for our analysis the version including more information concerning our study, if there were differences.

Applying the above criteria, our analysis concerned five textbooks published from 1970s:

Textbook-1: General biology courses for the 6th Grade of Gymnasium (Oikonomidis, 1970)—Grade 12.

Textbook-2: General biology courses for the 3rd Grady of Lyceum (Krimpas & Kalopisis, 1977)—Grade 12.

Textbook-3: Biology for the 3rd Grade of Gymnasium (Gkelti-Douka et al., 1982)—Grade 9.

Textbook-4: Biology for the 2nd and 3rd grade of Gymnasium (Mavrikaki et al., 2013)—Grades 8–9.

Textbook-5: Biology for the 2nd grade of General Lyceum-General Education (Adamantiadou et al., 2013)—Grade 11.

We selected the textbook as our unit of analysis rather than a smaller unit (such as chapters, paragraphs, or sentences) because our focus was on capturing the presence of specific concepts within the textbooks under analysis, rather than on finer details. Moreover, using the whole textbooks as units of analysis is considered a more suitable approach when examining changes over time within a book or series of books, or when comparing the content of different textbooks (Huang et al., 2022). The analysis of the chapters discussing environmental issues was driven by the objectives to present and discuss:

- the environmental issues mentioned
- the causes of environmental issues
- the solutions for environmental issues

Open coding (Cascio et al., 2019) was applied in order to create a set of data-driven categories. More specifically, the textual corpus of our study was coded under the categories "environmental issues", "causes of environmental issues", and "solutions for environmental issues" while the several sub-categories that emerged were organized into a coding scheme where every unit could be coded in more than one category (e.g. global warming, overgrazing, protection of wildlife for environmental issues, causes and solutions respectively—see Table 5.1). Both authors coded independently

Table 5.1 Analysis of textbooks in regards to environmental issues, causes of environmental issues and solutions for environmental problems

Categories	Subcategories	Textbooks
Environmental issues	Increase of chemicals in organisms	2, 3, 4, 5
	Photochemical smog	3, 4, 5
	Ozone hole	4, 5
	Acidic rain	4, 5
	Global warming	4, 5
	Biodiversity loss	1, 2, 3, 4, 5
	Desertification	5
	Eutrophication	2, 3, 4, 5
	Decrease of photosynthesis rate	3
Causes of environmental issues	Pollution of soil, water and atmosphere	1, 2, 3, 4, 5
	Deforestation	1, 2, 3, 5
	Overfishing	1, 2, 3
	Overgrazing	1, 2, 3, 4, 5
	Overhunting	2, 3
	Dewatering of ponds	1, 2, 3, 5
	Forest fires	1, 2, 4, 5
	Botanists' activity	1
Solutions for environmental issues	Establishment of protected areas	1, 2, 5
	Protection of wildlife	1
	Breeding of rare animals	1
	Reforestation	1
	Treatment of garbage/exhaust gases	3
	Regulation of pesticides' use	3
	Limitation of harmful human activity	3
	Ban of chlorofluorocarbons	4
	Recycling	4

the full textual corpus and the percentage agreement (Gisev et al., 2013) was calculated about 90%. The cases of disagreement were reviewed and discussed by the two coders until consensus was reached.

5.3 Results

Every textbook included one chapter which discussed environmental issues. The number of pages for textbooks-1 to textbook-5 were 2, 11, 6, 7, 22 respectively. As shown in Table 5.1, biodiversity loss is referred in all textbooks. For example, in textbook-2 we read: "*The extinction of a species not only impoverishes nature but also destabilizes it. The parts of the ecosystem are interconnected, much like the organs of a body. If a stone is removed from a stone structure, nothing may happen. However, if more stones are removed, the entire structure will collapse*" (Krimpas & Kalopisis, 1977, p. 211). Moreover, the authors of textbook-1 mention: "*The unlimited expansion of the human race results in the alteration of biotic communities through human intervention, which often progresses to their complete destruction"* (Oikonomidis, 1970, p. 198).

Some environmental issues are mentioned in most textbooks investigated (e.g. increase in the concentration of chemicals in organisms and eutrophication) (Table 5.1). An example of the former is the following excerpt from textbook-2: *"A toxic substance, such as an insecticide, can be transferred from species to species through food chains, increasing its concentration. For example, DDT enters lakes in small amounts and gets absorbed by phytoplankton. Zooplankton, which feed on phytoplankton, accumulate higher amounts of DDT because they need a much larger mass of phytoplankton for their food. Fish accumulate even more DDT, and by the time it reaches aquatic birds, the concentration is very high. The same happens with birds of prey in terrestrial ecosystems"* (Krimpas & Kalopisis, 1977, p. 210). Furthermore, an example of eutrophication mentioned in the textbooks investigated is the following excerpt from textbook-3: *"It is well known that fertilizers, usually phosphates and nitrates, are carried by water to rivers, lakes, and seas, providing nourishment to photosynthetic organisms. As a result, the flora there experiences significant growth and consequently consumes large amounts of oxygen dissolved in the water. This leads to the creation of suffocating conditions for heterotrophic organisms, which die off, leaving only a small number of them surviving"* (Gkelti-Douka et al., 1982, p. 112).

Other environmental issues are mentioned in only two or one textbooks (e.g. global warming) (Table 5.1). For instance, in textbook-4 we read: *"This phenomenon is called the greenhouse effect, as it operates in a way similar to a greenhouse. Thus, the average temperature on the Earth's surface is 15 °C, a fact that allows life to develop on it. However, in recent years, due to the release of large quantities of carbon dioxide into the atmosphere, mainly from industries and automobiles, more radiation is being trapped in the atmosphere. The result is an increase in the Earth's temperature beyond the normal levels. This abnormal rise in temperature can lead to the melting of the polar ice caps, resulting in a rise in sea levels, the loss of land areas, and a general change in the Earth's climate"* (Mavrikaki et al., 2013, p. 53).

Concerning the causes of environmental issues, pollution and overgrazing are very popular among the textbooks we researched (Table 5.1). The following excerpts are two examples of the pollution as shown in textbook-1 and textbook-5, and an example of overgrazing as shown in textbook-2: *"(Human) pollutes rivers and lakes with industrial and urban waste, making them unsuitable for the survival of animals and plants"* (Oikonomidis, 1970, p. 198), *"Volcanic activity, the decomposition processes of organic matter by soil bacteria, and primarily the burning of fossil fuels release various nitrogen oxides and sulfur dioxide into the atmosphere. These gases, after being converted into nitric and sulfuric acid respectively by the action of atmospheric water vapor, return to the Earth's surface dissolved in rain, snow, fog, or hail"* (Adamantiadou et al., 2013, p. 107), *"Overhunting, overfishing (with plastic nets and dynamite), excessive logging, and overgrazing pose significant threats to our animal and plant environment, as do pollutants from chemical substances (which poison biotic communities and are the most serious threats)"* (Krimpas & Kalopisis, 1977, p. 210).

On the other hand, there are harmful activities mentioned only in two (overhunting) or only one (botanists' activity) textbook (Table 5.1). What follows is an example of the former and the single example of the latter category from textbook-1: *"(Human) eliminates game and wild animals and radically alters the form of fauna and flora in various habitats. Even the drive to produce scientific knowledge becomes a cause for the extinction of various rare plant species. For example, foreign botanical collectors descend upon Greek territories annually to collect rare species of Greek flora, and some of these are on the brink of extinction"* (Oikonomidis, 1970, p. 198).

In regards to the solutions for environmental issues, the establishment of protected areas is mentioned in most textbooks investigated (Table 5.1). For example, in textbook-5 we read: *"The recognition of enormous wetland losses on a global scale has led to serious efforts to reverse this trend. The beginning was made in 1971 with the Ramsar Convention (named after the Persian city where the treaty was signed). Greece has also signed this convention, which aims to protect wetlands and their flora and fauna. Our country has included 11 wetlands (or aggregates of individual wetlands) in the Ramsar list of wetlands"* (Adamantiadou et al., 2013, p. 98). Other activities mentioned once are breeding of rare animals, recycling and regulation of pesticides' use (Table 5.1). Below are the relevant excerpts from textbook-1, textbook-4 and textbook-3 respectively: *"Efforts are being made to maintain the natural state of biotic communities as much as possible. In Greece, there is a Nature Protection Society which is involved in reforestation, protection of game animals, and breeding of rare native birds"* (Oikonomidis, 1970, p. 198), *"Divide into three groups and research information about recycling. What is the contribution of recycling to saving energy and natural resources? Is waste collected in your area for recycling purposes? Does your school participate in recycling programs? The first group will collect information about paper recycling, the second about glass recycling, and the third about aluminium recycling. Read your work to your class and choose ways you can contribute to paper, aluminium, and glass recycling"* (Mavrikaki et al., 2013, p. 59), *"(One way to deal with pollution is) the scientific control over the use of pesticides so that the appropriate ones are used in strictly defined doses"* (Gkelti-Douka et al., 1982, p. 113).

5.4 Discussion

The results of our analysis showed that the most popular environmental issue among the textbooks investigated is biodiversity loss. All textbooks mention the disappearance of animal and/or plant species and discuss its consequences for the environment. The relevant excerpts highlight a conceptualization of nature as balanced—see, for example in textbook-2: The extinction of a species not only impoverishes nature but also destabilizes it (Krimpas & Kalopisis, 1977, p. 211). Despite being scientifically discredited after a long history in the field of ecology and even earlier (for an overview of the history of the Balance of Nature idea see Ampatzidis & Ergazaki, 2021 and Ampatzidis & Ergazaki, 2023), the BON concept continues to be prevalent in school science curricula (Korfiatis, 2018; Westra, 2008) and remains widespread in public opinion (Ladle & Gillson, 2009). The Balance of Nature concept may hinder conceptual understanding by contradicting the notion of resilient nature, which is supported by current ecosystem research (Scheffer, 2009). According to the latter, nature is not seen as "balanced", with contingency being favoured over order and purpose. Ecosystems are thought to operate in multiple self-organized alternative states, and can shift between these states abruptly, in ways that may be reversible or irreversible (Pang et al., 2025; Scheffer, 2009). Moreover, the Balance of Nature concept may interfere with environmental awareness, as it may suggest that ecosystems possess an almost "magical" ability to return to their original state, potentially downplaying the importance of avoiding disturbances (Gunderson et al., 2010).

Global warming appears only in the two currently in use textbooks along with the ozone hole and acidic rain. It has been argued that educational topics and the time dedicated to them often shift in response to the evolving knowledge and demands of society (Ansari & Landin, 2022). For instance, it has been reported that the presence of insects in textbooks has been decreased after 1950s when their role in spreading diseases and their impact as agricultural pests was limited through prevention, availability of effective cures and the widespread use of modern synthetic insecticides (Gangwani & Landin, 2018). Similarly, it has been reported that after 1960s, textbooks began to expand their coverage of molecular biology topics in response to ground-breaking research publications on DNA and genetics (Gaster, 1990). Thus, the presence of global warming, ozone hole and acidic rain only in the textbooks currently in use arguably reflects the fact that during the last decades more knowledge has been built concerning these environmental issues and they have been grown more important for scientific research and society.

Pollution is a longstanding, key cause of environmental issues mentioned in biology textbooks from 1970s to date. In most cases pollution is caused by humans and non-human causes of pollution, such as volcano eruptions, are mentioned only in textbook-5. It is noted that humans are widely considered to cause the environmental issues mentioned in the textbooks we researched. Indeed, humans are largely considered as responsible for causing damage, but they also have the ability to restore and manage ecosystems. Even in cases that they are not able to fully restore the disturbances they cause, they are able of both stimulating and slowing down feedback mechanisms, thus they are responsible for maintaining the ecological balance (Westra, 2008).

These ideas reflect older views of ecosystems and their characteristics. The currently valid view of "resilient nature" supports that humans should be considered an integral part of the ecosystems and although they may monitor and create favourable conditions for the ecosystems, there is no guaranteed outcome from their efforts to preserve ecological balance (Ampatzidis & Ergazaki, 2014). Modern ecology views ecosystems as complex, dynamic systems that are characterized more by instability than by stability (Scheffer, 2009). Balance is not the norm; disruptions to this balance often arise not only from human or natural activities but can also be an intrinsic part of the ecosystems' own dynamics (Ampatzidis & Ergazaki, 2017).

We noticed that overgrazing is also mentioned in all textbooks as a cause of environmental issues, highlighting its persistence over the decades. Overgrazing largely concerns the Mediterranean region and Greece specifically since it is an important factor that contributes to loss of biodiversity (Fetzel et al., 2018), soil erosion and land degradation (Kairis et al., 2015). We argue that being one of the only two causes of environmental issues that are mentioned in all textbooks researched reflects the importance of this phenomenon for ecological research in Greece. At the same time, the omission or inconsistent reference to other ecological issues, whether specific to Greece (e.g., drought, freshwater scarcity, air pollution in large cities) or of a more global nature (e.g., climate crisis, depletion of natural resources), may illustrate how textbooks often act as barriers to knowledge rather than as facilitators (Isaksen et al., 2024).

Shifting to the solutions for environmental issues, we noticed that the most popular seems to be the establishment of protected areas. Greece is recognized globally as an ecological hotspot and is considered an important area for European conservation efforts. The practice of designating specific areas to protect biological values began in Greece in 1937 with the creation of the country's first two national parks. In later years five official categories have been introduced—National Parks, Marine Parks, Aesthetic Forests, Protected Natural Monuments, and Ramsar sites—while in 1992, the European Union adopted a more integrative approach with the establishment of the Natura 2000 network of protected areas which cover 18% of Greece's total land area (Papageorgiou & Vogiatzakis, 2006). As shown above, protected areas have a long history and cover a large part of Greece and this seems to be reflected in their presence in textbooks spanning from 1970s to present. On the other hand, their considerable coverage in textbooks seems to contribute to the building of knowledge and forming of attitudes towards them since research has shown that students in Greece generally express adequate knowledge and positive attitudes towards protected areas (Efthimiou et al., 2017; Tzaberis et al., 2012).

Greek education, both public and private, relies on a standardized set of textbooks mandated by the government. These books form the backbone of the curriculum across all grade levels, making them a reliable indicator of classroom content. Our analysis findings contribute to the literature on the content of school textbooks around the world in terms of environmental issues. Although educational systems vary across countries, textbooks remain a fundamental tool for providing students with knowledge. Analysing textbooks from different countries can offer valuable insights into various education systems worldwide. Analysis of a country's textbooks could be useful for identifying the strengths and weaknesses in other counties' textbooks, potentially guiding future improvements. Moreover, recognizing similarities and differences among textbooks from different nations can contribute to the development of universal textbooks, fostering a more inclusive and globally accessible education (Huang et al., 2022).

Our analysis findings also suggest that certain actions should be taken regarding textbooks currently in use in Greek secondary education, in order to improve teaching and learning about the environment and thus enhance students' environmental literacy. For example, we argue that the conceptualization of nature should be moved from a balanced to a resilient one. Students should be exposed to currently valid ideas about nature's function, such as the dynamic and complex nature of ecosystems, and the position of humans in it in order to build an adequate conceptual understanding of ecosystems. Moreover, the discussion of solutions for environmental problems seems to be rather limited. Pro-environmental actions that are extensively debated and practiced nowadays (e.g. recycling, limiting the use of plastic; forms of collective or political actions) are not widely discussed. We suggest that textbooks emphasize activities and actions aimed at addressing environmental issues to help students build understanding, adopt appropriate attitudes and behaviors, and develop citizenship skills along with a willingness to take action.

References

Adamantiadou, S., Georgatou, M., Giapitzakis, C., Lakka, L., Notaras, Δ., Florentin, N., Chatzigeorgiou, G., & Chantikonti, O. (2013). *Biology for the 2nd grade of general lyceum-general education.* CTI Diophantus.

Ampatzidis, G., & Ergazaki, M. (2014). Towards a learning environment for challenging the idea of the balanced nature: Insights from the first cycle of research. In C. P. Constantinou, N. Papadouris, & A. Hadjigeorgiou (Eds.), *E-book proceedings of the ESERA 2013 conference: Science education research for evidence-based teaching and coherence in learning. Part 3* (pp. 44–54). European Science Education Research Association.

Ampatzidis, G., & Ergazaki, M. (2017). Toward an "anti-balance of nature" learning environment for non-biology major students: Learning objectives and design criteria. *Natural Sciences Education, 46*(1). https://doi.org/10.4195/nse2017.07.0016

Ampatzidis, G., & Ergazaki, M. (2018). Challenging students' belief in the 'balance of nature' idea. *Science & Education, 27*(9), 895–919. https://doi.org/10.1007/s11191-018-0017-5

Ampatzidis, G., & Ergazaki, M. (2021). Can the history of the balance of nature-idea inform the design of narratives for highlighting general aspects of nature of science? *Review of Science, Mathematics and ICT Education, 15*(1), 77–88. https://doi.org/10.26220/rev.3574

Ampatzidis, G., & Ergazaki, M. (2023). How did Darwin prefer his tea? *Science & Education, 32*(1), 37–56. https://doi.org/10.1007/s11191-021-00305-z

Ansari, R. A., & Landin, J. M. (2022). Coverage of climate change in introductory biology textbooks, 1970–2019. *PLoS ONE, 17*(12), Article e0278532. https://doi.org/10.1371/journal.pone.0278532

Ardoin, N. M., Bowers, A. W., & Wheaton, M. (2023). Leveraging collective action and environmental literacy to address complex sustainability challenges. *Ambio, 52*(1), 30–44. https://doi.org/10.1007/s13280-022-01764-6

Cascio, M. A., Lee, E., Vaudrin, N., & Freedman, D. A. (2019). A Team-based approach to open coding: Considerations for creating intercoder consensus. *Field Methods, 31*(2), 116–130. https://doi.org/10.1177/1525822X19838237

Chiappetta, E. L., Koballa, T. R., & Collette, A. T. (2002). *Science instruction in the middle and secondary schools.* Prentice Hall.

Efthimiou, G., Ntouras, K., & Panagopoulos, T. (2017). Knowledge and attitudes of forestry students on nature and protected areas in Greece. *Journal of Spatial and Organizational Dynamics, 5*(1), Article 1.

Erdogan, M., & Ok, A. (2011). An assessment of Turkish young pupils' environmental literacy: A nationwide survey. *International Journal of Science Education, 33*(17), 2375–2406. https://doi.org/10.1080/09500693.2010.550653

Fang, W.-T., Hassan, A., & LePage, B. A. (2023). Environmental literacy. In W.-T. Fang, A. Hassan, & B. A. LePage (Eds.), *The living environmental education: Sound science toward a cleaner, safer, and healthier future* (pp. 93–126). Springer Nature. https://doi.org/10.1007/978-981-19-4234-1_4

Fetzel, T., Petridis, P., Noll, D., Singh, S. J., & Fischer-Kowalski, M. (2018). Reaching a socio-ecological tipping point: Overgrazing on the Greek island of Samothraki and the role of European agricultural policies. *Land Use Policy, 76*, 21–28. https://doi.org/10.1016/j.landusepol.2018.04.042

Gangwani, K., & Landin, J. (2018). The decline of insect representation in biology textbooks over time. *American Entomologist, 64*(4), 252–257. https://doi.org/10.1093/ae/tmy064

Gaster, B. (1990). Assimilation of scientific change: The introduction of molecular genetics into biology textbooks. *Social Studies of Science, 20*(3), 431–454. https://doi.org/10.1177/030631290020003002

Gisev, N., Bell, J. S., & Chen, T. F. (2013). Interrater agreement and interrater reliability: Key concepts, approaches, and applications. *Research in Social and Administrative Pharmacy, 9*(3), 330–338. https://doi.org/10.1016/j.sapharm.2012.04.004

Gkelti-Douka, E., Patargias, T. A., & Argyris, I. (1982). *Biology for the 3rd grade of gymnasium*. School Book Publishing Organization (OEDB).

Gunderson, L. H., Allen, C. R., & Holling, C. S. (Eds.). (2010). *Foundations of ecological resilience*. Island Press.

Hollweg, K. S., Taylor, J. R., Bybee, R. W., Marcinkowski, T. J., McBeth, W. C., & Zoido, P. (2011). *Developing a framework for assessing environmental literacy*. North American Association for Environmental Education.

Huang, R., Tlili, A., Zhang, X., Sun, T., Wang, J., Sharma, R. C., Affouneh, S., Salha, S., Altinay, F., Altinay, Z., Olivier, J., Jemni, M., Wang, Y., Zhao, J., & Burgos, D. (2022). A comprehensive framework for comparing textbooks: Insights from the literature and experts. *Sustainability, 14*(11), 6940. https://doi.org/10.3390/su14116940

Isaksen, M., Ødegaard, M., & Utsi, T.A. (2024). Science textbooks: Aids or obstacles to inquiry teaching? Science teachers' experiences in Norwegian secondary schools. *Science & Education*. https://doi.org/10.1007/s11191-023-00492-x

Kairis, O., Karavitis, C., Salvati, L., Kounalaki, A., & Kosmas, K. (2015). Exploring the impact of overgrazing on soil erosion and land degradation in a dry mediterranean agro-forest landscape (Crete, Greece). *Arid Land Research and Management, 29*(3), 360–374. https://doi.org/10.1080/15324982.2014.968691

Korfiatis, K. (2018). Ecology. In K. Kampourakis & M. Reiss (Eds.), *Teaching biology in schools: Global research, issues, and trends* (pp. 153–263). Routledge.

Krimpas, K., & Kalopisis, I. (1977). *General biology lessons for the 3rd grade of lyceum*. School Book Publishing Organization (OEDB).

Ladle, R. J., & Gillson, L. (2009). The (im)balance of nature: A public perception time-lag? *Public Understanding of Science, 18*(2), 229–242. https://doi.org/10.1177/0963662507082893

Mavrikaki, E., Gouvra, M., & Kampouri, A. (2013). *Biology for the 2nd and 3rd grade of gymnasium*. CTI Diophantus.

McBride, B. B., Brewer, C. A., Berkowitz, A. R., & Borrie, W. T. (2013). Environmental literacy, ecological literacy, ecoliteracy: What do we mean and how did we get here? *Ecosphere*, *4*(5), art67. https://doi.org/10.1890/ES13-00075.1

Negev, M., Sagy, G., Garb, Y., Salzberg, A., & Tal, A. (2008). Evaluating the environmental literacy of Israeli elementary and high school students. *The Journal of Environmental Education, 39*(2), 3–20. https://doi.org/10.3200/JOEE.39.2.3-20

Oikonomidis, I. (1970). *General biology courses for the 6th grade of gymnasium*. School Book Publishing Organization (OEDB).

Okeeffe, L. (2013). A framework for textbook analysis. *International Review of Contemporary Learning Research, 2*(1), 1–13.

Pang, Y.-Z., Li, L., & Jin, Z. (2025). Early warning signals of critical transitions in ecosystems: Entropy reduction in vegetation spatial patterns. *Nonlinear Dynamics*. https://doi.org/10.1007/s11071-025-10877-y

Papageorgiou, K., & Vogiatzakis, I. N. (2006). Nature protection in Greece: An appraisal of the factors shaping integrative conservation and policy effectiveness. *Environmental Science & Policy, 9*(5), 476–486. https://doi.org/10.1016/j.envsci.2006.03.003

Roth, C. E. (1968). On the road to conservation. *Massachusetts Audubon, 52*(4), 38–41.

Scheffer, M. (2009). *Critical transitions in nature and society*. Princeton University Press.

Şeker, M. (2024). A study on how environmental issues are discussed in social studies textbooks. *Environment, Development and Sustainability, 26*(8), 21325–21352. https://doi.org/10.1007/s10668-023-03532-2

Thiele, L. P. (2024). *Sustainability* (3rd ed.). Polity Press.

Tzaberis, N., Xanthakou, P., Papavasileiou, V., Matzanos, D., Hatzidiakos, D., & Papadomarkakis, I. (2012). A comparative investigation of the knowledge and attitudes of high school graduates in the topic of protected areas. *Procedia—Social and Behavioral Sciences, 69*, 404–413. https://doi.org/10.1016/j.sbspro.2012.11.427

Westra, R. (2008). *Learning and teaching ecosystem behaviour in secondary education*. Faculteit Betawetenschappen.

Yalçınkaya-Önder, E., Zorluoğlu, S. L., Timur, B., Timur, S., Güvenç, E., Özergun, I., & Özdemir, M. (2022). Investigation of science textbooks in terms of science process skills. *International Journal of Contemporary Educational Research*, 9 (2), 432–449. https://doi.org/10.33200/ijcer.1031338

Chapter 6
Save the Yellow-Bellied Toad (*Bombina variegata*)—The Impact of a Conservation Program on Students' Motivation and Relationship to Nature

Dorothee Benkowitz and Holger Weitzel

6.1 Introduction

The decline in biodiversity often goes unnoticed, and efforts to stop this trend have fallen short. Over a quarter of the species on the International Union for Conservation of Nature (IUCN) Red List are threatened with extinction. More than 44,000 species are officially listed, including 41% amphibians. Global human activities have caused biodiversity to decline rapidly over the past decades. The 2020 targets for halting biodiversity loss have not been met (United Nations, 2021). Amphibians are among the most affected groups on the UN Red List—especially the yellow-bellied toad (*Bombina variegata*). Due to river regulation, land reclamation, and climate-induced droughts, their habitats decreased. Paradoxically, human activities, e.g., forestry forwarders leaving trails that convert into ponds, foster habitats. This aspect of conservation highlights the complex nature of preservation.

Bombina variegata is a pioneering species native to Europe. It colonizes newly formed habitats, including those created by flooding in river and floodplain systems, open areas resulting from wind-throw, and areas affected by livestock in wetlands. The reproduction of the yellow-bellied toad is contingent upon the presence of small, newly formed water bodies, which provide a refuge from predators. However, floodplain systems with dynamic watercourses are uncommon in Germany, necessitating

D. Benkowitz (✉)
University of Education Karlsruhe, Karlsruhe, Germany
e-mail: benkowitz@ph-karlsruhe.de

H. Weitzel
University of Education Weingarten, Weingarten, Germany
e-mail: weitzel@ph-weingarten.de

O. Morin et al. (eds.), *Challenges of a Changing World in Biology Education*, Contributions from Biology Education Research,
https://doi.org/10.1007/978-3-032-05346-6_6

these animals' reliance on alternative habitats. Substitute habitats for the yellow-bellied toad include gravel, clay and sand pits, and quarries with temporarily unused post-mining areas (Gollmann & Gollmann, 2002). Their principal habitat is forests, wherein forestry operations predominantly shape habitat dynamics. Such activities entail the use of heavy machinery for the harvesting of trees and the disruption of soil (DBU, 2022).

Bombina variegata is included in Annexes II and IV of the Habitats Directive and is afforded strict protection under Germany's Federal Nature Conservation Act (LUBW, 2019). The species' current conservation status is unfavorable and inadequate regarding population, habitat, and prospects.

The yellow-bellied toad has the potential to serve as an appropriate subject for local conservation efforts and educational projects for students because its requirements for breeding waters are minimal. Students can create spawning ponds using spades in approximately two to three hours. Hence, a school class can establish a network of up to ten spawning ponds within a school morning. As the spawning grounds are not particularly large, maintenance measures with schoolchildren can already make an important contribution to protecting the species. This can ensure the importance of local nature conservation measures in schools, where teachers must address sustainable development issues.

Education for Sustainable Development (ESD) aims to equip individuals with the skills to actively shape the future, focusing on climate and environmental protection, resource conservation, and biodiversity preservation, among others (Kropp, 2019). Addressing environmental and developmental challenges requires more than just knowledge—it necessitates imparting skills that enable learners to translate their understanding of sustainable development into action (ibid.). In order to acquire sustainable skills in the sense of "Gestaltungskompetenz", the project can focus on the following goals: Thinking and acting proactively, gaining and acting on interdisciplinary insights, being able to plan and act together with others, participating in decision-making processes, motivating oneself and others to take action, reflecting on one's own and others' guiding principles, showing empathy and solidarity for the disadvantaged—in this case endangered species (de Haan, 2008).

The yellow-bellied toad project can be an excellent topic to discuss the ambivalence of nature conservation: Harvesters leave profound tracks in the forest floor, which can be used as spawning pools by toads. The tracks of the harvesting machines compact the soil and make it difficult to recolonize, and they are perceived by visitors as destruction of the forest. Foresters confront a multifaceted challenge, encompassing the imperative to facilitate economic utilization while minimizing forest degradation, the need to communicate forest utilization to visitors, and the preservation of biodiversity within forest ecosystems. Consequently, the issue of filling these tracks with gravel must be addressed through a deliberative process, wherein compromises must be sought to ensure the harmonious coexistence of various ecological interests. The multiple dilemmas faced by foresters are thus an excellent example of how to approach the problems of sustainable development in schools. Teachers face the challenge of addressing these complex issues while allowing students to develop sustainable competencies: On the basis of self-collected data they plan and execute actions to save an endangered species. This experience of self-efficacy plays a crucial role in initiating environmental action.

Schools represent a suitable starting point for the dissemination of knowledge pertaining to species and ready-to-use skills. Being part of an action oriented outdoor project leads to meaningful learning and fosters intrinsic motivation (Wilde et al., 2009). According to Deci and Ryan (2003), meeting the *basic needs* is considered a predictor for the development of internal motivation, an important factor for the development of personal interest.

Randler (2008) pointed out the importance of teaching species literacy as the basis of ecological understanding. The outdoor setting should be familiar to the students so that the novelty effect does not prevent learning (ibid.). Research highlights the importance of connecting students with nature to enhance their sense of belonging to nature and motivate proactive conservation efforts (Bogner, 2018). Lessons outside the traditional classroom encourage students to explore and discover on their own (Favre & Metzger, 2019; Köhler & Benkowitz, 2014). For example, under supervision, learners can observe living animals hands-on. Experiential learning and direct instruction have been shown to be effective in extracurricular settings (Favre & Metzger, 2019). Specifically, field trips that are integrated into a broader unit of study result in more significant knowledge gains than those that are not (ibid.). Additionally, students benefit from multiple visits to outdoor learning sites and opportunities for individualized learning experiences. Knowledge retention is also positively influenced when there is a follow-up lesson after the visit (ibid.).

Providing students access to authentic experiences with biodiversity will not only contribute to individual development but pave the way for an enduring connection between future generations and the natural world (Gebhard, 2020; Louv, 2008). This will promote the consciousness of, and responsibility towards, the ecological challenges facing our urbanized societies. Biodiversity plays a crucial role in the provision of ecosystem services that are essential for human well-being. Therefore, a more effective education in species literacy would not only meet one of the goals set out in the convention on biological diversity but also prepare students for the changes needed for sustainability (CBD, 2020).

The current disconnects between individuals, especially in highly industrialized societies, and the natural world emphasizes the need for direct interaction with plants and animals (Pilgrim et al., 2008). Especially young people, who, in the absence of direct contact with nature, struggle to develop a meaningful affinity towards it should be promoted e.g. by natural learning settings. Studies showed that students engaged in natural environments demonstrate an increased emotional connection to nature and inclination to support initiatives aimed at nature protection (Köhler & Benkowitz, 2014; Sellmann & Bogner, 2012; Soga et al., 2016). The degree in connectedness with nature serves as a positive predictor for environmental attitudes (Sellmann & Bogner, 2012).

Given the state of the research, we focused on the following questions:

1. To what extent does a local nature conservation project, which presents students with a complex dilemma situation, contribute to the enhancement of their

- knowledge,
- intrinsic motivation, and
- appreciation of nature?

2. What are the teachers' motives to take part in a nature conservation project?
3. What preparation do they need to participate in a nature conservation project?

6.2 Research Design and Method

6.2.1 Participants

The selection of project schools was based on the presence of toad populations in the vicinity, with the objective of facilitating access during the morning hours allocated for school activities. A request was extended to teachers at the selected schools to participate, on the condition that the subject of "amphibians and reptiles" was included in the curriculum of the grade.

Seventy-four students (47 males, 26 females, one diverse, aged 10 to 16 years) from six schools participated in the quantitative survey. Overall, 25 students (16 males, nine females, aged 10 to 16 years) from these six schools (four students from five schools each and five students from one school) participated in guided, problem-centered interviews. The students were selected based on their complete involvement in the intervention.

Furthermore, efforts were made to ensure the participants represented diverse attitudes towards nature. The selection of students was delegated to the participating teachers, as they were presumed to possess the most comprehensive understanding of the student's abilities and their active engagement in the practical conservation work.

To explore the teachers' motivations and affordances, one teacher from each school was asked for an interview. All interviewed teachers had participated in the project with their class. The teachers' professional experience ranged from three to 20 years.

6.2.2 Conservation Program

In preparation of the project, permits had to be obtained from the relevant nature conservation authority. To find suitable schools, it was important that these were located near *Bombina* habitats. The participating teachers went through a professional development program prior to the implementation of the project. The teacher training comprised three phases. In phase 1, the teachers acquired the scientific fundamentals regarding the toad that were essential for the project (habitat, occurrence, conservation status, etc.). In the subsequent phase, the rationale and methodology

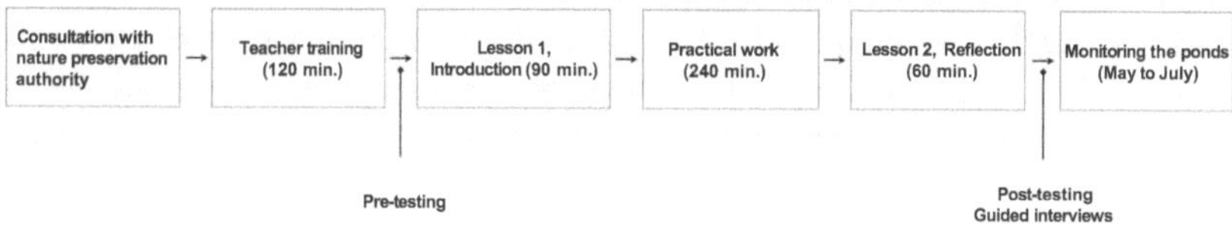

Fig. 6.1 Structure of the "Bombina" intervention

behind the utilisation of the yellow-bellied toad as a model to elucidate the intricacies of nature conservation measures and the challenges confronting nature conservation endeavours were elucidated. The third phase entailed a discussion on the project's course and the prerequisites for its successful execution. Additionally, a network of educators involved in the training programme was established, and the scheduling of subsequent activities was deliberated.

The theoretical and practical parts of the school project were divided (Fig. 6.1). The project began with a 90-min introduction to the topic. This theory lesson aimed to provide students with all relevant information about the strictly protected amphibians and ensure that the animals would not be endangered through the project. Students did some online research on the topic and used presentations and posters to inform each other about appearance and characteristics, development and migration, food and predators, habitat and distribution, threats, and protective measures. This was followed by a practical day, during which a field-trip was made to a nearby yellow-bellied toad population. Here, the students created small ponds which, when filled with rainwater and/or groundwater, could be used by the yellow-bellied toads to lay their eggs and connect habitats. The final step involved reflection on the practical activity conducted on-site and its contextualization within the framework of sustainable education.

Students from our university regularly monitored the ponds after the project was completed. Yellow-bellied toad spawns, larvae, and adults were counted, and the condition of the ponds documented. In addition, pH, light, and temperature measurements were taken in the water-filled ponds. This data and photos of the ponds were made available to the schools via a padlet. Schools were invited to regularly update the data of the development of the yellow-bellied toads and to upload new photos. Given the proximity of the ponds to the students' residential areas, some students also visited the ponds, facilitating further contact with the animals.

6.2.3 Data Gathering Tools

To investigate the research questions, we used an approach combining quantitative and qualitative instruments. A questionnaire contained 16 items in the pretest and 28 in the posttest. It included a multiple-choice test on knowledge about *Bombina variegata*. The questions were derived from the lessons and the outdoor activities. To investigate the appreciation of nature supported by the project we used the 2-MEV scale (Bogner, 2018), adapted by checking the items with Cronbach's alpha to ensure reliability. Additionally, in the posttest, the short intrinsic motivation scale (Wilde et al., 2009) was used, a shortened and time-efficient version of Deci & Ryan's Intrinsic Motivation Inventory (IMI, 2003). The questionnaire consisted of closed questions with predetermined response options, and 5-point Likert scales were used to answer the scales (Appendix).

The second measurement tool was a short semi-structured guideline interview to address two different areas:

1. Knowledge and knowledge gain about the yellow-bellied toad and its reproductive requirements as a pioneer species.
2. Attitudes toward Education for Sustainable Development (ESD) and its concretization through the experienced intervention.

The interview guidelines for the students were divided into the following subareas: perception of the project, knowledge about the yellow-bellied toad and conservation efforts, knowledge about and participation in local conservation efforts, perception of personal learning gains. The teacher interview guide focused on the same areas. In addition, we asked the teachers about the potential and feasibility of integrating the project into the regular curriculum.

The interviews with the students were conducted in person during school hours. Separate rooms were available for this purpose in the school buildings to create an atmosphere conducive to relaxed and confidential interviews. The interviews with the students lasted up to 12 min, the interviews with the teachers up to 15 min.

6.2.4 Data Analysis

The quantitative data were analyzed descriptively and inferentially using the statistical program SPSS (version 27). The qualitative data were examined through a content-structuring qualitative analysis (Kuckartz, 2018) using the software MAXQDA (2020). A deductive category system was developed for the evaluation. This system comprised the following main codes: knowledge about the yellow-bellied toad, ideas about nature conservation (e.g., its individual relevance, thematic areas, knowledge about and importance of local nature conservation measures), challenges in the project, potential of the project, and further comments on the project.

6.3 Results

The aim of the project was to investigate whether participation in a nature conservation project influences students' knowledge about yellow-bellied toads, their intrinsic motivation to participate in a nature conservation project, and their appreciation of nature. We also asked what motives the teachers have for participating in the project, what they need in preparation for the project and what potential the project has for teaching ESD.

6.3.1 Students' Knowledge and Intrinsic Motivation

The intervention significantly increased the participants' knowledge (Fig. 6.2), indicating a large effect size between the pretest and posttest ($t(74) = 10.33$, $p < 0.001$, $d = 1.19$). In addition, the school that students attended influenced the extent of knowledge improvement ($F_{1,69} = 2.966$; $p < 0.026$). One school where toads were found during the intervention experienced a greater increase in student knowledge, highlighting the positive impact of hands-on encounters with the species.

The project positively impacted the basic psychological needs of perceived choice (autonomy) and competence, which are predictors of increased interest (as shown in Table 6.1). The sense of felt pressure scored low.

Interest varied greatly between the different grades (Table 6.2), with sixth grade students showing more interest and enjoyment in the project than their ninth grade peers ($F_{1,72} = 5.124$, $p < 0.027$). Competence and perceived autonomy positively correlated with the knowledge gained during the intervention ($p < 0.001$).

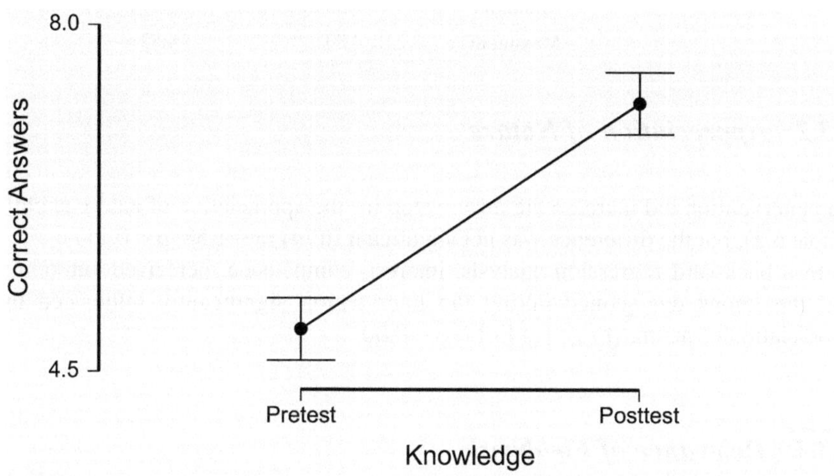

Fig. 6.2 Knowledge gained between pretest and posttest. Students answered a multiple-choice test consisting of ten items, where they selected the correct answers, n = 74

Table 6.1 Responses to the items measuring interest, autonomy, competence, and pressure, rated on a 5-point Likert scale, n = 74

	Interest	Competence	Autonomy	Pressure
Mean	3.734	4.018	3.595	2.032
Std. deviation	0.973	0.691	1.015	0.984
Minimum	1.000	2.667	1.000	1.000
Maximum	5.000	5.000	5.000	5.000

Table 6.2 Comparison of the answers on interest between grade 6 and grade 9, n = 74

	Grade 6	Grade 9
Valid	43	31
Mean	3.946	3.441
Std. deviation	0.945	0.948
Minimum	1.333	1.000
Maximum	5.000	5.000

Table 6.3 Appreciation of nature (APR) on a 5-point-Likert-scale in comparison from pretest to posttest, n = 74

	APR-pretest	APR-posttest
Mean	2.84	2.94
Std. deviation	0.84	0.94
Minimum	1.33	1.17
Maximum	4.83	4.83

6.3.2 Appreciation of Nature

The intervention did increase the mean value for the appreciation of nature slightly (Table 6.3), but the difference was not significant ($t(74) = -1.58$, $p < 0.117$).

In a backward regression analysis, interest, competence, perceived autonomy, and the knowledge gained during the intervention significantly influenced the appreciation of nature ($F_{1,72} = 143.12$, $p < 0.001$).

6.3.3 Relevance of Fieldwork

The interviews with the students showed that knowledge about yellow-bellied toad and their habitat was a particularly salient topic for them. A content analysis of the interviews revealed that the subjects "habitat of the toad" (number of mentions = 13), "approach to creating spawning waters" (11), and "endangerment of the animals" (8) were the most frequently mentioned aspects. Following the fieldwork, the students were able to recall specific details related to the creation of breeding ponds:

> The yellow-bellied toad requires a recently excavated pond that retains water for at least eight weeks. The pond should be situated in an area of direct sunlight rather than in the shade. The addition of branches within the pond allows the toads to attach their eggs securely. (S17)

> The depth of the pond should be between 10 and 20 cm, and the toads should have access to sticks to attach their eggs. (S25)

Most of the interviewed students had no prior experience with nature conservation measures before the commencement of the project (19 out of 25). Ten of the students were surprised to discover that nature conservation measures do not require significant effort:

> I was surprised to learn that even minor contributions can have a significant impact. It was remarkable to observe that our efforts were limited to excavation, yet we were able to create a valuable habitat for local fauna. (S7)

Overall, eight students (Interview number (IN) 4–7, 12, 13, 18, 20, 22) were so impressed by the simplicity of the action that they expressed a desire to alter their previous actions:

> We should protect the habitat of the yellow-bellied toad; we should simply take more interest in the topic because it is also important for our ecosystem. (S6)

In total eight students' (IN 4–9, 13, 14, 15) answers suggest that they value the conservation of the natural environment. However, they need a clearer understanding of the specific opportunities to get involved locally. One student's statement illustrates this perspective.

> I had assumed that undertaking such an initiative would necessitate travel to a more distant location. I was previously unaware of the presence of yellow-bellied toads in our vicinity, which piqued my interest. (S13)

The opportunity to interact with living animals proved to be a particularly motivating factor for the students (see Sect. 6.3.1), which helped to increase their motivation to engage in the physically demanding task of digging the ponds:

> It was a highly beneficial experience to observe the yellow-bellied toads and to recognize the potential of our assistance through the construction of ponds that would provide suitable habitat for them. (S22)

In total, 13 of the students (IN 4–8, 10, 12–14, 18, 20, 22–25) were mainly motivated by the practical work on-site in the forest, which involved the digging of ponds:

> The act of digging was beneficial and enjoyable, especially in terms of creating habitats. (S12)

> Furthermore, the process of creating ponds was especially gratifying when undertaken as a collective effort. (S13)

Only two students indicated that the autonomy afforded by independent group formation for challenging tasks contributed to their overall experience of autonomy:

> We were permitted to select our own groups, which enhanced the experience, even though it was physically demanding. (S7)

6.3.4 Teachers' Motivations and Project's Potential

The teachers articulated three primary motivations for their participation: their intrinsic motivation for the project's subject matter, their desire to provide students with tangible experiences of the natural world, and their confidence in the project team's expertise and meticulous organization. Similarly, the teachers were astonished to observe the yellow-bellied toads, thereby gaining first-hand insight into the practical value of their endeavors. All the teachers interviewed expressed admiration for the students' enthusiasm throughout the intervention, but especially during the fieldwork phase:

> On the day we venture into the woods, the physical exertion is considerable, and I was surprised at how much the class valued the experience and how much they enjoyed it. (L1)

The teachers were particularly impressed by the commitment of the more vulnerable students to the fieldwork, who often present behavioral challenges in the classroom:

> The students I find difficult in the classroom and the less able students were exemplary in this project, showing a side of themselves that I had not observed before. (L5)

In addition, we asked the teachers what benefits they derived from the project for their work in class. Four categories were identified: The teachers expressed frustration with the organizational burden of coordinating with the nature conservation authorities, foresters, and forest owners, which was a prerequisite for implementing a relatively straightforward nature conservation measure (1). For teachers, the application process for the conservation measure is a major obstacle to repeating the project. Secondly, the teachers expressed gratitude for the supplementary biological training by the project team (2), which gave them the necessary knowledge to develop the confidence to create spawning waters independently of the project team (3). All teachers were interested in repeating the project and expressed that they would like to integrate further nature conservation projects into their lessons (4). However, they also make it dependent on the resources made available, such as help obtaining applications.

Contextualizing the project within the Education for Sustainable Development (ESD) framework was a pivotal concern for the teachers. All teachers consider ESD to be a significant overarching topic in biology. However, they perceive challenges in teaching ESD, which pertain to the organizational effort and time required for the lessons and the content presented in often-used teaching materials, predominantly limited to topics such as energy conservation and waste separation. One teacher acknowledges a personal lack of experience with different methods to transmit ESD to their students.

Considering this rationale, the teachers identify multiple benefits associated with the project: The preliminary work with the nature conservation authorities by the supporting university provides a clear structure and organizational relief, allowing teachers to focus on the lesson content. The teachers perceive the fieldwork and its subsequent reflection to have the potential to bring about a change of perspective on nature in the students:

> The fieldwork at the very least engenders an awareness of perceiving things with greater consciousness, perhaps even with different eyes, on the subsequent visit to the natural environment. (L1)

The teachers consider the opportunity to gain insight into the interconnectedness of ecosystems as another benefit of the project. The yellow-bellied toad, which serves as a pioneer species, is dependent on the destruction of a supposed balance in nature:

> It is important to see an ecosystem as a whole and not just, as you hear everywhere, to plant more trees.... The project has contributed to this. (L5)

6.4 Discussion

Regarding research question 1, we showed that students acquire relevant knowledge: The lessons and the field trip in the forest led to higher scores in the knowledge test of the post-questionnaire. The results show a substantial improvement in understanding, reflecting the effectiveness of the intervention in enhancing students' knowledge. Specifically, the one school where toads were found during the intervention experienced a more significant knowledge increase among students, highlighting the positive impact of hands-on encounters with the species. Boeve-de Pauw et al. (2018) report a non-linear relationship between novelty and learning gains. A little novelty has a positive effect on learning growth, while too much novelty can inhibit learning. For this project, we can state that the yellow-bellied toad and the fieldwork were new to most students. Neither the quantitative data nor the interview results indicate that the encounter with the animal and the nature conservation measure could have negatively impacted the students' knowledge increase. It seems reasonable to conclude that the observed increase in knowledge is associated with the opportunity for practical involvement, as the students frequently mention both aspects together. However, we do not know whether the increase in knowledge is due to the classroom, the fieldwork, the reflection, or a combination of all of these, and whether the experiences in the forest are critical to the increase in knowledge. Seeing real yellow-bellied toads in the pond significantly increased students' knowledge. This first-hand experience allowed students to connect theory with reality, deepening their understanding of the species and its habitat. Such encounters are critical in fostering a greater appreciation for biodiversity and the importance of conservation efforts. Students realize (often for the first time) that conservation is a local issue.

The study indicates that students who feel a stronger connection to nature also exhibit greater interest and enjoyment in the yellow-bellied toad project. Those who value nature and feel connected to it are more inclined to engage actively in the project. Interest serves as a self-report measure of intrinsic motivation per se (Deci & Ryan, 2003), highlighting the importance of fostering a connection to the natural environment to enhance students' engagement and motivation in educational initiatives.

Interest, competence, perceived autonomy, and the knowledge gained in the project influenced the appreciation of nature. These factors collectively contribute to how students value and engage with their natural environment, highlighting the importance of fostering these attributes in educational settings to enhance environmental appreciation.

Importantly, the results indicated no significant differences between boys and girls regarding their engagement and outcomes in the project. This suggests that both genders benefited equally from the intervention, highlighting the program's effectiveness in fostering interest and knowledge about the yellow-bellied toad and conservation efforts among all students.

The motivations expressed by the teachers participating in the project—such as intrinsic interest, providing tangible experiences, and confidence in the project's organization—are consistent with the findings of Eliyawati et al. (2023), who emphasize the necessity of teacher training programs to integrate sustainability into both knowledge and practice. The teachers' observations of heightened student enthusiasm and engagement during fieldwork, particularly among students who typically encounter difficulties in traditional classroom settings, are consistent with research indicating that experiential learning can enhance student engagement and reveal hitherto unidentified strengths.

The project is a powerful example of the ESD slogan "act locally, think globally". By involving students in local conservation efforts, such as protecting the yellow-bellied toad, the project helps them understand the broader global importance of biodiversity and environmental protection. It emphasizes that small, local actions can contribute to global sustainability goals and fosters a sense of responsibility and empowerment in students to make a difference in their communities while considering the global impact of their efforts.

However, organizational challenges remain, as evidenced by teachers' frustration with coordinating conservation efforts. This is consistent with broader findings that highlight the need for structured support to alleviate these burdens. The structured support provided by the project facilitated a focus on teaching content.

6.4.1 Limitations

This study provides insights into the impact of a hands-on conservation programme on students' knowledge, intrinsic motivation, and appreciation of nature. However, it is important to consider the study's limitations. Firstly, the sample of students ($n = 74$) was relatively small and not randomly selected. Participants were drawn from six schools located near *Bombina variegata* habitats, and student selection for the interviews was delegated to the participating teachers based on their involvement in the intervention. While this purposive sampling approach allowed for the focus on students actively engaged in the conservation activities, it limits the generalizability of the findings to broader student populations. In order to achieve this objective, the study is currently being expanded to include a larger sample. Second, although teachers received professional development prior to the implementation of the project, this training was relatively short-term and consisted of a single afternoon activity. While teachers received ongoing organizational support from the university, sustained and repeated engagement in ESD initiatives is known to be more effective in increasing teachers' self-efficacy and integrating ESD principles into the curriculum (Boewe-de Pauw et al., 2022). Further research could explore the impact of more intensive and long-term teacher training programs on the sustainability of such conservation efforts in schools.

Finally, while the practical implementation of the project was successful, it required significant planning effort, coordination, and logistical support, highlighting the organisational and structural challenges associated with integrating such initiatives into the regular school curriculum, which may limit their scalability and long-term implementation without dedicated resources.

6.5 Conclusion

The local conservation project has successfully increased students' knowledge and motivation to become aware of endangered species. Many students learned for the first time about accessible nature conservation measures. Additionally, students took the initiative to visit the ponds with their parents to showcase the yellow-bellied toads (*Bombina variegata*). The project inspired teachers to incorporate conservation activities into their regular lessons, further promoting environmental awareness and action within the educational framework.

The teachers involved in the project were highly motivated and recognized the local nature conservation initiative as a valuable topic for teaching ESD competences. They saw the project as an opportunity to engage students in meaningful learning experiences that promote environmental awareness and sustainability.

The project provides an excellent opportunity to involve all stakeholders. A significant number of spawning areas for the yellow-bellied toad can be established quickly. This initiative helps to highlight conservation as a vital responsibility of civil society, and promotes community involvement and awareness of environmental protection efforts.

Acknowledgements We would like to extend our heartfelt thanks to all our students who collaborated with us on this project: Annemarie Eberhardt, Cathrin Hummel, Maximilian Ege, Karen Müller, and all that participated in our seminars. Together, we have taken meaningful steps toward protecting our environment.

The project was funded by Stiftung Naturschutzfonds, funding code SNR-8831.21/54691-2019G.

Appendix

Students Protect Yellow-Bellied Toads—Questionnaire

Knowledge test: One answer is correct for each question. Mark it with a cross.

1. How can different individuals of toads be distinguished from each other?

   By the warts on their upper side.
   The black and yellow pattern on the underside.
   By their calls.

2. In which way are the pupils of the yellow-bellied toad remarkable?

   They are heart-shaped.
   They are slit-shaped.
   They are star-shaped.

3. What are the original habitats of the yellow-bellied toad?

   Floodplains
   Lakes
   Marshland

4. Where do yellow-bellied toads hide?

   They hide under dead wood in the forest.
   They burrow into the ground a bit.
   They dive below the surface in the water.

5. What can yellow-bellied toads secrete through their skin when in danger?

   They secrete a foamy mucus.
   They secrete foul-smelling odors.
   They secrete a poison.

6. Why do male yellow-bellied toads call "uuh, uuh, uuh" during spawning season?

   They want to indicate where their territory is.
   They want to attract females who are ready to mate.
   They want to scare away other males.

7. Why are yellow-bellied toads highly endangered in Germany?

   They find no water to lay eggs.
   They find no winter hiding places.
   They find no partners.

8. Why should you put branches in the freshly dug ponds?

   That the toads can climb out of the pond more easily.
   That the toads can attach their eggs there.

That the toads can hide underneath.

9. What are the prerequisites for yellow-bellied toads to settle in constructed ponds?

    The ponds should be in the shade.
    The ponds should only be between 10 and 20 cm deep.
    The ponds should be exposed to sunlight.

10. The belly of the yellow-bellied toad is bright yellow-black. Why?

    The coloring is intended to deter potential predators.
    The coloring is intended to attract potential prey.
    The coloring is intended to attract females.

Motivation: How did you like participating in the yellow-bellied toad project?

Mark the scale: 1 = not true at all, 2 = not very true, 3 = partly true, 4 = fairly true, 5 = completely true

1. I enjoyed working on the yellow-bellied toad project.
2. I found working in the yellow-bellied toad project very interesting.
3. Working on the yellow-bellied toad project was entertaining.
4. I am satisfied with my performance in the yellow-bellied toad project.
5. I performed well in the yellow-bellied toad project.
6. I think I performed quite well in the yellow-bellied toad project.
7. I was able to control my activities in the yellow-bellied toad project.
8. During the activities in the yellow-bellied toad project, I was able to choose how I did them.
9. During the activities in the yellow-bellied toad project, I was able to do them the way I wanted.
10. During the activities in the yellow-bellied toad project, I felt pressured.
11. During the activities in the yellow-bellied toad project, I felt tense.
12. I was concerned about whether I could complete the tasks in the yellow-bellied toad project.

Education for Sustainable Development: What is your attitude towards the environment and nature in general? Adaptation of the 2-MEV-Scale (Bogner, 2018)

Mark the scale: 1 = not true at all, 2 = not very true, 3 = partly true, 4 = fairly true, 5 = completely true

1. I consciously watch for or listen to birds.
2. I take time to watch the clouds as they pass by.
3. I consciously take time to look at the stars at night.
4. I enjoy gardening.
5. I take care of plants myself.
6. Listening to the sounds of nature relaxes me.

Thank you for participating!

References

Boeve-de Pauw, J., Van Hoof, J., & Van Petegem, P. (2018). Effective field trips in nature: The interplay between novelty and learning. *Journal of Biological Education, 53*(1), 21–33. https://doi.org/10.1080/00219266.2017.1418760

Boeve-de Pauw, J., Olsson, D., Berglund, T., & Gericke, N. (2022). Teachers' ESD self-efficacy and practices: A longitudinal study on the impact of teacher professional development. *Environmental Education Research, 28*(6), 867–885. https://doi.org/10.1080/13504622.2022.2042206

Bogner, F. (2018). Environmental values (2-MEV) and appreciation of nature. *Sustainability, 10*(2). https://doi.org/10.3390/su10020350

CBD—Secretary of the Convention on Biologocal Diversity. (2020). *Sustaining live on earth.* https://www.cbd.int/doc/publications/cbd-sustain-en.pdf

DBU—Deutsche Bundesstiftung Umwelt. (2022). Entwicklung nachhaltiger Schutzkonzepte für die Gelbbauchunke (*Bombina variegata L.*) in Wirtschaftswäldern als Leitfaden zum angewandten Gelbbauchunkenschutz in der Forstwirtschaft. https://www.unkenschutz-bw.de/

Deci, E. L., & Ryan, R. M. (2003). Intrinsic motivation inventory. Self-determination theory, 267. https://selfdeterminationtheory.org/wp-content/uploads/2022/02/IMI_Complete.pdf

Eliyawati, E., Widodo, A., Kaniawati, I., & Fujii, H. (2023). The effectiveness of teacher training on environmental education: Challenges and strategy for future training program. *Jurnal Penelitian Pendidikan IPA, 9*(8), 6056–6066. https://doi.org/10.29303/jppipa.v9i8.3153

Favre, P., Metzger, S. (2019). Außerschulische Lernorte nutzen. In P. Labudde & S. Metzger (Eds.), UTB (3248) *Pädagogik. Fachdidaktik Naturwissenschaft* (3rd ed., pp. 167–182). Haupt Verlag

Gebhard, U. (2020). *Kind und Natur. Die Bedeutung der Natur für die psychische Entwicklung.* Springer Fachmedien Wiesebaden. https://doi.org/10.1007/978-3-658-21276-6

Gollmann, B., & Gollmann, G. (2002). Die Gelbbauchunke: Von der Suhle zur Radspur. *Zeitschrift für Feldherpetologie*: Beiheft (4)

Haan, G. de. (2008). Gestaltungskompetenzen als Kompetenzkonzept für Bildung für nachhaltige Entwicklung. In I. Bormann & G. de Haan (Eds.), *Kompetenzen der Bildung für nachhaltige Entwicklung: Operationalisierung, Messung, Rahmenbedingungen, Befunde* (pp. 23–44). VS Verlag für Sozialwissenschaften.

Köhler, K., & Benkowitz, D. (2014). Supporting awareness of urban biodiversity: School garden, schoolyard and school surrounding. In N. Kabisch, N. Larondelle, M. Artmann, & A. Reeve (Eds.), *Human-environmental interactions in cities—Challenges and opportunities of urban land use, planning and green infrastructure* (pp. 100–114). Cambridge Scholars Publishing.

Kropp, A. (2019). *Grundlagen der Nachhaltigen Entwicklung: Handlungsmöglichkeiten und Strategien zur Umsetzung* (1. Aufl.). Springer Fachmedien.

Kuckartz, U. (2018). *Qualitative Inhaltsanalyse. Methoden, Praxis, Computerunterstützung.* 4. Auflage. Beltz Juventa (Grundlagentexte Methoden).

Louv, R. (2008). *Last child in the woods: Saving our children from nature-deficit disorder.* Algonquin Books of Chapel Hill.

LUBW—Landesamt für Umwelt Baden-Württemberg. (2019). FFH-Arten in Baden-Württemberg. Erhaltungszustand 2019 der Arten in Baden-Württemberg. https://www.lubw.baden-wuerttemberg.de/documents/10184/258557/download_ffh_erhaltungszustand_arten_2019.pdf/aa4e94ed-5f90-4629-88ad-b031995b95d1

MAXQDA. (2020). [software]. VERBI Software. https://www.maxqda.com

Pilgrim, S., Cullen, L., Smith, D. J., & Pretty, J. (2008). Ecological knowledge is lost in wealthier communities and countries. *Environmental Science and Technology, 42*(4). https://doi.org/10.1021/es070837v

Randler, C. (2008). Teaching species identification—A prerequisite for learning biodiversity and understanding ecology. *Eurasia Journal of Mathematics, Science and Technology Education, 4*(3), 223–231. https://doi.org/10.12973/ejmste/75344

Sellmann, D., Bogner, F. X. (2012). Effects of a 1-day environmental education intervention on environmental attitudes and connectedness with nature. *European Journal of Psychology of Education, 28*(3), 1077–1086 (2013). https://doi.org/10.1007/s10212-012-0155-0

Soga, M., Gaston, K. J., Yamaura, Y., Kurisu, K., & Hanaki, K. (2016). Both direct and vicarious experiences of nature affect children's willingness to conserve biodiversity. *International Journal of Environmental Research Public Health* (13) 529. https://doi.org/10.3390/ijerph13060529

United Nations. (2021). Ziele für nachhaltige Entwicklung: Bericht 2021. https://www.un.org/depts/german/millennium/SDG%20Bericht%202021.pdf

Wilde, M., Bätz, K., Kovaleva, A., & Urhahne, D. (2009). Überprüfung einer Kurzskala intrinsischer Motivation (KIM). *Zeitschrift Für Didaktik der Naturwissenschaften, 15*, 31–45.

Chapter 7
Exploring the Educational Potential of Climate Change Exhibitions in Natural History Museums

Anna Pshenichny-Mamo, Maggie Demarse, Roberta Howard Hunter, and Dina Tsybulsky

7.1 Introduction

Climate Change (CC) constitutes one of humanity's most complex and multifaceted challenges, and has major implications for our planet and the wellbeing of its inhabitants. As an existential threat, it directly impacts critical areas of human life, including infrastructure, ecosystems, agriculture, health, and social networks, all of which are integral to survival and quality of life (Hamilton & Christian Ronning, 2020). However, this challenge is not solely a scientific issue; it is deeply intertwined with social and cultural dimensions and affects individuals differently based on their ideologies, values, and worldviews (Hulme, 2015). CC, which is classified as a "wicked problem", is characterized by intricate causes, a lack of straightforward solutions, and resistance to traditional problem-solving approaches (Achiam et al., 2021a, 2021b; Pellegrini, 2021). Addressing this issue requires a fine-grained understanding of the problem's complexity and the temporal scales involved (Achiam et al., 2021a, 2021b).

A. Pshenichny-Mamo (✉) · D. Tsybulsky
Faculty of Education in Science and Technology, Technion Israel Institute of Technology, Haifa, Israel
e-mail: panna@campus.technion.ac.il

D. Tsybulsky
e-mail: dinatsy@technion.ac.il

M. Demarse
Michigan State University, East Lansing, USA
e-mail: demarsem@msu.edu

R. H. Hunter
New Jersey Audubon, Bernardsville, USA
e-mail: robertahhunter@gmail.com

O. Morin et al. (eds.), *Challenges of a Changing World in Biology Education*, Contributions from Biology Education Research,
https://doi.org/10.1007/978-3-032-05346-6_7

CC education plays a crucial role in promoting awareness and equipping future generations to address environmental challenges. UNESCO white papers note that CC education fosters understanding, shifts perceptions and behaviors, and can instigate meaningful actions to address CC (UNESCO, n.d.). All types of formal, informal, or non-formal education are essential (Kondyli et al., 2024). Monroe et al. (2019) identified 5 key ways to convey the importance of CC in education: (1) Raising awareness and knowledge, (2) Fostering engagement and participation, (3) Addressing misconceptions, (4) Personal connection and relevance, and (5) Active and experiential learning.

Museums play a significant role in CC education. Museums influence public discourse and education on CC (Hamilton & Christian Ronning, 2020) since people tend to view museums as more reliable sources of information than mainstream media or government entities (Cameron et al., 2013; McGhie, 2020). Museums mediate public dialogue on CC and can encourage collective action by serving as platforms for public discussions, connecting diverse communities, and promoting scientific literacy (Hamilton & Christian Ronning, 2020; McGhie, 2020).

Museum staff plays a pivotal role in creating exhibitions and educational programs, while also facilitating the knowledge to the visitors. Museum staff aim to make science accessible to all, provide engaging and enjoyable learning experiences, and spark curiosity by connecting content to visitors' daily lives and classroom learning (Shaby et al., 2016; Tran, 2007). Some museums have put their beliefs into practice by embracing sustainable practices and green strategies to model environmental leadership to promote CC education (Goodger, 2019; Hamilton & Christian Ronning, 2020). Certain museums have set up collaborative networks to share support, experiences, and ways to develop CC communication and education (Nieroba & Geisler, 2024). Studies have found that educational exhibitions can convey scientific knowledge and create meaningful learning spaces (Kondyli et al., 2024) when they facilitate informal learning that resonates with visitors' emotions and values, and can enhance their understanding and concern for CC issues (Nieroba & Geisler, 2024). Museum staff also aim to provide opportunities for visitors to express their ideas and concerns by promoting discourse and critical thinking about CC (McGhie et al., 2018). By capitalizing on the emotional impact of CC, museums can transform visitors' feelings of despair into hope and action (Sutton, 2021). Exhibitions can emphasize human responsibility for this crisis while reframing the interrelationship between humans and nature as an interconnected system in which humanity is an integral part. Some exhibitions also emphasize the ties between CC and social injustice by showing that environmental effects are unevenly felt among different populations (Nieroba & Geisler, 2024). This positions these museums as agents of change in their communities (Sutton, 2021).

Natural History Museums (NHMs) are uniquely positioned to foster public understanding of CC due to their extensive and impressive collections of authentic natural specimens (Conn, 1998; Reiss et al., 2016). These collections are valuable biological data sources for scientists and educators, making them essential to biology education (King & Achiam, 2017). NHMs provide critical content, practices, and can enable discussions related to natural history by showcasing the historical aspects of nature and the scientific research associated with them (Marandino et al., 2015). Hence, NHMs have become central to learning and engagement in science (Mujtaba et al., 2018). As authentic environments, NHMs can emphasize the scientific process, which can help visitors understand the basis of CC research and policies. By offering an informal and engaging learning environment, NHMs can enhance public engagement with CC while tailoring their messaging to address specific environmental issues to local communities.

Given the unique role of NHMs, Dillon et al. (2016) identified five interconnected and interactive dimensions of wicked problems such as CC. The three core dimensions that set NHMs apart are:

- *Collections:* Utilized by scientists, educators, and the public to enhance knowledge and understanding of the natural world, these collections are valuable resources that reflect the institution's areas of expertise, values, and historical and geographical context. These collections, comprising fossils, plants, animals, and others, provide evidence of the impact of CC on biodiversity, making the subject more relatable and personally significant to visitors (Reiss et al., 2016).
- *Content:* NHMs explicitly connect their collections to natural history topics, including socio-scientific issues, wicked problems, and scientific processes. These connections are particularly important to CC education.
- *Expertise:* NHM experts, including scientists, curators, and educators, have specialized knowledge and skills in content, collections, and teaching. They work collaboratively within institutions or with external partners to engage audiences with the natural world. Some museums, in their capacity as research institutions, conduct research related to CC and can share their knowledge with visitors (e.g., Denney & Anderson, 2020).

These core dimensions are supported by facilitation and outreach to different audiences. Facilitation transforms collections, content, and expertise into tailored learning experiences that cater to the specific needs of diverse audience groups (e.g., through exhibitions, interactive displays, and self-guided activities). Elsewhere, we showed that facilitation in NHMs often taps visitors' cognitive experiences by integrating features related to scientific knowledge (Pshenichny-Mamo & Tsybulsky, 2024, 2025). Audiences encompass distinct groups who engage with museum content in various ways. NHMs not only serve existing visitors but also actively seek to engage new and underrepresented groups, thus broadening access to scientific knowledge and fostering authentic learning environments.

Given NHMs' unique role as informal learning spaces and their potential for addressing CC through exhibitions and educational programs, only scant research has examined the educational goals that NHM staff associate with CC exhibits. Understanding these perspectives is essential for maximizing the impact of CC education in museum settings and strengthening connections to formal science education.

The conceptual map proposed by Dillon et al. (2016) and the museum goals presented above all point to the potential for integrating scientific knowledge and knowledge about science in NHMs. In science education, 'knowledge about science' refers to Nature of Science (NOS), and encompasses fundamental aspects such as its definition, methodologies, scope, and limitations. Each dimension in the conceptual map can explicitly or implicitly incorporate various aspects of NOS.

One of the main frameworks for NOS science education is the Family Resemblance Approach (FRA) (Erduran & Dagher, 2014). The FRA provides a comprehensive picture of the interconnected cognitive-epistemic and social-institutional aspects of NOS. Each aspect is linked to categories that define the characteristics of science. The cognitive-epistemic aspects encompass practices, knowledge, aims and values, and methods and methodological rules. The social-institutional aspects encompass scientific ethos, professional activities, social certification and dissemination, social values, financial systems, political power structures, and social certification organisations and interactions. These categories are not discrete but interrelated, continuously shaping and influencing one another. Recent studies on NOS in NHMs (Pshenichny-Mamo & Tsybulsky, 2024, 2025) suggest they constitute an authentic environment where discussions of NOS can fruitfully take place. However, to date, no studies have examined which aspects of NOS are integrated into CC exhibitions. This study addressed this gap by analyzing which NOS aspects are incorporated into CC exhibitions at two NHMs, with a particular focus on both the cognitive-epistemic and social-institutional aspects.

Studies indicate that NHMs, as authentic informal educational environments, are well-suited to empowering visitors to engage with contemporary issues like CC (e.g., Knutson, 2018). Exposing learners to CC within the authentic environment of NHMs is thought to help enhance their understanding of ecological shifts, underlying causes, potential solutions, and ways to engage critically with scientific processes. These exhibits provide more informative and engaging discourse about CC, thus contributing to developing a more scientifically literate and environmentally-conscious society.

Thus, this study contributes to the scientific discourse about the role of museums worldwide in CC education. Its contribution lies in an initial exploration of the educational potential of these exhibitions, focusing on the goals the museum staff associate with these exhibitions.

7.2 Research Design and Method

This study employed a qualitative research design. Ethical approval for this study was obtained from the university ERB. Participants were informed of the study's purpose and provided their informed consent before participation. Anonymity and confidentiality were maintained throughout the research process.

7.2.1 Study Context

This study was conducted at two NHMs: the University of Michigan Natural History Museum (UMNHM) in the USA and the Steinhardt Museum of Natural History (SMNH) in Israel. These museums were selected for their accessibility, similar size, and similar-sized contemporary CC exhibitions: 'People and the Planet' (PATP) at UMNHM, and 'Global Warning: The Climate, the Crisis, and Us' (GW) at SMNH.

7.2.2 Data Collection and Analysis

To explore the goals of the museum staff associated with these exhibitions, the data were collected through semi-structured Zoom interviews with 15 museum staff members, including the head of exhibitions, the head of education, the scientific curator, the experience designer, and the museum guides. Examples of questions included:

- Why is it important to present exhibitions on CC in the museum?
- What are the challenges you have identified in conveying the complex science of CC to museum visitors?
- From your perspective, what is the museum's role in educating visitors about CC?
- What is the museum trying to teach visitors about CC?

The interviews were audio-recorded and transcribed verbatim. The first author conducted an inductive thematic analysis following the six-step process outlined by Braun and Clarke (2022), to identify key themes related to the educational goals of the exhibitions. In the first step, the data were transcribed and thoroughly read multiple times to facilitate familiarization, allowing an in-depth understanding and generating initial notes. During the second step, the data were coded to capture content specifically related to the educational goals of the exhibitions. The third step involved generating initial themes by identifying patterns among the codes. The themes were reviewed and refined in the fourth step to ensure they accurately represented the data. In the fifth step, the research team discussed and finalized the themes' labeling and definition, resulting in four themes that best captured the educational goals of the CC exhibitions. In the sixth step, a detailed narrative of the final analysis was written, providing a comprehensive interpretation of the data.

To explore the aspects of NOS presented in the signages, the data collection involved photographing the signages and displays of the exhibitions. The content of these exhibitions was then analyzed using deductive content analysis (Bingham & Witkowsky, 2022), guided by the FRA framework as an analytical tool (as in Pshenichny-Mamo & Tsybulsky, 2024, 2025). The FRA framework allows for the examination of the cognitive-epistemic and social-institutional aspects of NOS embedded within the exhibitions. The analysis followed a three-stage process. In the first stage, the lead author conducted a detailed reading of the exhibition signage to identify statements reflecting the NOS. In the second stage, each statement was examined within its context and assigned to one of the categories of NOS aspects based on the FRA framework. In the final stage, the research team convened to discuss the initial categorizations and reach a consensus on each statement's final classification.

7.2.3 Reliability

To ensure reliability, several measures were taken. Photographs, interview recordings, transcriptions, and detailed documentation of the data analysis process were preserved. External researchers independently coded approximately 30% of the data. A clear chain of evidence was maintained, including descriptions of the study setting and participants. Multiple quotes from museum staff were used to enhance credibility and contextual understanding, ensuring that the analysis stayed true to the intended meanings. This strengthened the validity of the findings and ensured confirmability (Shenton, 2004).

7.3 Findings

7.3.1 The Educational Goals of the CC Exhibitions

The findings revealed four key themes corresponding to the educational goals the museum staff associated with CC exhibitions: (1) Grounding knowledge in science, (2) Grounding knowledge in NOS, (3) Emotional connection, and (4) Personal activism. Below, we present each theme with supporting quotations from the interviews.

7.3.1.1 Grounding Knowledge in Science

The interviews with staff from the PATP and GW exhibitions revealed that the museum staff aimed to ground public knowledge about CC science by emphasizing what science already knows and ongoing research on urgent climate questions:

> [...] Not everybody agrees with the science. Not everybody accepts it as established science. So, we definitely took that into consideration [...] I mean, part of what we're trying to do in exhibits is not only show the problems but also show the way that researchers at the university are trying to get to solutions. (PATP)

Both exhibitions also trained their guides to be able to answer visitors' questions and clarify climate science:

> I teach our guides about common misconceptions among the public […] It's important to highlight posters and exhibits that provide the answers […] For example, we have a poster that compares natural disasters related to geophysics […] with natural disasters related to the climate crisis […] Or a poster that explains how we can claim the Earth is warming while still experiencing cold events. In that poster, we take the opportunity to explain the difference between weather and climate. […] There are many opportunities for dialogue, for question and answer. (GW)

7.3.1.2 Grounding Knowledge in NOS

In addition to scientific knowledge, the museum staff explained in the interviews that the exhibitions were intended to help visitors understand the NOS underlying CC research, in particular, how evidence is gathered and validated through the epistemic-cognitive aspects of NOS:

> […] talking about the nature of science. What a theory is, what a hypothesis is, what a fact is in science, how they differ from everyday interpretations, what a scientific law is, and how it differs from a theory. […] We talk about independent lines of evidence […]. (PATP)

At the 'Time Capsule' display in the GW exhibition, the museum staff sought to demonstrate how we know about past climate conditions despite the lack of direct measurements:

> […] We've been measuring greenhouse gases using instruments since 1957, which is about 70 years. So, how do we know that what's happening now is unprecedented? How do we know what was happening 1000 or 100,000 years ago? […] How do we even know what was happening here if we weren't measuring back then?' […]. (GW)

> […] Time capsules show real exhibits […] from so many different, real, physical sources […] there's a deep understanding that this is serious, complex, and well-founded science. (GW)

The museum staff also aimed to underscore science as a human endeavor by presenting the social-institutional aspects of NOS:

> […] We focus or talk about individual researchers and what they're doing. We also talk about why they're doing that research. So, visitors get to personal feelings or things that they see as wrong, that they want to fix […]. (PATP)

7.3.1.3 Emotional Connection

Fostering an emotional connection was identified by the museum staff as crucial for enhancing visitors' engagement with CC. This was addressed by pairing the presentation of problems with actionable solutions:

> I'm active, I play, I change things, I touch, and I feel it in my body. In contrast to a crisis that feels very distant, happening over time scales that I can't feel, very different from my everyday life—here, it was really important for us that people could touch, move, and feel things in relation to their own body dimensions and their everyday reality […] Optimism alongside urgency, because the situation is very critical, but there are things we can do about it. […]. (GW)

Another approach for fostering emotional connection was to focus on local climate issues:

> We wanted to include stories from our own space. In every part of the exhibition, there's always a local point, which we insisted on, even though it added a bit more text—something that was critical for us. (GW)

Writing exhibition signage in accessible language was also emphasized:

> When we're writing labels, we're always reading it to see: Did we translate this into the way people talk? We don't want it to be preachy. We don't want it to be too hopeless [...]. (PATP)

7.3.1.4 Personal Activism

One of the goals according to the museum staff was to encourage personal activism as a response to CC:

> The exhibition is divided into three parts [...] The third part presents solutions. Visitors are invited to take personal steps, such as choosing an action from a board of 24 pencils, each representing a different motivation. (GW)

The staff emphasized the importance of encouraging visitors to draw their own conclusions:

> [...] To provide information so that people can make their own judgments because those will be much more powerful than anything that we're hammering them with. (PATP)

7.3.2 NOS Aspects Integrated into the CC Exhibitions

In total there were 22 mentions of NOS aspects across the two exhibitions. In the PATP exhibition, six mentions were identified, with a greater emphasis on social-institutional aspects of NOS (4) than epistemic-cognitive aspects (2) (Fig. 7.1). The GW exhibition included 16 mentions, mainly focusing on epistemic-cognitive aspects (13). Social-institutional aspects were mentioned three times (Fig. 7.2).

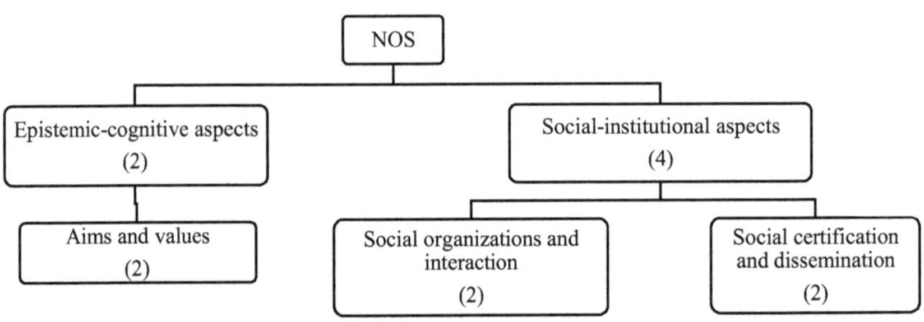

Fig. 7.1 NOS in the PATP exhibition (number of mentions)

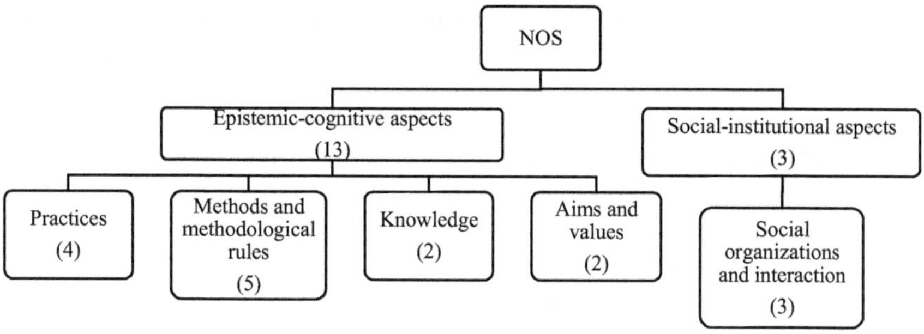

Fig. 7.2 NOS in the GW exhibition (number of mentions)

7.3.2.1 The PATP Exhibition

The findings showed that the social-institutional aspects of NOS were interwoven throughout the exhibition. For instance, images highlighted that science is conducted by individuals, men and women from diverse social, cultural, and ethnic backgrounds, thus emphasizing that science is a human endeavor. These mentions, along with descriptions of scientists as the ones who *'do'* science, reflect the social organizations and interactions of science. Including credits next to scientific information aligns with the social certification and dissemination aspect of NOS by acknowledging the contributions of multiple scientists and the collaborative effort involved in scientific publication. The PATP exhibition highlighted the epistemic-cognitive aspects of NOS by mentioning the aims and values of science and, in particular, by showing that scientific research has a purpose. The examples dealt with scientists' attempts to *'teach'* plants new features:

> Scientists are teaching plants to be better at photosynthesis by helping plants use the COs concentrating methods that cyanobacteria use, such as making carboxysomes and using specialized enzymes.
>
> Some scientists are trying to make faster growing plants by using carboxysomes to absorb CO_2 faster while others are using cyanobacteria to capture CO_2 from factories.

7.3.2.2 The GW Exhibition

In GW, most of the written signage related to NOS was in the 'Time capsules' display that began with:

> How do we know that the warming and the current rise in greenhouse gas concentrations are unprecedented and extreme? What do we know about the climate that existed before instrumental measurement? Glaciers, tree trunks, cave deposits, lake sediments and corals are built over thousands to millions of years. The composition of the layers is affected by the environmental conditions. By analyzing their composition and comparing the data with other findings we can reconstruct the local and global climate history

Below this text, a picture of the globe was presented, indicating (1) the *locations of ancient climate samples* and (2) the *locations of ancient climate exhibits from our* [museum's]*region.*

This signage was followed by a display showing different information sources in CC research: a model of an ice core, fragments of meteorites, trees' annual growth rings, the growth rings of stalactites and stalagmites, drill cores from the Dead Sea, coral growth rings, foraminifers, and pollen grains. Each was followed by an explanation of the ways in which scientists gather data, information, and evidence from the sources and pictures that add additional information, such as scientists at work, related graphs, etc. For example:

> The coral skeleton grows gradually in the form of annual growth rings. Chemical analysis of the rings allows us to determine the temperature and composition of the seawater during the coral's lifetime.

This signage was followed by objects presenting (1) a piece of a coral skeleton from Eilat, representing an 18-year growth period, and (2) a picture of a scientist conducting coral drilling. The epistemic-cognitive aspect of practices was integrated by presenting how observational, historical, and experimental data in research on CC are used to predict future CC outcomes in the real world. The different climate exhibits demonstrate that multiple data sources are used to find reasonable explanations. They illustrated the rigorous and systematic inquiry that scientists engage in by employing diverse methodologies to produce trustworthy evidence and establish theories, laws, and models within a particular field of study. These ideas reflect methods and methodological rules. These signages captured the NOS aspect of scientific knowledge by mentioning its empirical foundations and the dynamic nature of scientific understanding. The exhibition conveyed the idea that scientific information is subject to ongoing research and refinement as advancements in technology and the accumulation of knowledge shape and reshape scientific perspectives. The exhibition showed scientific inquiry's underlying aims and values, suggesting that a purpose and rationale guide scientific investigation and that the results can help account for CC. The social-institutional aspect of NOS was brought to the fore in the GW exhibition by two pictures of scientists during their research work, which were presented as a part of the 'Time Capsule' display and an additional mention during the introductory video presentation explaining that scientists are making predictions based on the information they had extracted from different sources. These mentions reflect social organizations and interaction and science as a human endeavor.

7.4 Discussion

This study investigated the educational potential of CC exhibitions at two NHMs in the USA and Israel, through examining the goals that museum staff associated with these exhibitions and the integration of NOS in the exhibition signages.

In general, the museum staff's educational goals were cohesive with their efforts to increase the visitors' understanding of the complexity of CC (Cameron et al., 2013) as a wicked problem that has no single simple solution (Achiam et al., 2021a, 2021b), thus nurturing personal connections, and inspiring activism (Monroe et al., 2019), and a deeper appreciation and comprehension of the scientific endeavor within the museum (Dillon et al., 2016; Mujtaba et al., 2018).

The staff of both NHMs identified four main goals for CC exhibitions: grounding knowledge in science, grounding knowledge in the NOS, fostering an emotional connection, and encouraging personal activism. The first goal, grounding knowledge in science, was emphasized by museum staff as an important aspect of CC education. During the interviews, the staff mentioned that they found it vital to expose visitors to what is already known about CC and new research being conducted to address issues scientists do not fully understand. They viewed this exposure as crucial for enhancing public awareness and deepening visitors' understanding of CC science. The second goal the museum staff mentioned during the interviews was grounding knowledge in NOS. This was concretized by demonstrating the process of CC research, including how knowledge is acquired, who conducts research, and how the evidence supports contemporary understandings of CC. NHMs provide access to authentic natural history specimens and scientific evidence and can show visitors tangible connections to historical and ongoing research on CC (Dillon et al., 2016; Reiss et al., 2016), which can lead to a more comprehensive understanding of the development of knowledge about CC. These two goals are interconnected. By providing visitors with scientific information alongside information on how scientific knowledge and assumptions have evolved and occasionally offering direct interactions with climate researchers, NHMs may foster a deeper understanding of scientific theories and an appreciation of the scientific enterprise. This approach underscores NOS's epistemic-cognitive and social-institutional aspects, as described in Erduran and Dagher (2014).

The museum staff's third goal associated with these exhibitions was to foster emotional connections. In both NHMs, the staff claimed that creating an emotional connection through the exhibitions was crucial for engaging visitors with CC's local and global impact. As Lin et al. (2014) noted, exhibitions designed with an emotional impact in mind, using relatable language, lead to positive audience experiences and lasting emotions. For example, the GW exhibition's interactive and experiential design allowed visitors to engage with the content physically. By incorporating play and hands-on interaction, visitors can *'feel'* the effects of CC. The staff also mentioned that the focus was on providing satisfying visitor experiences by informing and shaping visitors' expectations and preferences, which aligns with and supports previous research (Pekarik et al., 1999; Pshenichny-Mamo & Tsybulsky, 2024). The staff also mentioned fostering emotional connections by localizing the climate issues. The GW exhibition utilized the perception of the museum as a reliable source of information (Cameron et al., 2013) to demonstrate that CC affects numerous aspects of daily life in the museum region. By presenting CC as an immediate concern, these exhibitions sought to make the issue more tangible and meaningful for visitors, as Monroe et al. (2019) suggested. According to the PATP and GW staff, another way to promote an emotional connection was to use accessible language that allows visitors to understand the information easily. For instance, the GW exhibition featured signage in three languages, Hebrew, English, and Arabic, to cater to diverse audiences. Studies have shown that accessible language enhances comprehension and bridges the gap between scientific intent and public understanding, thus fostering empathy and emotional engagement (Chen et al., 2022).

The fourth goal was to enhance personal activism. The staff at both museums expressed their desire to encourage visitors to draw their own conclusions and consider potential actions based on knowledge acquired during their visit. This is consistent with the literature showing that as institutions that preserve and communicate the significance of natural history, NHMs play a vital role in encouraging personal activism and empowering visitors to take informed actions in response to the climate crisis. They present not only the problems but also the possible solutions and give visitors practical tools for action (McGhie, 2020).

The content analysis of the exhibition signages revealed differences in the emphasis on aspects of NOS between the two museums. The PATP exhibition at UMNHM highlighted social-institutional aspects of NOS, whereas the GW exhibition at SMNH focused more on epistemic-cognitive aspects. These findings suggest that the GW exhibition offered a deeper exploration of the systemic complexity of CC, which might enhance visitors' understanding of NOS, while the PATP exhibition underscored the human dimension of science, potentially fostering a personal connection through the portrayal of scientists and their work. Previous studies have explored museum guides' integration of NOS during guided tours (Pshenichny-Mamo & Tsybulsky, 2025) and the ways in which they integrate NOS while addressing visitors' experiences related to evolution and ecology topics (Pshenichny-Mamo & Tsybulsky, 2024). However, the focus on the integration of NOS into the content of CC exhibitions sheds new light on the integration of NOS in NHMs. Analyzing these NOS aspects through the theoretical framework of FRA also contributes to understanding its broad and comprehensive application in formal and informal educational settings (Cheung & Erduran, 2022).

The findings highlight the significance of NHMs as authentic educational environments for teaching CC (Mujtaba et al., 2018). Previous research suggests that nature-based museums providing specialized educational programs improve visitor knowledge, engagement, and climate activism (Swim et al., 2017). Thus, according to the museum staff's goals, these exhibitions may give students and the general public unique opportunities to engage with scientific concepts authentically. For biology educators, NHMs can complement classroom activities (Knutson, 2018; Swim et al., 2017), in abstract concepts such as ecological changes, carbon cycles, and biodiversity loss can become more relatable through interactions with physical scientific evidence. Thus, biology teaching on CC may benefit from incorporating visits to CC exhibitions into the curriculum. These exhibitions, which merge scientific knowledge with NOS components, offer students tangible, real-world examples of how scientific research is conducted, validated, and utilized to tackle global environmental issues. Interacting with these exhibitions can promote a deeper understanding of scientific inquiry, by covering processes such as data collection, hypothesis testing, and collaboration across disciplines. This approach creates a link between formal and informal education, thus reinforcing significant curricular themes while cultivating a broader understanding of science's role in society. It can help bridge theoretical knowledge and practical insight, and encourage the development of informed and environmentally aware citizens. Ultimately, positioning NHMs as vital allies in the collective effort to impart scientific knowledge regarding climate change effectively.

7.5 Research Limitations and Future Directions

This study focused on two NHMs in Israel and the United States. Future work may expand this study to other NHMs worldwide to identify other educational goals influenced by geographic location, target audiences, cultural, social, and political differences, and unique institutional characteristics. In this study, we relied on two main sources of data: interviews and a content analysis of exhibition signage. Studies could incorporate additional data sources such as observations, surveys, and other methodologies.

Further works should also explore the educational outcomes for NHM visitors, and biology students in particular, to determine how these exhibitions can support and enhance CC education in classroom settings. Another interesting research direction would be the development of an outreach program that integrates biology education with museum-based learning activities on CC. Studying how students' knowledge and perceptions of CC evolve throughout their learning could provide further insights into the impact of complementing biology education by visiting and learning in NHMs.

Acknowledgements Anna Pshenichny-Mamo is grateful to the Azrieli Foundation for awarding her an Azrieli Fellowship. The authors thank the staff of the University of Michigan Natural History Museum and the Steinhardt Museum of Natural History who collaborated with us and provided valuable insights for this study. The authors are also grateful to Prof. Tali Tal (Technion Israel Institute of Technology), Prof. Gail Richmond, and Irene Bayer (Michigan State University) for their guidance and support in the early stages of this research. Finally, the authors appreciate the constructive feedback from the anonymous reviewers, which contributed to the improvement of this chapter.

References

Achiam, M., Dillon, J., & Glackin, M. (2021a). *Addressing wicked problems through science education*. Springer.

Achiam, M., Glackin, M., & Dillon, J. (2021b). Wicked problems and out-of-school science education: Implications for practice and research. In M. Achiam, J. Dillon, & M. Glackin (Eds.), *Addressing wicked problems through science education* (pp. 229–237). Springer.

Bingham, A. J., & Witkowsky, P. (2022). Deductive and inductive approaches to qualitative data analysis. In C. Vanover, P. Mihas, & J. Saldaña (Eds.), *Analyzing and interpreting qualitative data: After the interview* (pp. 133–146). SAGE Publications.

Braun, V., & Clarke, V. (2022). *Thematic analysis: A practical guide*. SAGE Publications.

Cameron, F., Hodge, B., & Salazar, J. F. (2013). Representing climate change in museum space and places. *Wiley Interdisciplinary Reviews: Climate Change, 4*(1), 9–21. https://doi.org/10.1002/wcc.200

Chen, W., Bian, D., Shi, Q., & Wei, Z. (2022). On emotional expression of installation works in public art. *BCP Social Sciences & Humanities, 20*, 262–265. https://doi.org/10.54691/bcpssh.v20i.2326

Cheung, K. K. C., & Erduran, S. (2022). A systematic review of research on family resemblance approach to nature of science in science education. *Science and Education, 0123456789*. https://doi.org/10.1007/s11191-022-00379-3

Conn, S. (1998). *Museums and american intellectual life, 1876–1926*. University of Chicago Press.

Denney, D. A., & Anderson, J. T. (2020). Natural history collections document biological responses to climate change: A commentary on DeLeo et al. (2019), Effects of two centuries of global environmental variation on phenology and physiology of Arabidopsis thaliana. *Global Change Biology, 26*(2), 340–342. https://doi.org/10.1111/gcb.14922

Dillon, J., DeWitt, J., Pegram, E., Irwin, B., Crowley, K., Haydon, R., King, H., Knutson, K., Veall, V., & Xanthoudaki, M. (2016). *A learning research agenda for natural history institutions*.

Erduran, S., & Dagher, Z. R. (2014). *Reconceptualizing the nature of science for science education*. Springer.

Goodger, E. (2019). Environmental entrepreneurship: Adapting our museums for a greener future. In W. Leal Filho, B. Lackner, & H. McGhie (Eds.), *Addressing the challenges in communicating climate change across various audiences. Climate change management* (pp. 585–598). Springer. https://doi.org/10.1007/978-3-319-98294-6_35

Hamilton, P., & Christian Ronning, E. (2020). Why museums? Museums as conveners on climate change. *Journal of Museum Education, 45*(1), 16–27. https://doi.org/10.1080/10598650.2020.1720375

Hulme, M. (2015). Why we should disagree about climate change. In F. R. Cameron & B. Neilson (Eds.), *Climate change and museum futures* (pp. 9–15). Routledge.

King, H., & Achiam, M. (2017). The case for natural history. *Science and Education, 26*(1–2), 125–139. https://doi.org/10.1007/s11191-017-9880-8

Knutson, K. (2018). Rethinking museum/community partnerships: Science and natural history museums and the challenges of communicating climate change. In K. Drotner, V. Dziekan, R. Parry, & K. C. Schrøder (Eds.), *The Routledge handbook of museums, media and communication* (pp. 101–114). Routledge. https://doi.org/10.4324/9781315560168-9

Kondyli, C., Psychogiou, M., & Drinia, H. (2024). The museums of geology and paleontology as geoeducational learning environments for raising climate change awareness. *Sustainability, 16*(11), 4481. https://doi.org/10.3390/su16114481

Lin, Y.-J., Ma, J.-P., & Chen, J.-L. (2014). Cultural creativity in design exhibition: A case study of emotional effects experienced by the audience. In P. L. P. Rau (Ed.), *Cross-cultural design* (pp. 743–752). Springer International Publishing.

Marandino, M., Achiam, M., & de Oliveira, A. D. (2015). The diorama as a means for biodiversity education. In S. D. Tunnicliffe & A. Scheersoi (Eds.), *Natural history dioramas: History, construction and educational role* (pp. 251–266). Springer Netherlands. https://doi.org/10.1007/978-94-017-9496-1_19

McGhie, H. (2020). Evolving climate change policy and museums. *Museum Management and Curatorship, 35*(6), 653–662. https://doi.org/10.1080/09647775.2020.1844589

McGhie, H., Mander, S., & Underhill, R. (2018). Engaging people with climate change through museums. In W. Leal Filho, E. Manolas, A. M. Azul, U. M. Azeiteiro, & H. McGhie (Eds.), *Handbook of climate change communication* (Vol. 3, pp. 329–348). Springer International Publishing. https://doi.org/10.1007/978-3-319-70479-1_21

Monroe, M. C., Plate, R. R., Oxarart, A., Bowers, A., & Chaves, W. A. (2019). Identifying effective climate change education strategies: A systematic review of the research. *Environmental Education Research, 25*(6), 791–812. https://doi.org/10.1080/13504622.2017.1360842

Mujtaba, T., Lawrence, M., Oliver, M., & Reiss, M. J. (2018). Learning and engagement through natural history museums. *Studies in Science Education, 54*(1), 41–67. https://doi.org/10.1080/03057267.2018.1442820

Nieroba, E., & Geisler, R. (2024). Museums and climate change discourse: The selected case studies analysis. *Museum Management and Curatorship*, 1–21. https://doi.org/10.1080/09647775.2024.2393156

Pekarik, A. J., Doering, Z. D., & Karns, D. A. (1999). Exploring Satisfying Experiences in Museums. *Curator, 42*(2), 152–173.

Pellegrini, G. (2021). School sliding doors: Out-of-school activities and young people's engagement with climate change. In M. Achiam, J. Dillon, & M. Glackin (Eds.), *Addressing wicked problems through science education* (pp. 215–227). Springer.

Pshenichny-Mamo, A., & Tsybulsky, D. (2024). Museum guides' views on the integration of the nature of science while addressing visitors' experiences: The context of ecological and evolutionary issues. In K. Korfiatis, M. Grace, & M. Hammann (Eds.), *Shaping the future of biological education research. Contributions from biology education research* (pp. 79–91). Springer. https://doi.org/10.1007/978-3-031-44792-1_6

Pshenichny-Mamo, A., & Tsybulsky, D. (2025). Natural history museum guides' conceptions on the integration of the nature of science. *Science & Education, 34*, 511–529. https://doi.org/10.1007/s11191-023-00469-w

Reiss, M., Billingsley, B., Evans, E., Kissel, R., Lawrence, M., Mujtaba, T., Oliver, M., Pickering, J., Sheldrake, R., Shen, C., Stott, J., & Veall, D. (2016). *The contribution of natural history museums to science education.*

Shaby, N., Ben-Zvi Assaraf, O., & Tishler, C. E. (2016). The goals of science museums in the eyes of museum pedagogical staff. *Learning Environments Research, 19*(3), 359–382. https://doi.org/10.1007/s10984-016-9211-z

Shenton, A. K. (2004). Strategies for ensuring trustworthiness in qualitative research projects. *Education for Information, 22*(2), 63–75. https://doi.org/10.3233/EFI-2004-22201

Sutton, S. W. (2021). Museums as agents and settings for climate hope. *Nordisk Museologi, 30*(3), 14–28. https://doi.org/10.5617/nm.8627

Swim, J. K., Geiger, N., Fraser, J., & Pletcher, N. (2017). Climate change education at nature-based museums. *Curator, 60*(1), 101–119. https://doi.org/10.1111/cura.12187

Tran, L. U. (2007). Teaching science in museums: The pedagogy and goals of museum educators. *Science Education, 91*(2), 278–297.

United Nations Educational, Scientific and Cultural Organization (UNESCO) (n.d.). *Climate change education.*https://www.unesco.org/en/climate-change/education

Chapter 8
Plant Awareness Disparity in Secondary School Students' Nature Experiences

Marcus Hammann, Helene Kesseler, and Justin Wonnesch

8.1 Introduction

Sharing rich evolutionary and cultural relationships with plants, humans depend on plants (Schaal, 2019). Nevertheless, people tend to underestimate the personal relevance of plants and undervalue the interdependency. More specifically, plants are commonly viewed as an undifferentiated utilitarian resource, as insignificant background, and "nothing" (Schneekloth, 1989). Over the past 25 years, this phenomenon has been studied widely under the term plant-blindness, which was later renamed as plant awareness disparity (PAD) by Parsley (2020).

According to Wandersee and Schussler (1999), plant blindness is defined as

„(a) the inability to see or notice the plants in one's environment; (b) the inability to recognize the importance of plants in the biosphere and in human affairs; (c) the inability to appreciate the aesthetic and unique biological features of the life forms that belong to the Plant Kingdom; and (d) the misguided anthropocentric ranking of plants as inferior to animals and thus, as unworthy of consideration " (84).

Historically, the authors' intention to define plant-blindness in such a way was to call attention to the fact that animals are perceived as more important than plants and receive more attention in education. Similar to plant blindness, plant awareness disparity (PAD) is a relational construct explicitly comparing plants and animals in people's awareness, attitude, knowledge and interest (Parsley, 2020). The term PAD

M. Hammann (✉) · J. Wonnesch
Münster University, Münster, Germany
e-mail: hammann.m@uni-muenster.de

J. Wonnesch
e-mail: j_wonn03@uni-muenster.de

H. Kesseler
Amplonius Gymnasium, Rheinberg, Germany
e-mail: helene.kesseler@gmail.com

O. Morin et al. (eds.), *Challenges of a Changing World in Biology Education*, Contributions from Biology Education Research,
https://doi.org/10.1007/978-3-032-05346-6_8

places attentional aspects in the foreground insofar as “plants are not completely unseen, but rather placed in the background in the visual service of noticing other organisms,” namely animals (Parsley, 2020, 600). Knapp (2019, p. 164) argued that the deeper problem is not disregard for plants but for all organisms without “eyes like ours and backbones”, which includes invertebrates. Thus, human beings may not only visually disadvantage plants, but all organisms except vertebrates.

Whether inattentive to plants or to any non-vertebrate, people’s PAD is commonly investigated by using research designs which juxtapose plants and animals, for example, a rooster against a plant background (Jose et al., 2019). Another study focused on the extent to which images of plants and animals capture attentional resources by presenting cards with plants and animals, for example a seal, a hedgehog and a Macaw ara (Balas & Momsen, 2014). Other authors offered the finding that the absence of animals in an excursion to the Costa Rican rainforest led to the participants’ impression of having “seen nothing” after spending a day in one of the most biodiverse places on earth (Schneekloth, 1989). Similarly, young students participating in an excursion to a botanical garden were fascinated by the unexpected sighting of a fish under a lily pad which took away their attention from the plants (Tunnicliffe, 2001).

The aim of this study is to investigate plant awareness disparity in secondary school students’ authentic nature experiences. In particular, we tested the hypotheses from the plant awareness disparity literature, that secondary school students find nature experiences of plants less memorable and less personally relevant than nature experiences of animals.

8.2 Theoretical Background

A recent systematic review found support for the different dimensions of plant awareness disparity described by Parsley (2020), in particular the ranking of plants as inferior to animals, and hypothesized that a decline in relevant experiences of plants is the reason for plant awareness disparity (Stagg & Dillon, 2022). More specifically, the review revealed the following most common characteristics of PAD in educational studies: the ranking of plants as inferior to animals (30 studies), a lack of experience in observing and identifying pants (27 studies), a lack of knowledge or understanding in basic plant science (21 studies), a failure to see or notice plants (13 studies), and a lack of appreciation of plants or inability to recognize their importance (13 studies). The authors of the review discuss these findings by attributing PAD to a decline in relevant experiences of plants. More specifically, the authors of the review see the diminished experience of plants as the starting point of a cycle which leads to low knowledge, interest and attention for plants and a bias to animals in education.

The present study focuses on students’ experiences of plants and animals in authentic, self-reported nature experiences. It aims to add to the literature of PAD by investigating the role of plants and animals in students’ nature experience. In a previous study, we used the draw-and-write method to characterize the students’ nature experiences in terms of outer aspects (setting, type of interaction, entities encountered and social context) and inner aspects (emotions aesthetic judgment) (Hammann, 2024). As main findings, the following variables were associated with secondary school students' perceived relevance of their experiences: age, gender, setting of the nature experience, perception of beauty and frequency of the nature experience. The same study, however, did not find any association between the students' perception of personal relevance of the nature experience and the type of organism involved. This may have been due to small sample size because a considerable number of the 70 nature experiences studied involved neither plants nor animals. Furthermore, no statistically significant differences concerning perceived personal relevance were found between nature experiences with plants and/or animals ($n = 39$) and nature experiences without plants and/ or animals ($n = 29$). Sample size was too small to investigate possible differences between the perceived relevance of nature experiences of plants vs. nature experiences of animals. Such differences, however, can be expected from the argument found in the PAD literature that animals are valued over plants (Stagg & Dillon, 2022). Furthermore, sample size was too small to investigate the different roles plants and animals may have in exploratory and recreational experiences. Plants may feature more prominently as background in recreational nature experiences, whereas animals may be more often the focus of observations in exploratory nature activities. Therefore, to fill these research gaps, the present study seeks to use the draw-and-write method in a larger sample of secondary school students and compare larger numbers of experiences of plants and animals.

8.3 Research Rationale and Purpose

This study focuses on plant awareness disparity by comparing students' self-reported nature experiences of plants and animals. In particular, we asked the following research questions:

1. How frequent are experiences of plants and animals in this sample of secondary students?
2. How frequently are experiences of plants and animals represented in the different types of students' nature experiences, in particular the exploratory and recreational dimensions?
3. To what extent do students' experiences of plants and animals differ in terms of students' species knowledge in this sample?
4. How personally relevant are experiences of plants and animals in this sample of secondary students?
5. How connected to nature are secondary students in this sample who report experiences of plants and animals?

More specifically, we investigated the hypothesis (H 1) that students recall nature experiences with animals more frequently than nature experiences with plants. This hypothesis is based on the argument that PAD is commonly characterized by a lack of interest in plants and by an attention or memory advantage for animals compared to plants (Stagg & Dillon, 2022). Furthermore, because animals are commonly perceived as more interesting than plants, we expected (H 2a) animals to feature more prominently in exploratory nature experiences than plants. Likewise, we expected (H 2b) plants to feature more prominently in recreational experiences than animals. The latter hypothesis is based on the argument that plants often serve as the backdrop for human activities in nature (Schneekloth, 1989). For the third research question, we expected (H 3) students to display less knowledge of plants than of animals, as reflected by the specificity of terms used to denote plants and animals. This hypothesis is based on the argument about botanical education extinction (Stroud et al., 2022). Furthermore, we investigated the hypothesis (H 4) that students find nature experiences with animals more personally relevant than nature experiences with plants. This hypothesis is based on the argument that PAD stems from a decline in relevant experiences of plants (Stagg & Dillon, 2022). For the last research question, we investigated the hypothesis (H 5) that students who report nature experiences of animals are more connected to nature than students who report nature experiences of plants. This hypothesis is based on the argument that nonhuman animals have been shown to act as links to a caring relationship to nature (Myers & Saunders, 2002) and contribute to nature connection (Jacobs et al., 2023). More specifically, the core of the argument is that children care about animals as social others in their own right – not just as a result of projective anthropomorphism – which leads to caring about what that animal needs and the environmental conditions which affect the animal's well-being.

8.4 Research Design and Method

We first report on the participants and setting, before detailing the draw and write research method (Angell et al., 2015) which we used in combination with questionnaire items to elicit secondary school students' nature experiences, their perceived relevance of the nature experience and their nature connectedness. The questionnaire contained additional items and scales not described here because we used them to answer other research questions.

8.4.1 Participants and Setting

The participants were 814 secondary school students (395 female, 371 male, 15 diverse, 33 missing information about gender, aged 10–18 years, mean age 13.72, SD 2.25) from 20 secondary schools in the administrative district of Münster in North Rhine-Westfalia, Germany. 257 students were in grades 5–6 (mean age 11.25, SD 0.63), 388 students were in grades 7–10 (mean age 14.06, SD 1.32) and 169 students were in grades 11–13 (mean age 16.61, SD 1.30). Of the 814 students, 364 students

(44.75%) lived in cities with more than 100.000 inhabitants, 355 students (43.6%) lived in large towns with more than 50.000 inhabitants and 95 students (11.7%) lived in small towns with more than 5.000 inhabitants. The administrative district of Münster is 6.900 square kilometers large and inhabited by 2.6 million people. The region is dominated by agriculture (60.3%), which is above the average for North Rhine-Westfalia (47.3%). The percentage of woods was 15.3%, which is below the average for North Rhine Westfalia (24.9%).

8.4.2 Assessment Instrument and Data Collection

For eliciting the students' nature experiences, the assignment was: "Draw an image and write a text about a moment in nature, which you remember particularly well." For assessing the perceived personal relevance of the nature experience, we asked students to rate the item "How personally relevant are such moments in nature for you?" We used a 4-point Likert scale for this item (very relevant, relevant, not so relevant, not relevant at all). Furthermore, we used the NCI-items from the short nature connectedness scale by Richardson and colleagues (2019). The scale consisted of the following items: Being in nature makes me very happy. I find being in nature really amazing. I always find beauty in nature. Spending time in nature is very important to me. I always treat nature with respect and I feel part of nature. We added the following two negatively-keyed items: Nature does not play a role in my life, and I don't like to get out into nature. We asked the students to rate the items on a 4-point Likert scale (strongly agree, agree, disagree, strongly disagree). To assess students' nature connectedness, we calculated the mean of the scale. We investigated the reliability of the scale by calculating the scale's Cronbach's alpha, which was $\alpha = 0.79$ and, according to Streiner (2003), indicated a good reliability of the scale.

8.4.3 Data Coding

The main data sources for this study are students' drawings and texts. 763 students submitted both drawing and text so that the following analyses refer to this sample. Data coding focused on two aspects: secondary school students' type of interaction with nature and the absence/presence of plants and animals. For the first coding, our analyses were guided deductively by Bögeholz (1999, 2006) and Lude (2001) for the first five types of interaction and inductively for the last two types of interaction. The following categories were used to capture the students' types of interactions with nature:

- the aesthetic dimension – enjoying the beauty of plants, animals, landscapes
- the exploratory dimension – observing animals, looking closely at plants and exploring natural places

Dad and I went for a walk in the woods and there was a feeding trough. Dad and I waited for three minutes and then three deer came and ate. They didn't seem to be afraid because they watched us from about 25 meters away while eating.

Fig. 8.1 Example of a nature experience of animals (drawing and text in response to the draw-and write method described in the text)

- the instrumental dimension – cultivating and using parts of plants, as well as breeding or caring for domestic animals
- the recreational dimension – playing and doing leisure activities in nature
- the social dimension – having a relationship with a pet
- special experience – experiences not covered by the previous dimensions, like falling from a tree, burying a pet, flying in a hot air balloon
- no activity – student describes feelings, wishes, ideas about nature, etc., but no activity

Similar categories were used by Hammann (2024) in a study using the same draw-and-write method. Two independent coders coded 40 randomly chosen portrayals of nature experiences. Cohens Kappa was 0.83, which is almost perfect observer agreement, according to Landis and Koch (1977).

For the second coding, our analyses focused on the explicit reference to plants and animals in students' nature experiences. For nature experiences to be classified as nature experiences of animals, the students reported on animals or observations of animals in the texts (see Fig. 8.1). This category included both wild animals and farm animals. When students drew animals in their pictures, but did not mention them in their text, we excluded such nature experiences from this category.

In nature experiences of plants, students refer to plants, for example by reporting how they look at them, use plants for building a teepee in the woods or climb a tree (see Fig. 8.2). Statements in which plant organs, such as leaves, branches, etc., are reported also belong in this category. Plants that are drawn but not mentioned in the text are excluded from this category. For nature experiences without animals or plants, neither animals nor plants appear in the nature experiences depicted by the students (see Fig. 8.3). Animals and/or plants that are only drawn but not described in the text also belong to this category.

Fig. 8.2 Example of a nature experience of plants

Fig. 8.3 Example of a nature experience without plants and animals

8.4.4 Statistical Analyses

For investigating frequencies of experiences of plants and animals (RQ1) and frequencies of plants and animals represented in recreational and exploratory nature experiences (RQ 2), we used frequency counts as well as Chi-Square Goodness of Fit Tests for gender differences and differences between the different types of nature experiences. For reporting on students' perceived personal relevance (RQ 4) and nature connectedness (RQ 5), we calculated means and standard deviations, coding the categories 'very relevant' and 'strongly agree' as 4 and 'not relevant at all' and 'strongly disagree' as 1. Accordingly, high mean values signal high perception of personal relevance and a high degree of nature connectedness. For comparing students' ratings for nature connectedness and perceived personal relevance of nature experiences, we used Mann–Whitney U-Tests and Kruskal Wallis H-Tests because data for both variables were not normally distributed. Both tests compare mean ranks. For Mann–Whitney U-Tests, we report on effect size r, which is small for 0.10 – < 0.29, medium for 0.30 - < 0.49 and large for ≥ 0.5. For Kruskal Wallis H-Tests, we report on effect size f, which is small for 0.10 – < 0.24, medium for 0.25 - < 0.39 and large for ≥ 0.40. In non-parametric tests, the observed data is converted into ranks. Nevertheless, when we report on the significance level of the nonparametric tests, we also report on means and standard deviations to show differences between the groups of students we compared. This decision was made because reporting ranks does not reveal the differences. All statistical analyses were performed using SPSS statistical software version 28.

8.5 Results

Table 8.1 provides an overview of findings for the whole sample. The most frequent types of nature experiences portrayed by the students in this sample were recreational (50.6%) and exploratory (24.9%). Results are similar to Hammann (2024) who also reported that the majority of nature experiences were exploratory and recreational.

We used Kruskal Wallis H-Tests to investigate possible differences concerning the perceived personal relevance of the different types of nature experiences and the nature connectedness of the students reporting the different types of nature experiences. Both Kruskal Wallis H-Tests were significant with small effect sizes for personal relevance, $\chi^2(6) = 22.47$, $p \leq 0.01$, $f = 0.14$, and nature connectedness, $\chi^2(6) = 16.10$, $p = 0.013$, $f = 0.10$. Post hoc pairwise comparisons revealed that for the personal relevance of the nature experience, only the following three comparisons were statistically significant: instrumental – exploratory ($p = 0.05$), instrumental – social ($p = 0.009$) and social – no activity ($p = 0.02$). For nature connectedness, post hoc pairwise comparisons revealed no significances. This is attributable to the fact that for investigating post-hoc pairwise comparisons, we used adjusted p-values for multiple tests (alpha correction).

Table 8.1 Frequency of the different types of nature experiences (N = 763), personal relevance ratings for the types of nature experience and nature connectedness ratings of the person reporting the nature experience

Type of nature experience (coded for activity dimension)	Number (n)	%	Personal relevance Mean (SD)	Nature connectedness Mean (SD)
Aesthetic	53	6.9	3.09 (0.69)	2.91 (0.56)
Exploratory	198	24.9	3.17 (0.76)	2.95 (0.55)
Instrumental	11	1.4	2.36 (0.81)	2.39 (0.75)
Recreational	386	50.6	3.15 (0.78)	2.84 (0.56)
Social	38	4.9	3.37 (0.71)	2.91 (0.60)
Special event	24	3.1	2.92 (0.88)	2.88 (0.62)
No activity	53	6.9	2.83 (0.83)	2.71 (0.49)
Sum	763	100	3.11 (0.79)	2.86 (0.56)

8.5.1 How Frequent Are Experiences of Plants and Animals in This Sample of Secondary Students?

We used frequency counts to investigate whether or not secondary students recall nature experiences with animals more frequently than nature experiences with plants (see Table 8.2). Findings revealed a slight difference in frequency of nature experiences of plants ($n = 129$; 16.9% of 763 students who submitted text and drawing) and animals ($n = 155$, 20.3%), but the majority of students portrayed nature experiences involving neither plants nor animals ($n = 456$; 59.7%).

A Chi-Square Goodness of Fit Test was performed to determine whether the proportion of students reporting experiences either with animals (n = 149), or with plants ($n = 127$) or without animals and plants ($n = 218$) was equal between males ($n = 340$) and females ($n = 364$). The proportions differed by gender, χ^2 $(2, N = 704) = 9.93, p = 0.007$, with the main difference being that 16.2% males ($n = 55$) and 25.8% females ($n = 94$) reported experiences with animals. This was an unexpected finding, which we explored further. Therefore, we investigated whether the proportion of students who reported experiences with either pets (including horses) or wildlife was equal between males ($n = 55$) and females ($n = 90$). The same analyses were done for students who reported experiences with either plants as a utilitarian resource or plants as an object of observation. For experiences involving animals, the proportions differed by gender, χ^2 $(2, N = 145) = 28.8, p = .001$, with 9.1% males ($n = 5$) and 53.8% females ($n = 48$) reporting experiences with pets, and 90.9% male (n = 50) and 46.7% female students ($n = 42$) reporting experiences with wildlife. For experiences involving plants, the proportions did not differ by gender. Thus, the greater proportion of females than males reporting experiences of animals is related to the fact that pets featured more frequently in nature experiences reported by female students than wild animals.

Summing up the main findings for RQ 1, we found support for the hypothesis (H 1) that students find nature experiences of animals more memorable than nature

Table 8.2 Frequency of the nature experiences with plants, nature experiences with animals and nature experiences with neither plants nor animals ($N = 763$), personal relevance ratings for the types of nature experience and nature connectedness ratings of the person reporting the nature experience

Type of nature activity (coded for plants and animals)	Number (n)	%	Personal relevance *Mean (SD)*	Nature connectedness *Mean (SD)*
Neither plants nor animals	456	59.7	3.15 (0.75)	2.85 (0.55)
With animals	155	20.3	3.29 (0.75)	2.98 (0.56)
With plants	129	16.9	2.88 (0.84)	2.77 (0.61)
Other	23	3.0	2.78 (0.74)	2.89 (0.53)
Sum	763	100	3.11 (0.79)	2.86 (0.56)

experiences of plants, although the difference in frequency of nature experiences involving animals vs. nature experiences involving plants was small. Furthermore, we found an overwhelmingly large proportion of students who portrayed nature experiences involving neither plants nor animals.

8.5.2 How Frequently are Experiences of Plants and Animals Represented in the Different Types of Students' Nature Experiences, in Particular the Exploratory and Recreational Dimensions?

According to Stagg and Dillon (2022), the most common indicator of PAD is a lack of experience of observing and identifying plants. Therefore, our analyses focused on the question to what extent the secondary students' experiences of plants and animal were similar or different in terms of types of interaction with nature.

Plants featured primarily in recreational experiences (42.6% of the 129 nature experiences with plants) and less often in exploratory experiences (22.5%) (see Table 8.3). In contrast, animals featured primarily in exploratory experiences (46.5% of the 155 experiences with animals) and less often in recreational experiences (20.6%). Note that animals featured also in social nature experiences (24.5% of the 155 experiences with animals), which are defined as having a relationship with a pet, so that this category excludes plants. This detail is noteworthy because when the recreational and the social dimensions are combined, animals and plants feature equally often in these related categories. Furthermore, like nature experiences with plants, experiences without plants and animals were primarily recreational (64.9% of the 456 experiences without plants and animals) and less often exploratory (18.6%). A Chi-Square Goodness of Fit Test was performed to determine whether the proportion of students reporting experiences with either plants or animals was equal between the exploratory and the recreational dimensions, which were the main dimensions found in this study. The proportions differed by type of experience (either animals or plants), χ^2 $(1, N = 188) = 22.51, p \leq 0.001$, with plants underrepresented in the exploratory dimension and overrepresented in the recreational dimension.

Summing up the main findings for RQ 2, we found support for the hypothesis (H 2a) that animals featured more often than plants in exploratory nature experiences. Thus, animals were more often the focus of exploration than plants. Furthermore, when looking at the recreational dimension alone, we found support for the hypothesis (H 2b) that plants featured more prominently than animals in recreational activities. However, in light of the fact that animals also featured in nature experiences classified as social experiences, which often represented recreational activities with animals like walking a dog, hypothesis H 2b needs to be rejected because we found similar numbers for nature experiences with animals and plants when the recreational and the social dimensions were combined.

Table 8.3 Crosstable showing the relationship between the variables "type of nature experience coded for absence or presence of animals and/or plants" and "type of nature experience coded for the different activity dimensions"

Type of nature experience		Aesthetic dimension	Exploratory dimension	Instrumental dimension	Recreational dimension	Social dimension	Special event	No activity	Sum
With animals	Number	6	72	1	32	38	6	0	155
	Expected	10.5	39.0	2.3	80.2	8.0	4.8	10.3	155,0
	Percentage	3.9%	46.5%	0.6%	20.6%	24.5%	3.9%	0.0%	100.0%
	Standardized residuals	−1.4	5.3	−0.9	−5.4	10.6	0.5	−3.2	
With plants	Number	13	29	10	55	0	10	12	129
	Expected	8.7	32.4	1.9	66.8	6.6	4.0	8.5	129
	Percentage	10.1%	22.5%	7.8%	42.6%	0.0%	7.8%	9.3%	100%
	Standardized residuals	1.5	−0.6	5.8	−1.4	−2.6	3.0	1.2	
Neither plants nor animals	Number	31	85	0	296	0	7	37	456
	Expected	30.8	114.6	6.8	236	23.4	14.2	30.2	456.0
	Percentage	6.8%	18.6%	0.0%	64.9%	0.0%	1.5%	8.1%	100.0%
	Standardized residuals	0.0	−2.8	−2.6	3.9	−4.8	−1.9	1.2	
Sum	Number	50	186	11	383	38	23	49	740
	Expected	50	186	11	383	38	23	49	740
	Percentage	6.8%	25.1%	1.5%	51.8%	5.1%	3.1%	6.5%	100.0%

8.5.3 To What Extent Do Students' Experiences of Plants and Animals Differ in Terms of Students' Species Knowledge?

As an indicator of students' species knowledge, we analyzed the terms secondary students used to denote plants and animals in their descriptions of the nature experiences of plants and animals. Generally, students used less specific terms to denote plants than animals. When students denoted animals, they predominantly mentioned vertebrates and only few invertebrates. In detail, students used the following terms to denote plants in their nature experiences: tree ($n = 223$; 67,8% of $N = 329$ terms used in the whole sample), grass ($n = 27$; 8.2%), flower ($n = 23$; 7%), bush ($n = 13$; 4%), apple tree ($n = 11$; 3.3%), cherry tree ($n = 4$; 1.2%), palm tree ($n = 3$; 0.9%), plant ($n = 3$; 0.9%) and seaweed ($n = 3$; 0.9%). Furthermore, the students used the following terms once each ($n = 1$; 0.3%) to denote the following plants: corn, sunflower, tulip, rose, blueberry, stinging nettle, thorn bush, hazelnut, herb, moss, chestnut, dandelion, sequoia, shrub, hedge, fir tree, water plant, Christmas tree and grapevine.

For animals, the students used the following terms: dog ($n = 51$; 23.2% of the $N = 220$ terms used in the whole sample), bird ($n = 33$; 15%), deer ($n = 24$; 10.9%), fish ($n = 20$; 9.1%), horse ($n = 14$; 6.4%), stag ($n = 11$; 5.0%), animal ($n = 7$; 3.2%), fox ($n = 5$; 2.3%), bat ($n = 4$; 1.8%), hare ($n = 4$; 1.8%), butterfly ($n = 4$; 1.8%), frog ($n = 3$; 1.4%), swan ($n = 3$; 1.4%), shark ($n = 2$; 0.9%), chicken ($n = 2$; 0.9%), hedgehog ($n = 2$; 0.9%), mussel ($n = 2$; 0.9%), squirrel ($n = 2$; 0.9%) and wild boar ($n = 2$; 0.9%). Furthermore, the students used the following terms once each ($n = 1$; 0.5%) to denote the following animals: slow worm, buffalo, cow, duck, owl, mosquito, octopus, dew worm, shrimp, cricket, rabbit, beaver, insect, sea snail, goat, sheep, turtle, pig, woodpecker, skunk, stork, starfish, bald eagle, wild animal and wolf.

Summing up the main findings for RQ 3, we found evidence for the hypothesis (H 3) that secondary students used less specific terms to denote plants than animals and thus displayed less knowledge of plants. Evidence for this hypothesis is that 87% of the terms used to denote plants were fairly unspecific and above the genus level (tree, grass, flower, bush) whereas 27% of the terms used to denote animals were fairly unspecific and above the genus level (bird, fish, animal). Further support for this hypothesis comes from the frequency count of terms used at the genus and species level. In this sample, the students used twice as many terms at the genus and species level for animals (approx. 40 terms) than for plants (approx. 20 terms).

8.5.4 How Personally Relevant Are Experiences of Plants and Animals?

Secondary school students rated the perceived personal relevance of experiences involving plants (*M* 2.88; *SD* 0.84; $Z = 4.08, p \leq 0.001, r = 0.24$) significantly lower than experiences involving animals (*M* 3.29; *SD* 0.75) (see Table 8.2). Furthermore, students rated the personal relevance of experiences involving plants (*M* 2.88; *SD* 0.84; $Z = 3.13$; $p = 0.002$; $r = 0.12$) significantly lower than experiences involving neither plants nor animals (*M* 3.15; *SD*.75). The difference in relevance ratings between experiences of animals (*M* 3.29; *SD*.75) and experiences without animals and plants (*M* 3.15; *SD*.75; $Z = 2.11$; $p = 0.03$; $r = 0.08$) was statistically significant as well.

Summing up the main findings for RQ 4, we found evidence for the hypothesis (H 4) that students perceived experiences of plants as less personally relevant than experiences of animals. Furthermore, students perceived experiences without plants and animals more personally relevant than experiences with plants.

8.5.5 How Connected to Nature Are Secondary Students Who Report Experiences of Plants and Animals?

Secondary students reporting experiences with plants (*M* 2.77; *SD*.61) exhibited significantly lower nature connectedness than students reporting experiences with animals (*M* 2.98; *SD*.56; $Z = 2.77$; $p = .006$; $r = .16$) (see Table 8.2). Furthermore, students who reported nature experiences with animals (*M* 2.98; *SD*.56) rated their nature connectedness significantly higher than students who reported nature experiences involving neither plants nor animals (*M* 2.85; *SD* 0.55; $Z = 2.47$; $p = 0.014$; $r = 0.10$). There were no significant differences in nature connectedness between students reporting experiences with plants and students reporting experiences involving neither plants nor animals (*M* 2.85; *SD* 0.55; $Z = 1.09$; $p = 0.27$).

Summing up the main findings for RQ 5, we found evidence for the hypothesis (H 5) that students who report nature experiences of animals are more connected to nature than students who report nature experiences of plants.

8.6 Discussion

This study compared secondary students' authentic, self-reported nature experiences of plants and animals to investigate the different dimensions of plant awareness disparity (PAD). In this sample, secondary students reported less frequently on nature experiences of plants (17%) than animals (20%). However, most students reported on nature experiences involving neither plants nor animals (60%). Furthermore, plants featured less prominently than animals in exploratory nature experiences, in which animals were more often the object of exploration than plants. In their descriptions of the nature experience, secondary students referred to plants with less specific terms than to animals. Furthermore, they used fewer terms at the genus level and species-level for plants than for animals. Students who reported on nature experiences of plants perceived their nature experience less personally relevant and displayed less nature-connectedness than students who reported on nature experiences with animals.

Nature experiences of animals outnumbered nature experiences of animals. This finding lends support to the argument that experiences of plants are less memorable than experiences of animals, although the difference in numbers between nature experiences of plants and animals was surprisingly small and the percentage of students who reported on nature experiences without animals and plants was overwhelmingly large. Thus, when secondary students were instructed to portray only one memorable nature experience of their own choice, the majority of the students did not refer to plants and animals at all in their texts. This finding is new and needs to be investigated further by looking at the students' portrayal of nature experiences qualitatively. More specifically, qualitative analyses need to focus on the large group of nature experiences involving neither plants nor animals to understand what other factors make a nature experience memorable when references to animals and plants are missing in the text. Such analyses need to test the hypothesis that the students' extinction of experience (Soga & Gaston, 2016) is the loss of human-animal and human-plant interactions.

Despite the fact that the loss of human-animal and human-plant interactions is a larger problem than the plant-awareness deficit, this study yielded several findings which support the argument that students' plant awareness is deficient. Students who reported nature experiences of plants perceived these less personally relevant than students who reported experiences of animals. This finding is in line with the main argument in the literature that students lack personally relevant experiences of plants (Stagg & Dillon, 2022). Furthermore, when students wrote about their nature experiences of plants, they denoted plants with fairly unspecific terms, for example "tree", and used less specific terms than students who reported on nature experiences of animals. This study also found support for the argument that students are inattentive to all organisms without backbones (Knapp, 2019) because students denoted mostly vertebrates in their descriptions of nature experiences of animals.

This study yielded two entirely new findings relevant for the PAD literature. Plants featured less often in exploratory nature experiences than animals. Thus, an important difference seems to be that animals provoke curiosity so that they become the focus of an observation in an exploratory nature experience which students find memorable. Students, in contrast, seem to lack a similar interest in plants so that plants feature more prominently than animals in recreational activities, where the role of plants is more peripheral to the nature experience than the role of animals because plants are the background of the recreational activity. Furthermore, students reporting experiences with plants exhibited lower nature connectedness than students reporting experiences with animals. This finding lends support to the argument that animals act as links to a caring relationship to nature (Myers & Saunders, 2002) and contribute to nature connection (Jacobs et al., 2023). Further analyses need to look at these interesting new findings qualitatively to understand them more fully.

Because the students of this sample perceived experiences with plants as less personally relevant than experiences with animals and because students focused on plants less frequently as an object of exploration than animals, the main educational implications of this study are the need to focus on attitudinal and behavioral aspects of human-plants interactions. Essentially, the findings of this study point towards making human-plant interactions personally relevant and providing opportunities for interacting with and exploring plants. To foster positive attitudes towards plants, biology teachers should create interest in plants, for example by focusing on unusual aspects like plant-defense mechanisms (Nantawanit et al., 2012), and elaborate on the many ways in which plants are useful to human beings (Pany & Heidinger, 2014). To increase the personal relevance of plants, helpful teaching strategies are contextualizing plants and selecting relevant plant topics (Hammann et al., 2020). Furthermore, to foster positive attitudes towards plants, biology teachers need to avoid perpetuating false dichotomies like "loveable" mammals and "lifeless" plants (Lindemann-Matthies, 2005). Generally, biology educators should give plants and animals equal amount of coverage (Schussler & Olzak, 2008) and use plant and animal examples whenever possible (Wandersee & Schussler, 1999). In addition, biology educators need to focus on the many ways in which plants are central to human beings and ecosystems (Bowker, 2004), for example for food security and biodiversity issues (Drea, 2011). For improving human-plant interactions and for making plants the object of exploration, hands-on experiences with plants are important. Teachers need to provide interesting activities with plants, like growing plants. Furthermore, educators need to engage students with interesting plant science investigations like plant experiments and plant observations. Such activities should also be integrated into field trips so that students are encouraged to see, learn about and do activities with the plants in their environment. Such efforts provide the basis for an increased awareness of the personal relevance of plants and the willingness to engage in opportunities for interacting with and exploring plants.

References

Angell, C., Alexander, J., & Hunt, J. A. (2015). 'Draw, write and tell': A literature review and methodological development on the 'draw and write' research method. *Journal of Early Childhood Research, 13*(1), 17–28.

Balas, B., & Momsen, J. (2014). Attention "Blinks" differently for plants and animals. *CBE Life Sciences Education, 13*(3), 437–443.

Bowker, R. (2004). Children's perceptions of plants following their visit to the Eden Project. *Research in Science & Technological Education, 22*(2), 227–243.

Bögeholz, S. (1999). *Qualitäten primärer Naturerfahrung und ihr Zusammenhang mit Umweltwissen und Umwelthandeln.* [The Qualities of Primary Nature Experiences and their Relationship with Environmental Knowledge and Action]. Opladen: Leske+Budrich.

Bögeholz, S. (2006). Nature experience and its importance for environmental knowledge, values and action: Recent German empirical contributions. *Environmental Education Research, 12*(1), 65–84.

Drea, S. (2011). The end of the Botany degree in the UK. *Bioscience Education, 17*(1), 1–7.

Hammann, M., Jördens, J., & Büschgens, D. (2020). Students' situational interest in cultivated plants: The importance of contextualisation and topic selection. *International Journal of Science Education, 42*(16), 2765–2799.

Hammann, M. (2024). Personal relevance in secondary school students' nature experiences. In K. Korfiatis, M. Grace, M. Hammann (Eds.), *Shaping the future of biology education research: selected papers from the ERIDOB community* (pp. 261–276). Springer Nature.

Jacobs, T. P., Humphrey, B. T., & McConnell, A. R. (2023). Nature's best friend: Viewing pets as having greater emotional experience increases environmental engagement. *Anthrozoös, 36*, 625–639.

Jose, S. B., Wu, C.-H., & Kamoun, S. (2019). Overcoming plant blindness in science, education and society. *Plants, People, Planet, 1*(25), 169–172.

Knapp, S. (2019). Are humans really blind to plants? *Plants, People, Planet, 1*(25), 164–168.

Landis, J. R., & Koch, G. G. (1977). The measurement of observer agreement for categorial data. *Biometrics, 33*(1), 159–166.

Lindemann-Matthies, P. (2005). 'Loveable' mammals and 'lifeless' plants: How children's interest in common local organisms can be enhanced through observation of nature. *International Journal of Science Education, 27*, 655–677.

Lude, A. (2001). *Naturerfahrung und Naturschutzbewusstsein: eine empirische Studie. [Nature Experience and Environmental awareness: An empirical study].* Studienverlag.

Myers, O. E., & Saunders, C. P. (2002). Animals as links toward developing a caring relationship with the natural world. In P. H. Kahn & S. R. Kellert (Eds.), *Children and Nature: Psychological, sociocultural and evolutionary investigations* (pp. 153–178). The MIT Press.

Nantawanit, N., Panijpan, B., & Ruenwongsa, P. (2012). Promoting students' conceptual understanding of plant defense responses using the Fighting Plant Learning Unit (fplu). *International Journal of Science and Mathematics Education, 10*(4), 827–864.

Pany, P., & Heidinger, C. (2014). Students' interest in useful plants: A potential key to counteract plant blindness. *Plant Science Bulletin, 60*(1), 18–27.

Parsley, K. (2020). Plant awareness disparity: A case for renaming plant blindness. *Plants, People, Planet, 2*(6), 598–601.

Richardson, M., Hunt, A., Hinds, J., Bragg, R., Fido, D., Petronzi, D., Barbett, L., Clitherow, T., & White, M. (2019). A measure of nature connectedness for children and adults: Validation, performance, and insights. *Sustainability, 11*(12), 3250.

Schaal, B. (2019). Plants and people: Our shared history and future. *Plants, People, Planet, 1*, 14–19.

Schneekloth, L. (1989). "Where Did You Go?" "The Forest." "What Did You See?" "Nothing.". *Children's Environments Quarterly, 6*(1), 14–17.

Schussler, E. E., & Olzak, L. (2008). It's not easy being green: Student recall of plant and animal images. *Journal of Biological Education, 42*(3), 112–119.

Soga, M., & Gaston, K. J. (2016). Extinction of experience: The loss of human-nature interactions. *Frontiers in Ecology and the Environment, 14*(2), 94–101.

Stagg, B. C, & Dillon, J. (2022). Plant awareness is linked to plant relevance: A review of educational and ethnobiological literature (1998–2020). *Plants, People, Planet*, 1–14.

Streiner, D. L. (2003). Starting at the beginning: An introduction to coefficient alpha and internal consistency. *Journal of Personality Assessment, 80*(1), 99–103.

Stroud, S., Fennell, M., Mitchley, J., Lydon, S., Peacock, J., & Bacon, K. L. (2022). The botanical education extinction and the fall of plant awareness. *Ecology and Evolution., 12*, Article e9019.

Tunnicliffe, S. D. (2001). Talking about plants—comments of primary school groups looking at plant exhibits in a botanical garden. *Journal of Biological Education, 36*(1), 27–34.

Wandersee, J. H., & Schussler, E. E. (1999). Preventing plant blindness. *The American Biology Teacher, 61*, 84–86.

Chapter 9
Social Justice and School Biology Education: What Do Teacher Educators Think Can and Should Be Taught About Gender Issues?

Marian Mulcahy, Wilton Lodge, and Michael J. Reiss

9.1 Introduction

As citizens, the three of us hold that science education needs to play its part in helping to create a more socially just world. It is possible that this would have beneficial consequences but there is also a deontological argument for such a world. It would mean that all individuals would be accorded greater respect, their intrinsic worth would be recognised and they would not suffer indignities. We therefore need to consider what social injustice (the opposite of social justice) is, in order to clarify what it is we are striving for. Levy and Sidel (2013, p. 3) define social injustice as: 'a denial or violation of human rights … of specific populations or groups in society based on the erroneous perception of their inferiority by those with more power or influence. Populations or groups that suffer social injustice may be defined by racial or ethnic status, socioeconomic position, age, gender, sexual orientation, or other perceived population or group characteristics'.

During the course of high school biology, students encounter many topics that may directly or indirectly relate to social justice. For example, the current programme of study for 14–16 year-olds in England includes 'Health, disease and the development of medicines' and 'Evolution, inheritance and variation' (Department for Education, 2014). Upegui and colleagues (2022) argue that there are two ways in which students

M. Mulcahy (✉) · W. Lodge · M. J. Reiss
UCL Institute of Education, London, UK
e-mail: m.mulcahy@ucl.ac.uk

W. Lodge
e-mail: wilton.lodge@ucl.ac.uk

M. J. Reiss
e-mail: m.reiss@ucl.ac.uk

O. Morin et al. (eds.), *Challenges of a Changing World in Biology Education*, Contributions from Biology Education Research,
https://doi.org/10.1007/978-3-032-05346-6_9

can be helped to explore social inequities as they study biology: 'the pedagogical approaches adopted; and the actual biological content'. Their research participants highlighted how the same content differently delivered could either contribute to or disrupt notions of inequity (i.e., injustice). This point is well illustrated by the topic of sex and gender.

Currently, there is considerable debate around gender, gender (in)equality, stereotyping and the importance of language when talking about gender issues (e.g., Borger, 2023; Taasoobshirazi et al., 2019). In our experience as teacher educators, it is evident that beginning teachers often find topics such as these difficult to navigate. In a recent study, Long et al. (2021) noted that 45% of young people in the US recognise gender as being a spectrum, rather than a binary, and that 10% of TV characters in 2019–2020 were LGBTQ. However, in school science lessons only 2.4% of LGBTQ students identified any inclusive science content. This, combined with the paucity of literature regarding the preparation of new teachers to address gender issues in their teaching, shows that this is both a socially important but also an under-researched area prompting our study.

We therefore pose three research questions:

RQ1 Where in school biology do initial teacher educators feel that social justice issues might be addressed?

RQ2 What do initial teacher educators consider to be issues to do with social justice that are particularly relevant for people of different genders?

RQ3 What role do initial teacher educators think school biology could play in teaching about issues to do with social justice that are particularly relevant to people of different genders?

To answer these questions, we draw on a larger study that considered issues to do with social justice across the whole of school science, not just biology.

9.2 Theoretical Framework

Our theoretical framework has two related elements: the potential for school science education to promote social justice and the aims of science education. Within the school context, few educators would maintain that they are denying or violating the human rights of their students, but if we allow the cultural perpetuation of the status quo described by Freire (1996), it could be argued that we are complicit in doing just that. It is widely accepted that individuals from certain groups are discriminated against across the spectrum of society, economically, in the workplace, socially and by the expectations society has for them (Liou, 2023). In recent years there has been a growing emphasis on looking at how science education can address issues of social justice and hopefully thus contribute to the development of a world where each individual is valued whilst honouring their identity (e.g., Lesnefsky et al., 2023; Morales-Doyle, 2023; Rodriguez & Morrison, 2019).

In education, social justice is an overarching phrase encompassing a plethora of concepts, including welcoming and embracing diversity, promoting equity and ensuring that each individual feels valued and is both encouraged and enabled to achieve their potential. Because science has so often been misappropriated to endorse inequitable ideologies, perhaps most notably its use in 'legitimising' racism (e.g., Chatters et al., 2022), science education is especially well placed to challenge these misconceptions and incorporate social justice. Dimick (2012) examined how the science classroom could be used to empower students in social, political and academic arenas and could thus contribute to the development of critically aware adults with the potential, and ideally the desire, to be transformative.

There has long been a debate within science education as to the aims of school science (e.g., Mansfield & Reiss, 2020). Researchers have considered whether its primary role is economic, providing the scientists and engineers that society needs, the 'STEM pipeline' (Ahn et al., 2016), or whether it should be more focused on the needs and identities of individuals, being mindful of the multiple layers (personal, cultural and sociopolitical) that contribute to a person (Lee et al., 2021). It is clear that the pipeline (for all the shortcomings of this metaphor) is 'leaky' (Ahn et al., 2016), and that a disproportionately high proportion of those who leave it come from groups that experience discrimination (Nag Chowdhuri et al., 2023). Whichever of these two functions of school science education is prioritised – keeping more people in STEM for economic reasons or attending to needs and identities of individuals – there is a role for social justice to play in school science education. This is before we begin to think about our role as educators in contributing to shaping the society of the future, where social justice must surely be foregrounded.

Another approach to determining the aims of school science comes from examining three different visions of science education. Roberts and Bybee (2014) give a helpful conceptualisation of two contrasting visions – Vision I and Vision II. Vision I is all about 'mastering a sampling of the language, products, processes, and traditions of science itself' (p. 546), while Vision II 'is sometimes called science for citizenship, concentrating on matters of more obvious personal and social relevance to students than preparing to grasp more demanding science they might or might not study' (p. 546). More recently, Sjöström and Eilks (2018) have conceptualised what they term 'Vision III' to include 'critical scientific literacy' and 'socio-political action'. Vision III therefore builds on Visions I and II in that it requires a good understanding of science (Vision I) and is a science education for all (Vision II) but it extends these. Of these three visions, it is Vision III that most explicitly has a *focus* on social justice, though Vision II requires a *consideration* of social justice.

9.3 Research Design and Methodology

An important way in which the aims of science education are conveyed to new teachers is through initial teacher education (ITE). We therefore sought to gain the views of a sample of university tutors responsible for secondary science education about whether, and if so how, school biology might contribute to social justice.

We adopted a qualitative research design using individual, semi-structured interviews. Semi- structured interviews met our objectives in that they are highly purposeful tasks where the researchers serve as instruments and data are co-constructed by the researcher and the participants (Merriam & Tisdell, 2015). The questions were designed to be relatively open, allowing each research participant the opportunity to tell us about the issues around the intersection between social justice and school biology that were important to them. After an initial exploration of the participants' understanding of social justice, they were specifically asked whether they thought science education should address issues to do with social justice, where in school biology these issues might be addressed, what issues of social justice they considered of particular relevance to people of different gender, and what role school biology could play in teaching about such issues.

Twelve university tutors (lecturers or senior lecturers) at universities with secondary science ITE cohorts within England participated in the interviews from March to June 2022. The sampling of interviewees was largely purposive, though we ensured that four had a specialism in biology, four in chemistry and four in physics (the three disciplines into which the National Curriculum for science in England is divided), and that half were female and half male. Each came from a different university, with a range of universities chosen with respect to location, academic standing, history and student diversity. Table 9.1 provides a description of the interviewees with respect to their self-declared specialism, gender, age and ethnicity.

Table 9.1 Description of the research participants

Research participant	Specialism	Gender	Age	Ethnicity
RP 1	Chemistry	Female	64	White British
RP2	Physics	Female	68	White British
RP3	Chemistry	Female	39	White
RP 4	Physics	Male	51	African
RP5	Physics	Male	49	European
RP6	Biology	Male	48	White British
RP7	Biology	Female	45	White British
RP8	Physics	Male	55	White British
RP9	Biology	Male	44	White British
RP10	Chemistry	Female	60	European
RP11	Biology	Male	69	White British
RP12	Chemistry	Female	53	White

Interviews were undertaken, after receiving research ethics approval from our university, by the same person (the third author) using Microsoft Teams and were recorded and professionally transcribed. Each interview lasted about 30 min. The transcripts were repeatedly read, and inductive thematic analysis used to identify major themes. The initial analysis was undertaken by the third author with subsequent analysis being undertaken by the first. This helped ensure that the themes identified were indeed those arising from the data, increasing the validity of their identification. Using Braun et al.'s (2016) framework, we adopted a semantic focus, concentrating on what the research participants explicitly said, allowing us to identify some of the commonalities underpinning their self-reported viewpoints. It was from these commonalities that we identified the themes discussed below, with themes being identified for each of our three research questions.

9.4 Findings

In this section we report on what teacher educators considered to be areas of biology teaching where social justice issues could be addressed (RQ1), the issues they considered were particularly relevant to people of different genders (RQ2), and how they thought school biology could help teach about these (RQ3).

9.4.1 Opportunities to Address Social Justice Issues in the Biology Curriculum Identified by Teacher Educators

The research participants identified a number of areas within biology where they thought that biology education could play a role in addressing issues to do with social justice (Table 9.2).

Table 9.2 Areas identified by research participants where school biology could address issues to do with social justice

Biology areas that could address social justice issues	Number of responses
Reproduction	6
Genetics	5
Evolution	4
Environment	3
Health and disease	2
Respect for indigenous knowledge	1

Each of these six areas – 'Reproduction', 'Genetics, 'Evolution', 'Environment', 'Health and disease' and 'Respect for indigenous knowledge' – can be considered as a theme. The biological curricular area that was most frequently identified as having potential to incorporate social justice issues was reproduction, which includes sex education. Research participant (RP)3 noted that 'Anything to do with relationships, sex education sort of stuff, social justice is important there, things like abortion'.

This participant recognises that addressing social justice can involve moving beyond teaching the biological aspects of reproduction, to also examining the relational, sociological and/or cultural implications. This was a view echoed by others, for example RP8: 'We did sex education with science departments. And we addressed quite a lot of those sorts of issues then in terms of sexual politics, saying yes, no, peer pressure'.

In general, the teacher educators highlighted that there are opportunities to engage with what some may consider to be sociological and cultural issues whilst teaching about reproduction.

The teacher educators also noted that genetics can be (mis)used to either promote or to combat racism. RP1 advocated that we teach 'how science was misappropriated to justify IQ and racism', whereas RP12 argued that genetics highlights our similarities rather than differences, stating: 'We should be aware that genetically in terms of our DNA and the DNA, for example, of a fly very comparable, so why wouldn't the DNA of someone who is, for example, male and female, be very similar or a BAME person and a white person?'.

When teaching evolution, the dissonance between what is scientifically accepted and many students' beliefs was cited as an area that should be addressed in the classroom. This is a social justice issue in terms of students feeling that they and their cultures are valued: 'I think that the danger of excluding people because their values do not subscribe to, you know, a Darwinian evolutionary model is purely wrong. And I think that people's understanding and views should be accepted' (RP2).

Teaching about environmental issues was identified as another important aspect of biology that intersects with considerations of social justice. The specific areas indicated were climate education, care for the environment and inequitable access to nature: 'I guess a greater embedding of a climate environmental focus education would require a great focus on social justice, would it not?' (RP8). 'We teach about ecosystems and care for creation, what I would understand is creation but care for the environment' (RP3). 'There's a big disparity between children whose parents can afford to take them outside, are National Trust members, that kind of thing, compared to our inner-city children who think that they have a very small world compared to some of those other children' (RP7).

These responses indicate some of the areas encompassed by environmental biology that can be used or addressed with respect to teaching for greater social justice.

Human health and disease was another area where opportunities to look at inequities in the biology curriculum were identified. One teacher educator pointed out that the biology curriculum in England relies on western ways of knowing, thus implicitly exemplifying a differential value system within the biology classroom: 'We do the human body in biology, but we discount oriental knowledge on the human body. We don't teach things like acupuncture' (RP4).

9.4.2 Issues to Do with Social Justice Particularly Experienced by People of Different Genders, as Identified by Teacher Educators

In this section we look at the responses when we asked our research participants about what social justice issues they considered to be particularly relevant to those of different genders. Three themes were identified (Table 9.3): 'Sexism'. 'Gender diversity' and 'Lack of representation'.

9.4.2.1 Sexism

In considering issues to do with gender, the most common theme that emerged was to do with sexism. Research participants identified a power imbalance at multiple levels. They asserted that men hold positions of power and are more likely to progress in their careers. RP12 pointed out that the world is 'very male-dominated and … most young women may see the world as not being something where they have particular power'. Similarly, RP6 noted that 'there is injustice in the system where women aren't paid as much for the same jobs, and they are less likely to be promoted'.

9.4.2.2 Gender Diversity

Looking at the theme of gender diversity, teacher educators focused on the growing recognition that gender is both fluid and non-binary: 'The first thing that strikes me is around gender diversity and not pigeon-holing into two genders and so particularly non-binary people and trans people … So that's a massive issue for social justice because we know that that has real devastating and deadly impact for people in their lives' (RP3).

Recognising sex as a social construct was a related issue that emerged. This is closely linked to considering gender as not being binary.

Table 9.3 Social justice themes of particular relevance to people of different genders identified by teacher educators

Social justice theme	Number of responses
Sexism	8
Gender diversity	5
Lack of representation	4

For a number of our interviewees, issues of gender diversity were mentioned alongside issues to do with sexualities and therefore the diversity of ways in which families exist, beyond the 'standard' heterosexual family: 'So, the social constructs we have of men and women have to be carefully thought about. Also, when we're talking about reproduction and humans, babies, the different ways, but also, assuming heterosexuality as well, there's that, that's a big issue, much better addressed now, and especially when it comes to sexual health' (RP6).

This research participant appreciates the complexities and constrictions of heteronormativity. In this quotation there is recognition of the implicit assumptions that are often made about men and women.

Some of the teacher educators pointed out the importance of cultural differences regarding gender; for example, RP12 reported: 'We have quite a few schools in our local areas, a lot of Asian girls who have got a real interest in science, but culturally are not allowed to develop that. So, I do think that there's a lot to learn. I do think we start to, you know, we should be using, or working with these children in our local context and our broader context to enable them to see themselves in science but also the parental involvement to see that science is an option. Because, you know, a lot of these children, the male aspect, the cultural male aspect will go into science but not the female. And that, to me, is limiting'.

This quotation demonstrates an appreciation of the changing role of a teacher if they want to adopt a socially just stance. The need for a deeper understanding of school students' contexts was acknowledged as being desirable if we are to help children and families consider science as an option regardless of the student's gender. As the above suggests, 'working with', rather than challenging families, is potentially a route to removing some of the barriers to a more equitable future for children regardless of their gender.

9.4.2.3 Lack of Representation

A number of the interviewees commented on how there was a lack of representation of females in the curriculum. For example, RP12 said: 'The curriculum that we use is very much a male-dominated curriculum. Any scientists that we talk about are male, in general, apart from Marie Curie, really, who's the only female scientist that anyone can name'.

9.4.3 Role of School Biology in Addressing Social Justice Issues to Do with Gender Identified by Teacher Educators

Two principal themes were identified: 'Classroom climate' and 'Tackling injustice'.

9.4.3.1 Classroom Climate

Teacher educators expressed the importance of the classroom climate in helping to promote a social justice agenda. Valuing every member of the class, representing them where possible during teaching, recognising that we are helping prepare students for the rest of their lives were some of the non-subject-specific suggestions made (e.g., by RP2, RP4, RP12).

It was questioned whether 'we can explore relations of power within school science' (RP4), with RP4 couching this specifically as 'looking at people like women'.

When teaching sex education, the importance of including all students when teaching about relationships, and of fostering a nurturing, not only a tolerant, but an affirming, ethos, was highlighted: 'For biology colleagues, it is the way that we teach relationships and sex education that can have a massive influence on how people feel about themselves and their right to exist and the right to be in love and all those sorts of things' (RP3).

9.4.3.2 Tackling Injustice

The importance of attention being paid by biology teachers to the context in which the work that they are presenting in lessons was originally carried out was pointed out. This can help paint a richer picture than the standard one in which science is only successfully done by 'privileged western white men' (RP5). This quotation highlights the prevalence of socio-economic, geographic, ethnic and gendered bias encountered by our students as they study science. If we allow these biases to pass unchallenged, we tacitly reinforce what for many may be an unconscious bias about who can 'do' science.

The issue of low numbers of girls choosing to study physics was commented on by many of our research participants, but two were concerned that fewer boys choose to study biology, with RP1 specifically commenting that 'we should worry that not enough boys are doing biology'.

9.5 Discussion

Our findings have identified key areas where initial teacher educators feel that social justice issues may be addressed within school biology education (Sect. 4.1, RQ1), the specific social justice concerns related to people of different genders that initial teacher educators consider particularly relevant (Sect. 4.2, RQ2), and the role they believe school biology could play in addressing these issues (Sect. 4.3, RQ3).

A prominent issue identified was the underrepresentation of women in biology-related fields and the historical invisibility of female biologists in school curricula. Although biology has a relatively high proportion of female students, systemic gender biases persist in representation, career progression, and access to leadership roles (Driessen et al., 2024). Like other STEM disciplines, biology education has often marginalised the contributions of female scientists, reinforcing the perception that key discoveries and advancements have primarily been made by men (Ross et al., 2022). This reflects Schiebinger's (2000) critique of the historical exclusion of women in science, where significant contributions – such as those of Rosalind Franklin in genetics or Barbara McClintock in cytogenetics – have been undervalued or omitted from mainstream narratives.

Beyond gender disparities, the issue of intersectionality further compounds inequalities in biology education. Representation is not solely shaped by gender but also by factors such as socio-economic status, race, and other identity markers that influence participation and recognition (Crenshaw, 2019). Many of the biologists traditionally featured in curricula are not only male but also from privileged backgrounds, reinforcing a narrow and exclusionary view of scientific achievement (Henri et al., 2023). Women of colour, in particular, have been systematically excluded from scientific narratives, despite their significant contributions to biology and related fields (Nash & Moore, 2022). Their absence from curricula limits the diversity of role models available to students and perpetuates the misconception that scientific advancements have been driven predominantly by white men. Addressing these inequalities requires a comprehensive restructuring of the curriculum to include a more representative spectrum of scientists, aligning with calls to challenge historical omissions and promote a more inclusive perspective on scientific contributions (Charlesworth & Banaji, 2019).

For biology education to contribute meaningfully to social justice, it must actively challenge these historical omissions. Research suggests that increasing the visibility of female biologists in teaching materials can positively influence students' perceptions of gender roles in science and boost female students' confidence in pursuing biology-related careers (Clark Blickenstaff, 2005). This is particularly significant given that gender disparities persist at higher academic levels and in research leadership (Charlesworth & Banaji, 2019). However, representation alone is insufficient. A truly inclusive curriculum must also address the structural barriers that have historically excluded women and other marginalised groups from scientific recognition.

Freire's (1996) concept of education as a site of 'true dialogue' is relevant here. Rather than treating inclusion as a superficial exercise in diversifying content, biology education should create spaces for critical engagement, where students explore the historical and structural factors that have shaped the discipline. Encouraging open discussions about the systemic exclusion of women and other marginalised groups in science enables students to develop a deeper understanding of these inequities. Such an approach not only challenges exclusionary narratives in science but also empowers students to actively question and reshape the discipline's future.

Another concern that emerged in the findings relates to gendered subject choices within science education. Traditionally, discussions around gender disparity in science have focused on the underrepresentation of girls in fields like physics, engineering, and computer science (Charlesworth & Banaji, 2019). However, an ongoing issue is that fewer boys than girls opt to study biology (Smith & White, 2024). These patterns in science education reflect the reality that subjects are often gendered, based on societal expectations.

One contributing factor is the not infrequent perception of biology as a less 'rigorous' or 'prestigious' science compared to physics or chemistry (Scantlebury, 2014). Cultural attitudes, reinforced by media portrayals, often associate certain scientific fields with specific gender norms, shaping students' perceptions of what is considered an appropriate subject choice. As a result, biology is often perceived as more suitable for females, while physics and chemistry continue to be linked with traditional notions of masculinity. This reinforces existing gender divides in subject selection and career aspirations.

Addressing these issues requires a broader re-evaluation of how science is framed within education. While efforts to increase female participation in male-dominated fields remain crucial, there is also a need to challenge gendered assumptions about biology and other sciences. A more inclusive and socially just science curriculum would not only highlight the contributions of women and marginalised scientists but also work to dismantle stereotypes that shape subject choices. By fostering critical discussions about gendered narratives in science, educators can encourage all students to engage with the subject based on interest and ability rather than societal expectations.

The argument that school science should contribute to social justice has gained considerable traction in recent years (Rodriguez, 2015). Our findings align with this growing recognition, as the majority of interviewees acknowledged that science education can play a role in promoting social justice. However, there was significant variation in how participants understood the concept, and the discussions often lacked theoretical depth. Key areas such as the aims of education, the nature of social justice, diverse worldviews, pedagogy, and curriculum design were rarely explored in a structured or reflective way. Instead, the discourse tended to focus on blanket endorsements or rejections of specific approaches to science education.

For example, there was noticeable resistance to what is sometimes referred to as 'girl-friendly' science – efforts to tailor science content to stereotypical gendered interests. This reflects broader concerns within the literature about the risks of reinforcing gender stereotypes through the curriculum (Brotman & Moore, 2008). While some argue that such approaches could make science more appealing to girls (Saini, 2012), others, as reflected in our findings, contend that these efforts can be patronising and counterproductive (Scantlebury, 2014). This resistance also aligns with the work of Archer et al. (2012), who argue that the marginalisation of certain topics in the curriculum can inadvertently contribute to the gendered division of knowledge. In this view, emphasising 'feminine' topics may unintentionally undermine the status of science, presenting it as less rigorous or prestigious, and can limit students' broader engagement with the subject.

Overall, there was consensus that biology education presents lots of opportunities to raise social justice issues in the classroom though there was a range of opinions about how much school biology should engage. A number of areas of the biology curriculum were highlighted as being suitable to teach from a social justice stance.

In addition to curricular areas, interviewees frequently talked about issues to do with discrimination and access.

We would argue that there is a synergy between good biology teaching and teaching for social justice when teaching about gender, sex and reproduction. For example, the issue of gender diversity illustrates a number of points. In particular, its recognition does not fit with a narrow, essentialist reading of sex and gender in which there is a one-to-one correspondence between a person's sex (understood as a binary) and their gender (also understood as a binary). While most humans are unambiguously male or female, not all are. If we just focus on sex determination, the term 'intersex' is widely used to identify individuals who cannot unambiguously be assigned (whether at birth or later) as male or female. While some intersex individuals find the term helpful, others do not. It can be read as reifying the notion of two 'correct' sexes with 'intersex' stuck in between in a sort of no-man's (!) land. There are many types of intersexuality and it is difficult to determine its overall extent. If one includes conditions like Klinefelter syndrome (XXY individuals), Turner syndrome (XO individuals, with a total of 45 chromosomes as they have a single sex chromosome) and late-onset adrenal hyperplasia, a figure of 1.7% has been calculated. Adopting much narrower criteria, a figure of 0.02% has been calculated. A recent review noted that if we use a definition of intersexuality that focuses on genitalia that make classification at birth as a boy or girl difficult, within certain populations a figure of about 0.6% is reasonable (Reiss, 2023).

Good biology teaching needs, when teaching about gender, sex and reproduction, to draw on anatomy, on genetics, on physiology and on endocrinology. This particular topic illustrates well the reality is that as humans, we are neither independent of our biological heritage nor reduced to it. Good biology teaching can help students appreciate this. Indeed, when we think of the whole of the natural world, including but also beyond humans, it is clear that there are a large number of standard school biology topics where biology teaching that engages with issues of social justice also entails (requires) the teaching of good biology.

In conclusion, the school science initial teacher educators we interviewed generally felt that social justice issues could and should be addressed in school biology. The area of the biology curriculum that they most often mentioned as being one where it was particularly appropriate to tackle issues of social justice was reproduction, and they had a range of suggestions as to how school biology could play a role in teaching about issues to do with social justice that would be particularly relevant to people of different genders.

References

Ahn, J., Luna-Lucero, M., Lamnina, M., Nightingale, M., Novak, D., & Lin-Siegler, X. (2016). Motivating students' STEM learning using biographical information. *International Journal of Designs for Learning, 7*(1), 71–85.

Archer, L., DeWitt, J., Osborne, J., Dillon, J., Willis, B., & Wong, B. (2012). Science aspirations, capital, and family habitus: How families shape children's engagement and identification with science. *American Educational Research Journal, 49*(5), 881–908.

Borger, J. G. (2023). From sex to biology: The case for gender-neutral language in science education. *Immunology & Cell Biology, 101*(8), 690–692.

Braun, V., Clarke, V., & Weate, P. (2016). Using thematic analysis in sport and exercise research. In *Routledge handbook of qualitative research in sport and exercise* (pp. 213–227). Routledge.

Brotman, J. S., & Moore, F. M. (2008). Girls and science: A review of four themes in the science education literature. *Journal of Research in Science Teaching, 45*(9), 971–1002.

Charlesworth, T. E., & Banaji, M. R. (2019). Gender in science, technology, engineering, and mathematics: Issues, causes, solutions. *Journal of Neuroscience, 39*(37), 7228–7243.

Chatters, L. M., Taylor, R. J., & Schulz, A. J. (2022). The return of race science and why it matters for family science. *Journal of Family Theory & Review, 14*(3), 442–462.

Clark Blickenstaff, J. (2005). Women and science careers: Leaky pipeline or gender filter? *Gender and Education, 17*(4), 369–386.

Crenshaw, K. (2019). The marginalization of Harriet's daughters: Perpetual crisis, misdirected blame, and the enduring urgency of intersectionality. *Kalfou (Santa Barbara, Calif.), 6*(1), 7–23.

Department for Education (2014) *The national curriculum in England: Key stages 3 and 4 framework document.* Available at https://www.gov.uk/government/publications/national-curriculum-in-england-secondary-curriculum

Dimick, A. S. (2012). Student empowerment in an environmental science classroom: Toward a framework for social justice science education. *Science Education, 96*(6), 990–1012.

Driessen, E. P., Walker, K. E., Hallman, T., Casper, A., Eddy, S. L., Schneider, J. R., & Lane, A. K. (2024). "It's been a process": A multiple case study of biology instructor efforts to reform their sex and gender =curriculum to be more inclusive of students with queer genders and intersex students. *CBE Life Sciences Education, 23*(4),

Freire, P. (1996). *Pedagogy of the oppressed (revised edn).* Continuum.

Henri, D. C., Coates, K., & Hubbard, K. (2023). I am a scientist: Overcoming biased assumptions around diversity in science through explicit representation of scientists in lectures. *PLoS One, 18*(7)

Lee, V. R., Wilkerson, M. H., & Lanouette, K. (2021). A call for a humanistic stance toward K–12 data science education. *Educational Researcher, 50*(9), 664–672.

Lesnefsky, R. R., Kirk, E. A., Yeldell, J., Sadler, T. D., & Ke, L. (2023). Socioscientific modelling as an approach towards justice-centred science pedagogy. *London Review of Education, 21*(1), 30.

Levy, B. S., & Sidel, V. W. (Eds.). (2013). *Social injustice and public health.* Oxford University Press.

Liou, D. D. (2023). Expectations as property of white supremacy: The coloniality of ascriptive expectations within the racial contract. *Race Ethnicity and Education, 26*(4), 478–496.

Long, S., Steller, L., & Suh, R. (2021). Gender-inclusive biology: A framework in action. *The Science Teacher, 89*(1), 27–33.

Mansfield, J., & Reiss, M. J. (2020). The place of values in the aims of school science education. *Values in science education: The shifting sands*, 191–209.

Merriam, S. B., & Tisdell, E. J. (2015). *Qualitative research: A guide to design and implementation* (4th ed.). Jossey Bass.

Morales-Doyle, D. (2023). Putting science education in its place: The science question in social justice education. *Cultural Studies of Science Education, 18*(1), 81–94.

Nag Chowdhuri, M., King, H., Godec, S., & Archer, L. (2023). Towards justice-oriented science teaching: Examining the impact of the science capital teaching approach on teachers. *London Review of Education, 21*(1), 37.

Nash, M., & Moore, R. (2022). In/visible: The intersectional experiences of women of color in science, technology, engineering, mathematics, and medicine in Australia. *Gender, Work & Organization, 29*(5), 1571–1590.

Reiss, M. J. (2023). A biologist looks at sex and gender. *Education in Science, 293*, 10–12.
Roberts, D. A., & Bybee, R. W. (2014). Scientific literacy, science literacy, and science education. In N. G. Lederman & S. K. Abell (Eds.), *Handbook of research on science education* (Vol. II, pp. 559–572). Routledge.
Rodriguez, A. J. (2015). Managing sociocultural and institutional challenges through sociotransformative constructivism: A longitudinal case study of a high school science teacher. *Journal of Research in Science Teaching, 52*(4), 448–460.
Rodriguez, A. J., & Morrison, D. (2019). Expanding and enacting transformative meanings of equity, diversity and social justice in science education. *Cultural Studies of Science Education, 14*, 265–281.
Ross, M. B., Glennon, B. M., Murciano-Goroff, R., Berkes, E. G., Weinberg, B. A., & Lane, J. I. (2022). Women are credited less in science than men. *Nature, 608*(7921), 135–145.
Saini, A. (2012). Making science more female friendly. *MRS Bulletin, 37*(6), 547–548.
Scantlebury, K. (2014). Gender matters: Building on the past, recognizing the present, and looking toward the future. In N. G. Lederman & S. K. Abell (Eds.), *Handbook of research on science education* (pp. 187–203). Routledge.
Schiebinger, L. (2000). Has feminism changed science? *Signs: Journal of Women in Culture and Society, 25*(4), 1171–1175.
Sjöström, J., & Eilks, I. (2018). Reconsidering different visions of scientific literacy and science education based on the concept of *Bildung*. In Y. J. Dori, Z. R. Mevarech, & D. R. Baker (Eds.), *Cognition, metacognition, and culture in STEM education: Learning, teaching and assessment* (pp. 65–88). Springer.
Smith, E., & White, P. (2024). Gender, participation and attainment in STEM: A comprehensive overview of long-term trends in the United Kingdom. *British Educational Research Journal.*
Taasoobshirazi, G., Puckett, C., & Marchand, G. (2019). Stereotype threat and gender differences in biology. *International Journal of Science and Mathematics Education, 17*, 1267–1282.
Upegui, D., Coiro, J., Battle, S., Kraus, R., & Fastovsky, D. (2022). Integration of the topic of social justice into high school biology curricula. *Science & Education, 31*(4), 923–941.

Part II
Scientific Inquiry, Scientific Explanation and Nature of Science

Chapter 10
Metrological Concepts in Biology Courses in Higher Education

Myriam Régent-Kloeckner, Clément Maisch, and Christophe Daussy

10.1 Theoretical Background

Uncertainty is inherent in science, as it is in new societal issues, and it challenges biology education, both for teaching scientific and socio-scientific topics. Previous studies have defined different types of uncertainties without resulting in a consensus. For instance, Gustafson and Rice (2019) refer to four types of uncertainty: deficient uncertainty, technical uncertainty, scientific uncertainty and consensus uncertainty. In contrast, Blauza et al. (2024) subsume deficient and scientific uncertainty into epistemic uncertainty, rename technical uncertainty as measurement uncertainty, and add two other types: structural uncertainty and aleatory uncertainty (this last type corresponds to the definition of risk according to Neth & Gigerenzer, 2015). Most studies addressing uncertainty in science (as seen in the symposium "Making sense of uncertainty in biology education" in ERIDOB24) focus on one or all of these types of uncertainty but not on measurement uncertainty. This key concept of metrology is overlooked or reduced to merely communicating a margin of error (Gustafson & Rice, 2019). However, the evaluation of measurement quality and significance (including

M. Régent-Kloeckner (✉)
Sciences Et Société; Historicité, Éducation Et Pratiques, Université Claude Bernard, Lyon 1, Lyon, France
e-mail: myriam.regent-kloeckner@univ-lyon1.fr

C. Maisch
Laboratoire de Didactique André Revuz, CY Cergy Paris Université, Université Paris Cité, Université Paris Est Creteil, Université Lille, Université Rouen, Paris, France
e-mail: clement.maisch@cyu.fr

C. Daussy
Laboratoire de Physique Des Lasers, Université Sorbonne Paris Nord, Centre National de La Recherche Scientifique: UMR7538, Paris, France
e-mail: christophe.daussy@univ-paris13.fr

O. Morin et al. (eds.), *Challenges of a Changing World in Biology Education*, Contributions from Biology Education Research,
https://doi.org/10.1007/978-3-032-05346-6_10

uncertainties) should be a key component of science practice to prevent misinterpretation. One of the most striking examples remains the famous debate surrounding the alleged discovery that neutrinos travel faster than the speed of light, as initially reported in 2011 by the Opera experiment (Adam et al., 2011). Recently, Gaume and Desquilbet (2024) reported numerous issues in the InsectChange ecological database, which underlie certain meta-analysis. Some of these issues include "*errors in insect counts, sampling biases, inclusion of noninsects driving assemblage trends, omission of drivers investigated in original studies and inaccurate assessment of local cropland cover*" (p. 1). The authors also explain that the use of disparate and non-standardised units of measurement (which could be linked to metrology concepts) is a "*general issue of the database*", leading to incorrect data transformation and flawed trend estimations in the meta-analysis. They conclude: "*[…] InsectChange conveys unsubstantiated information to scientists, decision-makers and the general public*" (p. 17).

Metrology is considered as the science of measurement (Perdijon, 2020) and pertains to all experimental sciences. According to the French Network of Metrology, its goals are to design methods, to formulate observing conditions of a phenomenon, to build and maintain instruments and to establish the meaning of the results obtained. More precisely, metrology defines a methodological framework to help experimenters ensure a certain level of confidence in the results they obtain when conducting measurements. Thus, Perdijon (2020) describes the measurement process as consisting of four key components: the definition of the quantity to be measured, the taking of a measurement, the quality of the measurement, and its statistical processing. The metrological framework encompasses all these components. However, in this study, we will consider two levels of experimenter practice:

- A common level involves conducting an experimental process aimed at obtaining a single qualitative or quantitative result for each set of experimental conditions. Interpretation is typically based on comparisons of results or comparison of data with a model.
- An advanced level that considers confidence in measurements by incorporating more specific metrological concepts, such as the definition of measurement uncertainties linked to statistical processes due to repeatability or reproducibility, as well as the evaluation of uncertainties throughout the measurement process. Interpretation is enhanced by including feedback on the measurement process.

Biology, based on an experimental approach, deals increasingly with quantitative data, and could thus benefit from the metrological framework. Unfortunately, these concepts and methods are obscure to most biology researchers or are restricted to statistics (Coxon et al., 2019; Houle et al., 2011). Lange (2000) pinpoints a key question about the meaningfulness and reliability of a measurement in biology. This question is linked to the Nature of Science (NoS) and to the representational measurement theory which discriminates between the different kinds of attributes, or scales type of measurement (Houle et al., 2011; Lange, 2000). Regarding biology education, this reflection is central to teach how a discrepancy between measurements and its meaning could lead students to make mistakes in the analysis and the interpretation of data and to misunderstand biological concepts and more generally the NoS. Therefore, we suggest that an acculturation to the key concepts of metrology, such as measurement errors or uncertainties, simple statistical data treatment (standard deviation), significant digits, instrumental calibration, etc. might improve the students' confidence in scientific results and their significance and thus future researchers' competences.

Currently, a first approach to metrology is introduced in the French high school physics and chemistry curriculum. However, references to measurement are sparse in the biology curriculum which focuses more on the common level of practice in an experimental approach. For instance, there are only two references in the senior year curriculum: one dealing with the variability of results and the other concerning the measurement process. Thus, related skills to measurement such as "*identifying error sources to calculate uncertainties for an experimental result*" are mentioned in the French national standards for bachelor's degrees of experimental sciences (2015). Despite these recommendations, it seems that metrology concepts are rarely included in biology course programmes.

Some science education studies deal with the teaching and learning of quantities and measurement concepts. Many studies in mathematics education focus on these concepts at primary school. However, they question the concept of quantity (expressed by integers, and later by decimal numbers) and limit the view of measurement to the exact value of an "ideal" quantity (Chesnais & Munier, 2015). Physics education researchers have studied measurement issues through curricula, uncertainties issues, measurement processes at primary, secondary school or university level (Caussarieu & Tiberghien, 2017; Lubben et al., 2001; Maisch et al., 2008, Passelaigue & Munier, 2015; Pollard et al., 2020; Séré, 2008). Buffler et al. (2009) studied the relationship between the students' views of the NoS and their understanding of the Nature of Scientific Measurement (including the issue of measurement uncertainties). Despite the support from these studies and the new initiatives in higher education curricula in the USA and high school physics curricula in France, none of them indicate a genuine incorporation of metrological concepts into everyday teaching practices. However, when metrology is restricted to statistics processes, some curricular intervention has been shown to improve biological variation understanding (Dewey et al, 2022). But when regarding metrology as the whole process of measurement, only one comparative research study dealing with biology education and measurement during laboratory sessions in biology, chemistry, and physics, was conducted by Fondère et al. (1998). The authors noticed some characteristics of these disciplines and underlined the significance of a decision on the quality of the measurement values. Although this study deals with uncertainties in the measuring systems, larger issues based on the teaching of metrology are not discussed.

Some studies (Coxon et al., 2019; Houle et al., 2011) suggest that concepts of metrology could be transferred to biology research but they might need an adaptation to the disciplinary characteristics dealing with the study of living systems. A lack of adaptation may explain why metrology seems to be rarely a clear goal for university biology teaching (Bachelor degree) but it might also be the result of a lack of knowledge of the concepts and methods by biology teachers and researchers.

In this context, this study investigates to what extent teaching goals related to measurement are rare, and which concepts and methods of metrology are present in the first years of university degree of biology in France. In this study, we analyse the content and objectives of standard curricula. First, a grid of analysis based on 5 dimensions has been defined to identify the concepts and methods dedicated to metrology formulated in the documents of tutorial and laboratory sessions. Teachers' perspectives on the role of measurement in biology research and its place in teaching have also been explored through interviews. Finally, a questionnaire submitted to a large number of teachers and researchers responsible for biology training programs at universities (programme coordinators in the rest of this paper) throughout France, helps us to specify teachers' views about measurement in biology teaching.

10.2 Research Design and Methodology

10.2.1 Analytical Framework

A framework has been defined to characterise measurement in teaching. Firstly, we developed a grid based on the work of Fondère et al. (1998) and the four components describing metrology (Perdijon, 2020) leading to the classification reported in Fig. 10.1. The following 5 dimensions have been defined: 'type of data and measurement', 'implementation', 'management', 'results and values processing' and 'interpretation'.

Each of these dimensions is characterised by items which deal with experimental processes in general (solid-line boxes in Fig. 10.1) or with more specific concepts of metrology (dotted-line boxes in Fig. 10.1). Only items in the dimension 'interpretation' cannot be directly associated with the concepts of metrology or experimental processes (dash-line box in Fig. 10.1).

10.2.2 Course Materials

Subsequently, biology and physics course materials from the first-year university degree of biology were collected in the Paris area. These materials included worksheets for tutorial sessions and written instructions for laboratory sessions that were handed out to students. We analysed 18 worksheets (out of over 40 for the first year) covering a broad range of biology domains (biology of animals, plants or cells, molecular biology, genetics, biodiversity, ecology, biochemistry, biomodelling and scientific methods). Worksheets from the "Physics for Biologists" tutorial sessions

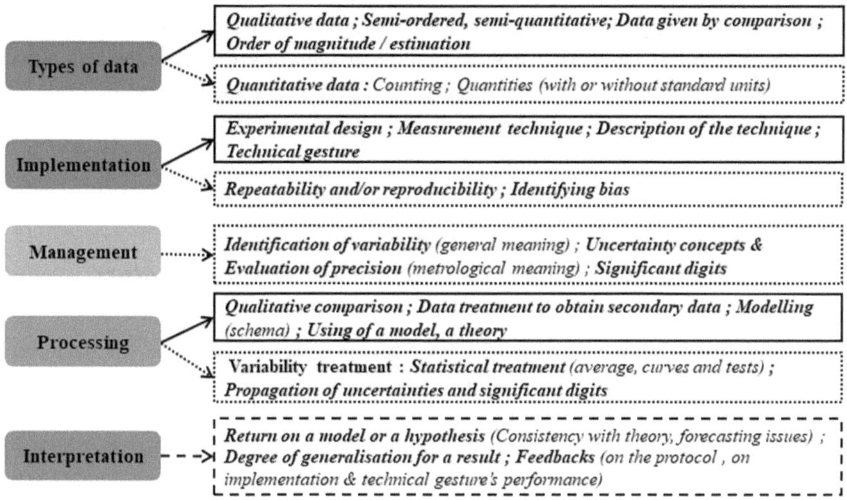

Fig. 10.1 Classification grid: 5 dimensions and associated items to characterise measurement in teaching

were also included in the analysis. Additionally, we analysed 2 out of 3 written instructions for laboratory sessions.

Based on our classification grid, the coding of the documents was performed collectively to reach mutual agreement. Each paragraph, sentence, or part of a sentence corresponding to an item on the grid is counted as a 'hint', representing a single coding unit. An example of analysis is presented in Appendix 1.

10.2.3 Teachers' Interviews

The analysis of documents described in the previous section does not provide information regarding the intentions of teachers when using the documents with students. Therefore, we interviewed four biology teachers, two full-time university teachers without research activities (in animal and plant biology), and two assistant professors engaged in research (in cell biology and genetics). These teachers were responsible for the lectures, authors of the documents and involved in the tutorial sessions. The interview guideline (Appendix 2) was developed to explore (1) the intended purposes dealing with measurement in the teaching sessions and (2) their conceptions (as scientist) of measurement in research, especially in their own research field as assistant professors. The interviews lasted between 25 and 50 min and were audio recorded and transcribed. Thus, the interview transcripts were analysed in order to allow for a better understanding of what constitutes measurement in biology for researchers and for university teachers using our framework of analysis. The aim was to get deeper insights into their pedagogical and practical objectives (mainly through explicit questions), and, finally, to gain an overview of the learning progression in the tasks given to students regarding measurement during the first three years of the bachelor degree in biology at the university. A sample analysis is presented in Appendix 3.

10.2.4 Views on Measurement in Bachelor's Degree Programs: Insights from Programme Coordinators

Except for the guideline provided by the French national standards for bachelor's degrees (MENESR, 2015), French universities do not have a common curriculum. Consequently, to investigate whether or not our results reflect a trend in most university degrees, we developed a questionnaire addressed to programme coordinators, teachers responsible for managing biology training programs (either 1 year or the whole 3-year formation) for 30 biology degrees across French universities. This questionnaire has been structured in 5 parts: (1) practical information about their responsibilities and status; (2) the domains of biology studied in the current and subsequent years of their study program, as well as their connection to other scientific courses included in the bachelor's programme such as physics, mathematics, chemistry or computer science; (3) the status of measurement in programmes; (4) special interest in the different items of the classification grid within biology domains or other scientific fields; (5) the level of competency targeted for each of the 5 dimensions of the grid. We sent the questionnaire to 144 targeted emails of 18 French universities (mainland and French islands). 24 questionnaires were returned and analysed, representing 14 biology bachelor degrees from 13 universities.

10.3 Results

10.3.1 Analysis of Course Materials

Our analysis of biology worksheets of tutorial session revealed several pieces of information. First, we identified several situations in the documents that allow for the exploration of measurement, presenting various pedagogical opportunities. When the text addresses a type of data, it is often presented in a descriptive manner (i.e. intended to solve exercises). However, when the texts focus on the implementation of a measurement (or experiment), they are mostly informative (i.e. not intended to solve exercises). Few are descriptive, and those are solely intended to illustrate procedures and technical principles. For example, in the exercise "*A solution of NaCl is prepared by weighing 10 g of this compound using a precision scale and a weighing pan. What is the molarity of the solution if the final volume is 250 ml?*", the mentioning of the precision scale and weighing pan is unnecessary for solving the

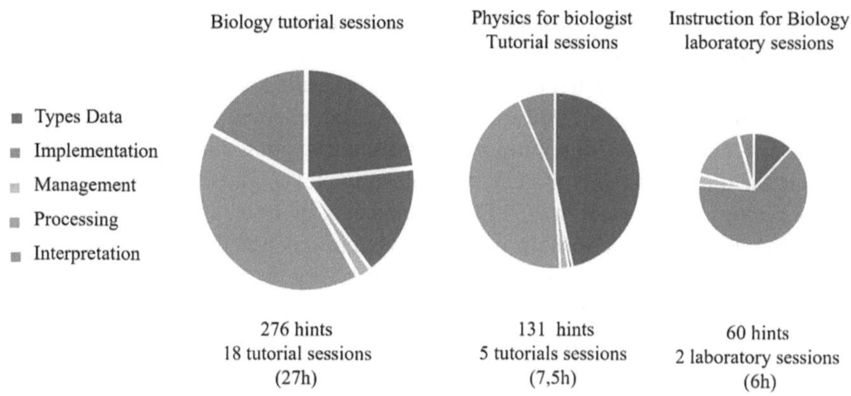

Fig. 10.2 Percentage distribution of hints by dimension in the three types of learning sessions (biology tutorial sessions, 'physics for biologists' tutorial sessions, and biology laboratory sessions)

problem. So, this technical information is considered informative. In contrast, in an exercise focusing on the molecular structure of insulin, students are asked: "*Why are there two protein bands in the presence of SDS and β-mercaptoethanol [and only one in the presence of SDS alone]?*" The introduction to this exercise includes a detailed explanation of key aspects of electrophoresis and the molecular actions of SDS and β-mercaptoethanol, along with a schematic diagram of the electrophoresis setup. To correctly answer this question, students must use the provided information about the molecular effects of β-MSH, and the principles of electrophoresis to deduce that insulin is a dimer composed of two subunits linked by a disulphide bond. They must also verify that the molecular mass of the entire protein is equal to the sum of the molecular masses of its subunits. In this case, the technical background provided is essential for solving the exercise: it is considered descriptive.

Figure 10.2 shows the percentage distribution of hints fitting with items of the grids by dimension.

Each of the 5 dimensions of the grid is represented by at least one hint for every type of session. Depending on the discipline or the type of teaching situations, the percentage distribution of the dimensions is different. More specifically, the dimension "types of data" was all linked to quantities in physics tutorial session (59 hints), but was more diverse in biology tutorial sessions, mixing quantitative and qualitative data (34 quantities, 17 qualitative data, 9 counting data and 8 other types of data). The major goal for tutorial sessions in the worksheets (in biology and physics) focused on the processing of data provided to students. In biology, students had to calculate secondary data with physical laws but also often needed to compare data in order to interpret two or more situations. In physics, students usually only had to calculate secondary data. The implementation dimension was present in biology documents both in tutorial sessions (43 hints) and in laboratory sessions (43 hints) such as naming a technique or describing a procedure, whereas this dimension was hardly ever addressed in physics written instruction (one informative hint, not used to solve problem). Although the majority of hints found in the written instruction for the laboratory sessions dealt with this dimension, students were not engaged in questioning the relevance of this information. The dimension dealing with the management of variability was scarcely covered in all the analysed teaching situations (9 hints all together). Finally, hints of the processing and interpretation dimensions are mainly found in the questions rather than in the explanation texts. The interpretation dimension in biology was present in 2/3 of the worksheets analysed and mainly dealt with discussion about the relationship between assumptions and the consistency of the models (40 hints). Students were required to work very little on feedback on experimental design and its implementation (3 hints) or generalisation (3 hints). In physics tutorial session worksheets, this last dimension was rare (9 hints) and dealt with a return to a model (revising or rejecting a model) or a generalisation of the results.

When examining the items within the different dimensions, more specific concepts of metrology were rare in all situations: there was no item dealing with measurement management and decision-making about confidence in values—the quality of data (reliability, validity) was never questioned—, no item questioned the compliance of the model, and items about reflective feedback on the protocol were rare. The variability of data (caused by measurement or by biological process) was almost missing. Students did not appear to be involved in an advanced level of measurement, for example, by being encouraged to question the quality of the data or the overall measurement process.

10.3.2 Analysis of Teachers' Interviews

When asked whether or not they address measurement with their students, the teachers' first responses were quite varied: from "*I don't directly address the concepts of measurement*" (genetics teacher) and "*I make allusions more than I speak about it*" (animal biology teacher) to "*There's a lot to speak about, as early as my first lecture, similar to what my colleagues are doing [the animal and plant biology teachers], we give data about [cell] dimensions ...*" (cell biology teacher). However, all the interviewed teachers pinpointed the issues of scales and order of magnitude and the main features of some experimental techniques during their teaching sessions. The data mentioned as measurement results were mainly quantitative (quantities or countable), which is consistent with the fact that their definitions of a measurement in biology research was linked to quantification (sometimes restricted to quantities) and often linked to a technique of measurement. For example, the link to a technique of measurement was obvious in the cell biology teacher's discourse: "*[a measure] is the determination of a value, a quantity with various types of units, so, for me, it's necessarily linked to a technique designed to quantify something*". On the other hand, the genetics teacher put forth the understanding of the situation: "*A measurement, it's a value which gives information, hum, about a situation we are analysing to be able to understand, hum, about a mechanism*".

The interviewed teachers did not spontaneously mention issues of reliability and validity of measurement when discussing teaching aspects but one of them spontaneously addressed these issues while explaining her views on measurement in biological research "*[...] essential, whatever. For me, it's a clue for reliability, the validity of a result*" (*Plant biology teacher*). The issue of variability was not raised or was implicit (through mentioning of means or standard deviations) during the teaching part of the interview, but it came out during the second part and was linked to statistics: "*Variability issues are often linked to statistics and can be linked to living matter*" (*Plant biology teacher*); "*I can give a precise measurement if we work with techniques [...] for which you can calculate a statistical mean on a set of molecules*" (*Cell biology teacher*). The dimensions of data processing and interpretation were rarely spontaneously present, either explicitly or implicitly.

Despite the fact that three teachers considered measurement and its quality as important for biology research, measurement education is mostly not addressed or delegated to physics and mathematics lectures without questioning the specificity of measuring biological features: "*I consider these notions [mean, standard deviation, graphical representation...] already known, they were studied elsewhere, they [the students] had maths lectures [in high school]*" (Plant biology teacher). Also, they characterise specificities of measurement in biology by making comparisons with other scientific domains: "*Measurement in biology is considered extremely precise whereas in physics it is more about obtaining an average*" (Cell biology teacher).

Regarding teaching difficulties about measurement concepts, none of them name a specific concept of metrology (such as measurement error, significant digits...). Some teachers talked about the complexity of teaching about the meaning of a specific measure (including the meaning of quantities such as a molecular concentration and the corresponding orders of magnitude in a given context) (as shown in the interview extract presented in Appendix 3).

10.3.3 Analysis of Programme Coordinators' Views About Measurement in Bachelor Degrees

We analysed 24 responses corresponding to general biology bachelor degrees, mostly designed for cohorts of over 100 students (21 responses out of 24).

Regarding the biology disciplines included in their 3-year programme, there exists a diversity of biology domains studied (see Appendix 4). Four of them represent a common core of biology domains (cell biology, biochemistry, molecular biology, biology of organisms) whereas the others could represent local specificities (such as immunology, pharmacology). The course materials analysed fit with the six most represented ones.

The main mandatory courses in other scientific domains are chemistry, statistics or biostatistics, physics (including biophysics), and mathematics. These courses are generally mandatory at least one semester or one year of the formation but rarely over the 3 years. Finally, 15 coordinators described specific work on experimental and/or scientific methods (in which we suggest could be existing hints to measurement education). This dedicated work can be included in a specific biology course (such as ecology) or organised as an interdisciplinary course with other scientific domains or as a transversal course on general skills like communication.

Regarding measurement education, 15 of them declared that measurement is explicitly included in the programme of biology lectures and 5 expressed that it was also sometimes in the programme of other scientific lectures. However, 6 (over 24) of them did not describe any associated goals, and 3 coordinators, who declared that measurement was not a specific focus of their programme, nevertheless described some teaching goals related to measurement. This suggests that these goals might be implicit for students or just locally addressed by some teachers.

The different measurement-related elements identified by the teachers in their courses highlighted certain key items and specificities. To further explore what was taught about measurement, programme coordinators were asked to declare whether each item within our dimensions was covered (or not) in the biology course programme and/or in other scientific courses (see Appendix 5). For all dimensions, most items were not exclusively delegated to other scientific domains and were at least present in some biology courses. All items related to the implementation dimension were declared to be present in their training, whereas other dimensions contained items that were frequently not addressed or that raised issues for coordinators, particularly those related specifically to metrology in the management and processing dimension. The management dimension was associated with the lowest level of skill expected at the end of formation (see Appendix 6). This is consistent with the limited number of hints related to the specific concepts of metrology found in the analysed course materials, except for repeatability and the identification of bias (in the implementation dimension). Even when associated with a partial level, one coordinator explained that "*the concepts can be partially mastered by students at the end of the second year of formation, but the fundamentals of metrology are often left aside, wrongly so, because of a maths phobia among most of our biology students and because of the choices we have to make regarding teaching contents, leading to give little attention/place to metrology. As a result, no students in graduate formation know how to use these concepts.*" The dimension interpretation and the item 'description of techniques' (within the implementation dimension) seemed to be the most specific issues in biology courses. Along with the dimension of processing, the dimensions of implementation and interpretation were associated with the highest expected skill level.

More surprisingly, the expected skill level for the dimension type of data was only partial, despite it appearing to be an important dimension in the course materials and in the discourse of the interviewed teachers.

10.4 Discussion

The framework we defined was useful for classifying the course materials of biology and physics education. We identified two types of items: those related to measurement from a metrological perspective, and those that focused more on an experimental approach. The grid allowed us to distinguish between items related to the variability of measurements and those related to more conventional methods of handling experimental processes.

The data obtained through three methodologies (analysis of course materials, interviews and a questionnaire) revealed common points regarding how measurement is taught in higher education biology in France (in ordinary teaching practices).

First, measurement appears to be a stated issue for biology research (cf. the interviews) but also for biology teaching (interviews and questionnaire). However, the results from the documents, a deeper analysis of the interviews and of the questionnaire responses, using the different items from our grid, showed that the variability and quality of data obtained through measurement processes or techniques are not a systematic issue in teaching biology at bachelor's level.

Second, the documents analysis and the interviews revealed a specific focus on 'technique of measurement', especially for laboratory sessions, cell biology and genetics sessions and their associated teachers (more than for animal or plant biology). The emphasis on measurement techniques may be linked to the common level of experimenter practice (described in the first part) and could be related to their field of expertise, as the emergence of cell biology and genetics is associated with technological advancements during the twentieth century and/or with their roles as researchers. In the questionnaire, the item 'description of techniques' also appears to be a focus in biology programmes, although it is mentioned as an explicit goal by only four programme coordinators.

Third, teachers emphasised 'scales and order of magnitude' as a major goal, especially in the first year of the programmes. We found exercises both in biology and physics that address this goal, and some programme coordinators explicitly corroborated this statement, suggesting that it is a common goal across all scientific disciplines (including biology). However, it is unclear how teachers relate these magnitudes to measurements. Indeed, we found very little evidence of work or discourses on the concepts of variability and reliability. Some interviews indicate a tendency to devolve variability (and metrology) to other science courses. However, the analysis of documents from physics tutorial sessions does not reflect this.

Although tutorial sessions present different types of data, reflexive questions about this dimension are rare. Teachers and questionnaire responses reveal that this dimension is less emphasised compared to skills such as implementation, processing, or interpretation. These observations corroborate Houle's critique (2011), where he explained how a lack of attention to the type of data can induce errors during the measurement processing, leading to misinterpretations.

Finally, the available data provide a preliminary view of how bachelor's degree biology teaching at universities addresses measurement issues, including aspects related to variability (which can be linked to metrology). They recognize its importance for understanding experimental processes and the nature of biology more broadly; however, it appears to be a "hidden" issue for teachers and supervisors, and thus likely for students as well.

10.5 Conclusions

The results of this study shed light on the poor role of measurement in biology education in the first years at French universities. They reveal certain deficiencies, particularly in the critical evaluation of data quality (reliability, validity), which could be an opportunity to teach how confidence in scientific results can- or could-be assessed.

While recent authors have focused on uncertainty issues from a societal perspective, they often overlook the issue of uncertainty directly related to the scientific processes through the measurement procedures. Our results suggest that this issue is also neglected by university biology teachers, even though they acknowledge its potential significance. For example, one programme coordinator's goal is "*to raise awareness that measurement is a source of uncertainty.*"

However, as one programme coordinator rightly pointed out, the issue is not only due to "a lack of time to delve deeper into these concepts" but also due to the fact that "teachers are unfortunately not well (or not at all) alert about metrology. These concepts are poorly addressed in biology labs."

Therefore, we suggest making the various aspects of measurement more explicitly visible in both programmes and course materials, according to the dimensions and items provided to highlight the specificity of biology studies. This can be achieved at least informatively; however, we recommend integrating these explicit descriptive goals into specific sections such as reflective questions at the end of exercises. To achieve this goal, it seems crucial to focus on the management dimension but also to encourage students to interpret data variability, thereby placing them in a more authentic researcher position. Beyond this, a specific curriculum focusing on measurement could be developed in collaboration with teachers to address these issues. The curriculum intervention presented by Dewey et al. (2022) can be used as a pattern to initiate such collaboration. Therefore, we present the argument that biology teachers should improve their acculturation to metrological issues. Furthermore, they will have an additional pedagogical tool for debating recent and ongoing scientific findings with students.

Appendix 1: Example of a Coded Worksheet Using Our Framework of Analysis

(Bold words show the key elements for classification; Colours used in the "Coding 'hint'" feature colours used in Fig. 10.1).

Course material (translated from French)	Coding 'Hint'	Explanation
"During your last laboratory session, you measured by **spectrophotometry, absorbance values** of a toluidine blue solution, a coquelicot solution, and ONPG solution in order to **find the wavelength that has a maximum of absorption** and then **determine the concentration** of these solutions.	Implementation	Description of the experimental context, naming the measurement technique and the purpose of these measurement processes. *No information about technical bias, or any repetition process.*
In the attached table, you will find some measurement results from various groups of students.	Type of Data (quantitative)	The table contains for each substance **absorbance measurements** for different **wavelengths (in nm)**.
1. Which **averages** can be extracted from these results? Calculate them and the corresponding **standard deviation**.	Management & Processing	Choice of averaging Variability treatment
2. Indicate the **type of data** present in the table.	Type of Data	Reasoning about the type of data (quantitative)
3. **Draw a graph** that represents data in order to show that absorbance depends on both the chemical substance and wavelength.	Processing	Data treatment and comparison to determine which wavelength has a maximum absorption for each substance
4. For each substance, estimate the **real concentration of the solution** used for in the laboratory session.	Management & Processing	To carry out this data treatment, they should mobilise the Beer-Lambert's law, which relates absorbance to concentration [using a model], and apply this model on the average value [management] calculated from the measurements obtained at the wavelength of maximum absorbance [data treatment]
Supplementary data: chemical information for each substance: molecular formula, **molecular mass, and molar absorption coefficient** (for the wavelengths of maximum absorption)."	Type of Data (quantitative)	

(continued)

(continued)

Course material (translated from French)	Coding 'Hint'	Explanation
	No interpretation marks	The exercise does not propose any reflection on the validity of the model, nor on generalisation (such as the absorbance of any substance depends on the wavelength according to an absorption spectrum) used, or explicit feedback on why measurement results in the table are fluctuating.

Appendix 2: Interview Guideline

Part of the interview	Planned questions	Associated objectives and hypothesis
Personal information and teaching experience	About their university status and their professional background About their past and current teaching activities: the sub-discipline they teach, their teaching responsibilities, if any, level of education…	Obtaining information about their education and teaching profile
Measuring in their teaching activities	When you are in class, do you talk about measuring? (Often/rarely/always/whenever the opportunity arises/…) In what type of teaching sessions (lecture, laboratory session, tutorial session…)? For which level of education (first/last year of bachelor or graduate students…)?	Investigating the importance of measurement in teaching goals Is the measurement issue addressed in all types of teaching sessions or dedicated to one of them? Is the measurement issue addressed in all students whatever the level of education or restricted to higher levels?

(continued)

(continued)

Part of the interview	Planned questions	Associated objectives and hypothesis
	Do you have different objectives according to these teaching levels? What are your teaching goals associated with the measurement issue? Can you give me some examples of theses teaching moments? What do you exactly talk about measuring/measurement? (generally speaking, then with a special focus for first year of bachelor degree)	Investigating the intended purposes of measuring, while developing more precise information in order to determine if they will discuss uncertainties, how to obtain measurements, techniques, interpretation… (**key ideas of our analyse framework**)
	Is it difficult to discuss the notion of measurement with students? What are the students' difficulties?	Another way to explore measurement dimension (**key ideas of our analysis framework**)
	If you had more time with students, what would you like to discuss about measuring with them?	Another way to explore measurement dimension: for example, is uncertainty education a dreaming world?
Measuring in their research activities	According to you, what is "a measure" in biology research? What is the place of "the measure" in your research activity (assistant professor interview) / in research in general (full-time university teachers' interviews)?	Catching a quick definition of measure: does it focus on the result values? On the process to obtain that/these value(s)? (**key ideas of our analysis framework**) Are measurements central to scientific research for them? Are measurements linked to decision-making, a question, a model? Why and when do they need to measure something?
Measurement's characteristics (in their research context)	In your research, do you work with different types of measures, of quantities? If any, can you give me some examples? (Assistant professor interview, and Ph-D full-time teachers)	Obtaining the **diversity of types of data**, they work with

(continued)

(continued)

Part of the interview	Planned questions	Associated objectives and hypothesis
	How do/did you check your measurements are workable? (Assistant professor interview, and Ph-D full-time teachers) How does a researcher check measurements are workable outcomes? (full-time university teachers' interviews)	Investigating whether they will discuss repeated measurements, statistics and/or measurement uncertainties (**management and handling**), feedback processes on theory and on procedure conception (**interpretation**), degrees of confidence, and decision-making
	Do you make a difference between a measurement protocol and an experimental protocol?	

Appendix 3: Example of an Interview (Extract) Rephrased and Coded Using Our Framework

Extract of one response (translation) to the question: *"Do you experience any difficulties discussing the notion of measurement with students, or do you feel them have trouble with this notion?"*	***Rephrasing and interpretation***
*"**I don't directly address the concepts of measurement** with them. So afterwards there's... I think about a problem with scales (which means that the smaller it is, the harder it is for them to understand) and putting them into any kind of context, and therefore integrating the results in relation to an experiment afterwards. Typically, for example, one of the first experiments we do in L2 [second year of bachelor degree] is a bacterial culture, where we extract a plasmid using a specific protocol, and I calculate my concentration. Often, it's a value, when they do their calculation, with a power value and they don't ask themselves whether the power relates to the real quantity: afterwards if you go by 1.10 minus something gram [1.10 -... g] or 10 to the power of something gram [10... g] that can result in tons of DNA, they sometimes don't even ask themselves the question so...[...] if my result is coherent with what can be expected from that type of experiment."*	Through this response, we learn that: – Measurement and associated competencies are mostly an implicit goal. – Some quantitative data are worked with students (type of data). – The measured values are obtained through a specific protocol. – The measured value is not directly a useful or meaningful result; they need to performed a calculation to obtain a biological or chemical measurement (processing step). – During the calculation, students are confronted with a scale problem (type of data). – Students should perform an interpretation of the calculated results to check the coherence of their process according to the experiment context (feedback), but they usually do not spontaneously do this feedback step. In all the part "Measuring in their teaching activities" of the interview, we do not find any mention of variability. So, the calculation competencies associated with making sense of scale (scientific notation) seem to be more essential (or a prerequisite) than an education on uncertainties.

Coding description
Bold: Highly informative ideas regarding teaching **goals** and **justificative**.
Measurement characteristics: Type of data, Implementation, Management, Processing, Interpretation.

Appendix 4: Frequency Distribution of Scientific Domains (Top) and Biology Domains (Down) in the Bachelor Programmes Described by Programme Coordinators

Programme coordinators were asked to indicate which disciplines are included in the biology degree curriculum. The disciplines were presented as a checklist including a space for additional entries if necessary.

The course materials analysed in the first part correspond to sessions dealing with the domains labelled with an asterisk (*).

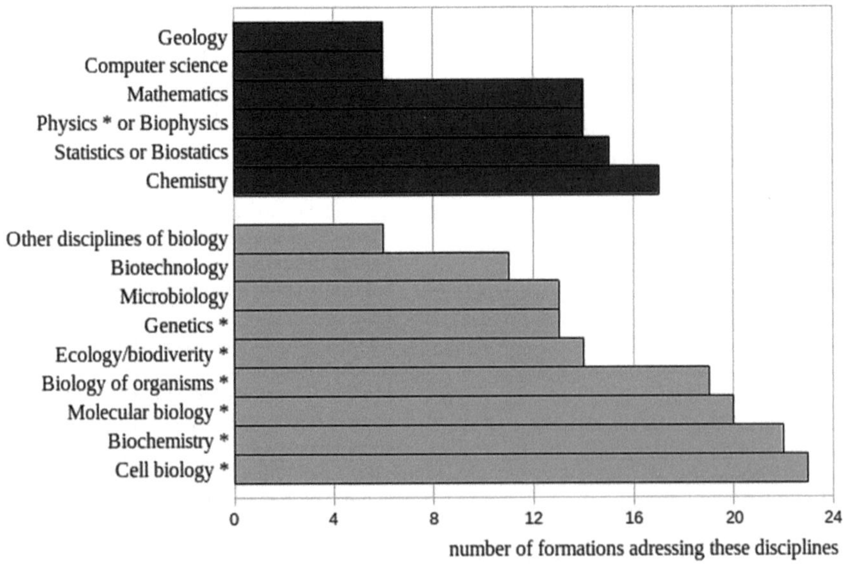

Appendix 5: The Programme Coordinators' Declarations Indicating Whether Each Item of the Grid is Present in the Biology Course Programme and/or in Other Scientific Courses

For each item, programme coordinators were asked to indicate whether it was covered in biology courses, in other science courses, in biololgy and others science courses, not covered at all or if they don't know (button list).

		Biology courses	Other science courses	Biology and other science courses	Not covered / I don't know / No answer
Type of data	Counting (eg., number of drosophila)	9	0	9	6
	Quantitative attribute with unity	10	0	13	1
	Quantitative attribute without unity (ratio, percentage...)	12	0	10	12
	Qualitative observations or descriptive ones with words	12	0	8	4
	Semi-quantitative data (such as grayscale, size as small / medium / large	8	0	8	8
	Order of magnitude	9	0	14	1
Implementation	Experimental design	10	0	12	2
	Description of a measurement technique (PCR, electrophoresis, biodiversity survey...)	16	0	7	1
	Work on specific technical gestures (pipetting, device setting...)	11	0	10	3
	Identifying bias and uncertainties origins (choosing of devices, inappropriate gestures...)	9	0	11	4
	Work on repeated measurements (by a student or results pooling in a group)	13	0	9	2
Management	Qualitative study on the variability of results	13	0	8	3
	Quantitative determination of uncertainties	8	1	9	6

(continued)

(continued)

		Biology courses	Other science courses	Biology and other science courses	Not covered / I don't know / No answer
	Work or awareness given on significant digits	7	0	12	5
Processing	Calculus of average and /or standard deviations (descriptive statistics)	13	0	9	2
	Calculus to obtain secondary results (slope, concentrations, biodiversity index ...)	12	0	7	5
	Taking into account uncertainties in calculus	7	1	7	9
	Statistical treatment (comparison statistical tests, slope linearization…)	15	0	5	4
	Qualitative comparisons (descriptive) of measurement	11	0	8	5
Interpretation	Argumentative feedbacks on the handling: analysis of technical gesture	16	0	5	3
	Argumentative feedbacks on the experimental design	16	0	4	4
	Validation or invalidation of a model or initial assumptions	14	0	6	4
	Interpretation generalisation of results (until the development of an explanatory law/model/ mechanism)	11	0	5	8

Appendix 6: The Skill Level Expected for Each Dimension

For each dimension, programme coordinators should choose one skill level in the following list: high/perfect; medium/partial; awareness; unexpected.

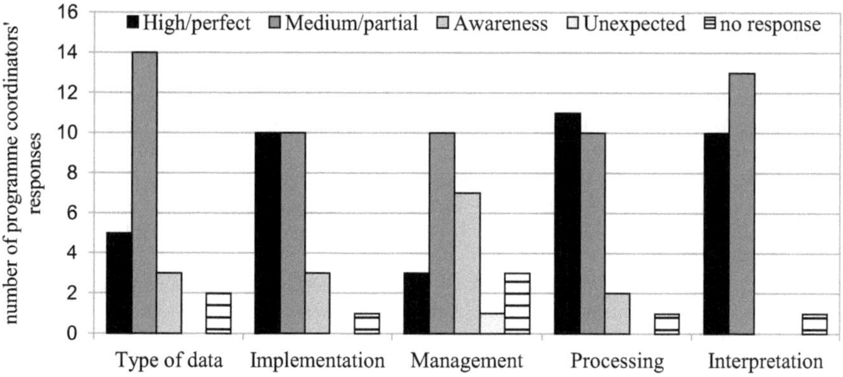

References

Adam, T., et al. (OPERA Collaboration). (2011). *Measurement of the neutrino velocity with the OPERA detector in the CNGS beam*. arXiv:1109.4897

Blauza, S., Kremer, K., & Heuckmann, B. (2024). *Delphi study on the nature of uncertainty in science education and competencies to navigate uncertainty*. 14th Conference of European Researchers in Didactics of Biology, ERIDOB.

Buffler, A., Lubben, F., & Ibrahim, B. (2009). The relationship between students' views of the nature of science and their views of the nature of scientific measurement. *International Journal of Science Education, 31*(9), 1137–1156.

Caussarieu, A., & Tiberghien, A. (2017). When and why are the values of physical quantities expressed with uncertainties? A case study of a physics undergraduate laboratory course. *International Journal of Science and Mathematics Education, 15*, 997–1015.

Chesnais, A., & Munier, V. (2015). Mesure, mesurage et incertitudes: Une problématique interdidactique mathématique/physique. *Proceedings of the Annual Conference of the Association De Recherche En Didactique des Mathématiques, 2015*, 212–237.

Coxon, C. H., Longstaff, C., & Burns, C. (2019). Applying the science of measurement to biology: Why bother? *PLOS Biology, 17*(6), Article e3000338.

Dewey, J., Hicks, J., & Schuchardt, A. (2022). Improving student's understanding of biological variation in experimental design and analysis through a short model-based curricular intervention. *CBE—Life Sciences Education, 21*(1), ar11.

Fondère, F., Pernot, C., & Richard-Molard, C. (1998). Analyse comparative de la gestion du mesurage en TP de DEUG à Orsay (biologie, chimie et physique). *Didaskalia, 12*, 63–92.

Gaume, L., & Desquilbet, M. (2024). InsectChange: Comment. *Peer Community Journal, 4*, Article e97.

Gustafson, A., & Rice, R. E. (2019). The effects of uncertainty frames in three science communication topics. *Science Communication, 41*(6), 679–706.

Houle, D., Pelabon, C., Wagner, G. P., & Hansen, T. F. (2011). Measurement and meaning in biology. *The Quarterly Review of Biology, 86*(1), 3–34.
Lange, J.-M. (2000). Les relations biologie/mathématiques interrogent l'enseignement des sciences de la vie. *Aster, 30*, 123–142.
Lubben, F., Campbell, B., Buffler, A., & Allie, S. (2001). Point and set reasoning in practical science measurement by entering university freshmen. *Science Education, 85*(4), 311–327.
Maisch, C., Ney, M., & Balacheff, N. (2008). Quelle est l'influence du contexte sur les raisonnements d'étudiants sur la mesure en physique? *Aster, 47*, 43–70.
Ministère de l'Éducation Nationale, de l'Enseignement Supérieur et de la Recherche. (2015). *Référentiels de compétences des mentions de licence*. https://www.enseignementsup-recherche.gouv.fr/sites/default/files/content_migration/document/Referentiels_de_competences_licence_formatMESR_2014_12_29_ssblancs_380001.pdf
Neth, H., & Gigerenzer, G. (2015). Heuristics: Tools for an uncertain world. In R. Scott & S. Kosslyn (Eds.), *Emerging trends in the social and behavioural sciences: An interdisciplinary, searchable, and linkable resource* (pp. 1–18). Wiley. https://doi.org/10.1002/9781118900772.etrds0394
Passelaigue, D., & Munier, V. (2015). School teacher trainee's difficulties about the concepts of attribute and measurement. *Educational Studies in Mathematics, 89*, 307–336.
Perdijon, J. (2020). *Pour faire bonne mesure. Entre faits et réalité*. EDPscience.
Pollard, B., Werth, A., Hobbs, R., & Lewandowski, H. J. (2020). Impact of a course transformation on students' reasoning about measurement uncertainty. *Physical Review Physics Education Research, 16*, 020160.
Séré, M.-G. (2008). La mesure dans l'enseignement des sciences physiques. Évolution au cours du temps. *Aster, 47*, 25–42.

Chapter 11
Complexity and Non-knowledge as a Challenge: Is There a Tendency to Prefer Linear Explanations?

Christina Ehras and Arne Dittmer

11.1 Introduction

Biological phenomena are often complex and therefore explanations of those phenomena are characterised, among other things, by multi-causality, interactions between factors, and non-knowledge or uncertain knowledge (Mainzer, 2008; Mitchell, 2008; Potochnik, 2013; Wehling, 2008). Non-knowledge and uncertain knowledge differ in the degree of knowledge that is missing: If there is uncertain knowledge about phenomena, scientists have (data-based) assumptions, whereas if there is non-knowledge about a phenomenon, assumptions are completely missing.

In contrast, teachers' explanations for their students (instructional explanations) tend to focus on the presentation of known facts and on linear, single-cause-effect explanations that may suggest certainty and closure (Braaten & Windschitl, 2011; Jacobson & Wilensky, 2006; Potochnik, 2013). Given that instructional explanations not only communicate content about biological phenomena but also introduce students to adequate explanations of the discipline (Larreamendy-Joerns & Muñoz, 2010), the contradiction between a teaching culture that is often dominated by fact-oriented and complexity-reduced instructional explanations on the one hand and scientific practice characterised by complex explanations from biologists on the other hand requires attention from educational researchers and biology teachers. Especially with current global challenges (e.g. climate change, covid-19-pandemic), the ability to deal with complexity and non-knowledge seems crucial for the education of our students and their participation in democratic processes (Dittmer & Ehras, 2024). There is evidence that communicating non-knowledge and uncertain knowledge can

C. Ehras (✉) · A. Dittmer
University of Regensburg, Regensburg, Germany
e-mail: christina.ehras@ur.de

A. Dittmer
e-mail: arne.dittmer@ur.de

O. Morin et al. (eds.), *Challenges of a Changing World in Biology Education*, Contributions from Biology Education Research,
https://doi.org/10.1007/978-3-032-05346-6_11

be disorienting to people outside of science (especially when opinions are controversial; Gustafson & Rice, 2020; Varwig, 2020), which reinforces the urgency of integrating these aspects of scientific knowledge into the classroom.

In biology didactics, the topic of non-knowledge and uncertain knowledge are usually dealt with under the broad heading of uncertainty (Lübke & Heuckmann, 2024) and is also commonly associated with the assessment of situations (Christensen & Fensham, 2012). However, if non-knowledge and uncertain knowledge are linked to the assessment and consideration of possible actions, ethical complexity is added to the factual complexity, which already includes a variety of cognitive challenges (e.g. Jacobson, 2000; Mehren et al., 2015).

In this study we take a step back and focus on the perception of factual complexity and, in particular, on the communication of non-knowledge and uncertain knowledge in instructional explanations. The study is part of a larger, transdisciplinary research project on criteria of good explanations at the University of Regensburg (Schilcher et al., 2021).

11.1.1 Complexity as a Characteristic of Biological Phenomena

Biological objects and phenomena such as the function of the human brain (Mainzer, 2008) and ecosystem dynamics (Bar-Yam, 1997) serve as prototypical examples for researchers to elaborate characteristics of complex phenomena and complex systems. In the following, common characteristics are presented with reference to the philosophers Kuhlmann (2007), Mainzer (2008) and Mitchell (2008) as well as the system scientist Bar-Yam (1997):

- Complex systems are usually open systems
- For the most part, many elements are involved
- There is a high degree of interconnectedness. One element can be influenced by several other elements and feedback loops are a typical feature
- Individual elements can form networks; the result is a "system of systems"
- There can be non-linear cause-effect
- When critical points are reached, unexpected states of the system can occur at the macroscopic level
- The system is not always an additive aggregate of its elements (emergence)

As a result, complex phenomena exhibit a high degree of variability and dynamics.

Scientists are frequently confronted with non-knowledge and uncertain knowledge when reconstructing complex systems. The borders of the systems are not clear, and elements of the system are sometimes not recognised as such (Kampourakis & McCain, 2020; Mitchell, 2008). Although non-knowledge and uncertain knowledge do not exist exclusively in association with complex systems, they are particularly evident in this context.

11.1.2 Non-knowledge as an Inherent Part of Scientific Knowledge

Considerations relating to the phenomenon of non-knowledge and uncertain knowledge particularly come from the disciplines of philosophy, linguistics, sociology of knowledge and increasingly from the didactics of science. With different emphases, they focus on *subjects*, the characteristics of different *states of knowledge* and their respective *effects* on communication, decision-making, as well as the identification and realisation of actions (see Fig. 11.1).

In terms of the *subjects*, a distinction can be made as to 'who' possesses a knowledge status (non-knowledge, uncertain knowledge, knowledge). The knowledge status pertains either to individual persons, groups of persons, or to all persons equally. When it applies to all persons, knowledge is not available to any person—no (scientific) knowledge has yet been created.

The different possible *knowledge statuses* can be arranged on a quantitative continuum from non-knowledge (meaning a completely blind spot), to uncertain knowledge, to knowledge about a phenomenon (Wehling, 2008). According to Wehling (2008), non-knowledge refers to the complete absence of ideas about factors or interrelationships, whereas in the case of uncertain knowledge there already exist assumptions and ideas about the phenomenon in question. For the classification presented here, the discriminating criterion between the knowledge statuses is not an individual perception (e.g. concerning fake news), but the scientific evidence available. Over time, knowledge can prove to be incorrect or to be valid scientific knowledge (Wehling, 2008). As the amount of data in supporting the explanation increases, the probability of the existence of incorrect knowledge merely

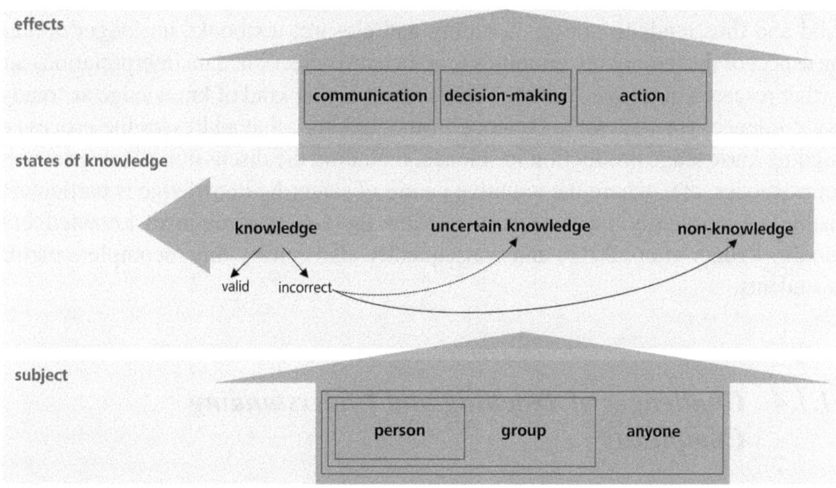

Fig. 11.1 Knowledge-non-knowledge-continuum—central concepts and distinguishable terms (Ehras, 2024)

decreases (Kampourakis & McCain, 2020). The tentativeness of scientific knowledge is also taken into account in approaches of 'nature of science' (Lederman, 2007; McComas & Olson, 1998).

An exemplary analysis of the Bavarian curriculum shows that non-knowledge and uncertain knowledge in biology classes are mainly understood as tentativeness of scientific knowledge (Ehras, 2024). Tentativeness as part of the nature of scientific knowledge is widely discussed in concepts of the 'nature of science' (Lederman, 2007; McComas & Olson, 1998)—but the phenomenon of non-knowledge is considerably wider.

The last level of the figure refers to the *effects* of the knowledge status. This particularly concerns social phenomena arising from how non-knowledge and uncertain knowledge are communicated, perceived, evaluated, and influence decision-making processes. The focus is commonly on uncertainty (Gustafson & Rice, 2020; Kampourakis & McCain, 2020; Lübke & Heuckmann, 2024).

11.1.3 Relationship Between Scientific Knowledge and School Knowledge

In the process of didactic reduction, scientific knowledge is transformed by teachers for the purpose of teaching. The degree of difficulty and the amount of information included in explanations are adjusted in relation to the students' knowledge and abilities (Kattmann, 2008). This process contains critical points with regard to the adequate presentation of subject-specific phenomena: Teachers often use textbooks for preparation that only represent parts of the characteristics of scientific knowledge (Härtig et al., 2012). Textbook knowledge is what is considered to be valid and thus tends to convey certainty and closure; textbooks no longer contain the aspect of discussing uncertainties (e.g. in data collection, data interpretation) and further research questions. Latour (1987) refers to that kind of knowledge as 'ready-made-science' (in contrast to 'science-in-the-making', that addresses the process of ongoing knowledge production in science, including the discussion of uncertainties, controversies, etc., where the tentative nature of scientific knowledge is particularly visible). Accordingly, teachers tend to view the school's canon of knowledge as certain (Krüger et al., 2013) and consequently also convey this incomplete picture to students.

11.1.4 Challenges of Teaching and Understanding Complexity

There are several challenges in dealing with complexity for both explainers and recipients of instructional explanations. Regarding the understanding of complex systems, a tendency towards a 'centralized mindset' is described by Jacobson (2000): Novices tend to reduce the factors of a system towards a central cause, to reconstruct a step-by-step sequence, and to think primarily in linear cause effect relationships. For complex systems, however, a 'decentralized mindset' is required, in which many factors and their interactions bring about the phenomenon (Jacobson & Wilensky, 2006). While novices assume complete predictability of effects, experts also think of probabilistic relationships (Jacobson, 2000). The knowledge and the level of expertise about complex phenomena respectively are therefore important predictors for understanding these phenomena.

In addition, when dealing with complex phenomena, the 'cognitive load' (CL) can be much higher than it is with linear explanations (Grotzer et al., 2017; Sweller, 2011). The term 'cognitive load' originates from the 'cognitive-load theory' (Sweller, 2011). It is assumed that the working memory has limited capacity, and learning involves three types of cognitive load that can fill it. The CL on the working memory in a learning situation results from an *intrinsic* CL (ratio of complexity of the learning content and prior knowledge) and the *extraneous* CL (load due to the favourable or unfavourable design of the learning material). The *germane* cognitive load refers to the part of the cognitive load that is beneficial for productive learning, which arises through the application of higher-level cognitive processes. The *intrinsic* CL and the *extraneous* CL must therefore leave room for learning (the *germane* CL).

Furthermore, there are considerations that expressing non-knowledge can increase a loss of trust in science (Kampourakis & McCain, 2020; Varwig, 2020); a review study by Gustafson and Rice (2020) partially mitigates this assumption, since negative effects are only observed when uncertain knowledge is manifested as a lack of consensus among scientists' positions.

11.1.5 Perspectives on the Quality of Teaching Situations from Different Status Groups

It can be assumed that people perceive, and as a result rate, teaching situations differently due to different experiences, knowledge, aims and needs (Krauss et al., 2020). From various reasons, multi-perspective research approaches are recommended for research on good instructions. In addition to validity aspects, the focus is mainly on a deeper understanding of the research subject such as teacher activities (Fielding & Fielding, 1986). So, a contrastive approach of status groups can lead to new ideas, identification of discrepancies, and a better understanding of the research subject.

The following groups are relevant in the context of educational research on good instructions:

- Students as addressees of the instructions
- Teachers in various stages of professionalisation (teacher students and practicing teachers) as those who design and implement instructions

- Didacticians who train student teachers and can thus make a significant contribution to their knowledge on providing good instructions

11.1.6 Research Objectives

Regarding the relevance of understanding complexity and non-knowledge as characteristics of biological phenomena, the explorative study focuses on the perception of explaining complex phenomena in biology classes to reconstruct difficulties. The following research questions are addressed:

(1) Are linear and complex instructional explanations rated differently?
(2) How are complexity, non-knowledge, and uncertain knowledge perceived in instructional explanations?
(3) Are there differences between status groups in terms of their rating and reasoning?

11.2 Research Design and Method

We developed a computer-based questionnaire containing six video vignettes of instructional explanations. The explanations address different biological phenomena, varying systematically in their degree of complexity. Four different status groups (students, student teachers, teachers, didacticians) were asked to rate the explanations and give reasons for their ratings afterwards. The perspectives of the status groups are brought together to triangulate perspectives for a thorough understanding of the research objective (Fielding & Fielding, 1986).

11.2.1 Instrument

Online questionnaire: Throughout the questionnaire, the participants are shown six video vignettes of instructional explanations. After each explanation, they initially give a global rating with a school mark. In Germany, the scale of school marks ranges from 1 (best mark) to 6 (worst mark)—the likert scale was designed accordingly (Item: "What school mark would you give this explanation overall?"). The mark can additionally be made more precise with a + or − (Item: "You can also indicate a tendency in this section."). The respondents were then asked to provide a reason for this rating. The reasons were stated in an open text field (Item: "You are welcome to briefly provide reasons for your grade here.").

Video vignettes: The vignettes were planned or re-enacted, i.e. they are classified as 'staged videos' (Seidel & Thiel, 2017). To conceptualise and operationalise the degree of complexity (independent variable [IV]), two aspects were chosen: the

Table 11.1 Topics of the six video vignettes

Subject area/Version	Linear explanation	Complex explanation
Genetics	Cause of cystic fribrosis (G_L)	Causes of multiple sclerosis (G_C)
Ecology	Cause of population fluctuations (snowshoe hare) (E_L)	Causes of population fluctuations (common toad) (E_C)
Physiology	Cause of a vitamin A deficiency (P_L)	Causes of a folate deficiency (P_C)

number of factors involved and the inclusion of non-knowledge and uncertain knowledge. Multi-causality as well as non-knowledge and uncertain knowledge are central to the concept of complexity (Bar-Yam, 1997; Kuhlmann, 2007; Mainzer, 2008; Mitchell, 2008); they can be regarded as "core concepts" (Jacobson & Wilensky, 2006, p. 21). Furthermore, they can be communicated within instructional explanations. Thus, the implemented vignettes of linear explanations provide only one cause to explain the phenomenon and do not explicitly mention non-knowledge and uncertain knowledge. Complex explanations, in contrast, include several causes, show interactions between factors, and explicitly refer to non-knowledge and uncertain knowledge. The selected topics are shown in Table 11.1.

The topics explained in the vignettes are selected based on the prior knowledge of the 9th grade of the Bavarian secondary school; the topics themselves are to be found in the curriculum of the 10th or 11th school year. In addition to their placement in the curriculum, there were two other considerations in the selection process: two similar phenomena had to be suitable for the explanation in a linear and complex variant and the explanation had to be feasible in about three minutes.

To increase reliability and validity, the order in which the six video vignettes were presented was altered systematically and interviews were conducted with two students, two student teachers, and two didacticians to ensure the validity of the IV realisation.

11.2.2 Data Analysis

11.2.2.1 Quantitative Analysis of the Global Ratings

Mixed ANOVAs are used to investigate differences in ratings of linear and complex instructional explanations as well as between the four status groups (dependent variable: global rating; independent variables: status group, degree of complexity). Regarding the research questions, the main effect on the degree of complexity and the interaction effect between degree of complexity and status group are of particular interest.

For the analysis, the global rating was combined with the optional specifications (+ or −): a + improves the given rating by 0.3 (2 + = 1.7) and a−lowers the given rating by 0.3 (2− = 2.3).

11.2.2.2 Qualitative Analysis of the Given Reasons

The responses in the open text fields are used to answer the research question addressing reasons for the ratings and differences between status groups. They were analysed by using structuring and evaluative content analysis (Kuckartz, 2018). The category system and the coding process of the responses are described below.

Category system: The category system was developed deductively and inductively. The final category system contains six main categories that summarise criteria for good instructional explanations: addressee orientation, structuredness, linguistic comprehensibility, teacher's speech and body expression, representations, and subject-specific criteria. Each main category has several subcategories. The category for subject-specific criteria contains nine subcategories with six directly addressing the perception of complexity, non-knowledge, and uncertain knowledge: inclusion of non-knowledge, perceived vagueness, feeling of (un-)certainty, orientation towards the phenomenon including the complexity of the natural phenomenon, and orientation towards an underling principle. Each subcategory has three valences to identify good and poor implementations in the explanations: a positive, a negative, and a neutral.

Coding process: In a first step, the main- and subcategories were coded. The entire material was analysed by two trained coders and discrepancies were discussed. In a second step, valences were coded; 25% of the material was analysed by two trained coders (interrater agreement 89.7%).

11.3 Results

11.3.1 Participants

In total, the sample (see Table 11.2) consisted of 250 participants. The participating teachers mostly have a role at their school in which they rate colleagues' lessons, provide further training for colleagues or train student teachers. Teachers were recruited through personal contacts and during various teacher traning courses. Didacticians ($n = 31$) were contacted by e-mail; they work at German universities in the field of biology education and are professors ($n = 6$), have a PhD ($n = 8$) or are assistants for teaching and research ($n = 17$). The teacher students ($n = 51$) are enrolled at the University of Regensburg to study biology as a secondary school teacher (German expression: Gymnasium) and were initially contacted by e-mail. Due to low response rates, the survey was then integrated into various seminars.

Table 11.2 Sample

Status group	*N*	Gender f/m/d	Grade/Semester/Work experience *M* (*SD*)	
Students	134	–	Grade	10
Student teachers	51	33/18/–	Sem.	5.33 (2.17)
Teachers	34	19/15/–	Work.	15.63 (9.48)
Didacticans	31	15/15/1		–
$\sum$	250	67/48/1		

Note N: number of participants; f/m /d: female, male, diverse; *M*: mean, *SD*: standard deviation; grade: school grade; sem.: number of semesters studied; work.: work experience in years

The students surveyed ($n = 134$) were in the 10th grade of secondary school. Five classes were surveyed at three Bavarian schools; contact was established via the school headmasters and teachers were able to voluntarily register for participation with their class.

11.3.2 Ratings of Linear and Complex Explanations

Table 11.3 shows the descriptive values and the results of the mixed ANOVA.

Genetics: There is a significant interaction effect for the genetics explanations—the linear variant is rated better depending on the status group, didacticans are the only group to rate the complex variant better. Post-hoc tests (Bonferroni corrected t-tests for dependent samples) for differences between the rating of linear and complex explanations within a status group show significant differences for students, student teachers and teachers. So, for didacticians there is a tendency to rate the complex explanation better.

Ecology: A significant main effect is found for the ecological explanations—all groups prefer the linear version. Post-hoc tests show that there are significant differences between the groups of student teachers and teachers.

Physiology: There are no significant differences in the ratings of the physiological explanations—only students tend to prefer the linear explanation, while all other status groups prefer the complex explanation.

11.3.3 Reasons for the Ratings of Linear and Complex Explanations

Along with numerous specifics for each video vignette, some tendencies can be identified in the reasons stated in the open text field regarding the ratings of complex and linear explanations. The following examples from the data material, printed in

Table 11.3 Mean ratings (incl. standard deviations) differentiated by status group and topic of explanation, and analysis of variance with a mixed design

<table>
<tr><td>Topic</td><td colspan="3">Genetics</td><td colspan="3">Ecology</td><td colspan="3">Physiology</td></tr>
<tr><td>Degree of complexity</td><td colspan="2">Linear</td><td>Complex</td><td colspan="2">Linear</td><td>Complex</td><td colspan="2">Linear</td><td>Complex</td></tr>
<tr><td>Status group</td><td colspan="2">M (SD)</td><td>M (SD)</td><td colspan="2">M (SD)</td><td>M (SD)</td><td colspan="2">M (SD)</td><td>M (SD)</td></tr>
<tr><td>Students ($n_{\max} = 134$)</td><td colspan="2">2.01 (0.76)**</td><td>2.30 (0.74)</td><td colspan="2">1.84 (0.81)</td><td>2.00 (0.85)</td><td colspan="2">2.04 (0.72)</td><td>2.15 (0.79)</td></tr>
<tr><td>Student teachers ($n_{\max} = 51$)</td><td colspan="2">1.88 (0.81)**</td><td>2.22 (0.86)</td><td colspan="2">1.87 (0.89)*</td><td>2.14 (0.75)</td><td colspan="2">2.09 (0.81)</td><td>1.88 (0.68)</td></tr>
<tr><td>Teachers ($n_{\max} = 34$)</td><td colspan="2">1.98 (0.58)*</td><td>2.38 (0.98)</td><td colspan="2">2.20 (0.77)*</td><td>2.64 (0.95)</td><td colspan="2">2.28 (0.75)</td><td>2.06 (0.88)</td></tr>
<tr><td>Didacticains ($n_{\max} = 31$)</td><td colspan="2">2.01 (0.87)</td><td>1.88 (0.77)</td><td colspan="2">1.99 (0.69)</td><td>2.33 (1.05)</td><td colspan="2">2.06 (0.82)</td><td>1.90 (0.75)</td></tr>
<tr><td>ANOVA</td><td>df</td><td>F</td><td>η^2</td><td>df</td><td>F</td><td>η^2</td><td>df</td><td>F</td><td>η^2</td></tr>
<tr><td>Complexity</td><td>1</td><td>8.51**</td><td>0.04</td><td>1</td><td>11.96**</td><td>0.05</td><td>1</td><td>2.47</td><td>0.01</td></tr>
<tr><td>Status group</td><td>3</td><td>1.39</td><td>0.02</td><td>3</td><td>5.25**</td><td>0.07</td><td>3</td><td>0.72</td><td>0.01</td></tr>
<tr><td>Complexity × status gr.</td><td>3</td><td>2.71*</td><td>0.03</td><td>3</td><td>0.74</td><td>0.01</td><td>3</td><td>2.08</td><td>0.03</td></tr>
<tr><td>Residuals</td><td>229</td><td></td><td></td><td>226</td><td></td><td></td><td>222</td><td></td><td></td></tr>
</table>

Note n_{max}: maximum number of participants; *M*: mean; *SD*: standard deviation; ANOVA: analysis of variance with mixed design; *df*: degrees of freedom; *F*: *F*-value; η^2: partial Eta squared; level of significance (in the upper halb of the tbale for Bonferroni-corrected post-hoc t-tests comparing ratings between linear and complex explanations within a topic and one status group and in the lower half of the table for the F-test of the ANOVA): *: $p \leq 0.05$; **: $p \leq 0.01$

grey boxes, provide illustrations; the language of the survey was German, for the purpose of this article the statements have been translated.

Students indicate with their statements that they expect explanations to be specific in the form of findings or a take-home message, which is sometimes not achievable due to gaps in knowledge. Reasons for the ratings of complex explanations therefore relate negatively to a perceived vagueness.

> *"I really liked the notes" ($Ö_L$)*
> *"no clear result" ($Ö_C$)*

What is missing in students' statements are references to the implemented non-knowledge. In other words, non-knowledge as a feature of complex explanations does not initially seem to be globally remarkable or disturbing.

Teachers rate complex explanations worse if they perceive vagueness. Students also cite vagueness negatively, but the causes of the perceived vagueness differ: Teachers evaluate negatively that only possible reasons for the phenomenon are presented.

> *"But with limited substantial content: the immune system attacks nerve cells. All causes appear rather arbitrary, without any mechanism" (G_C)*

Additionally, they refer to non-knowledge exclusively in a negative way. From the reasons for the inappropriateness of non-knowledge, a connection with the function of instructional explanations can be reconstructed: The assumed goal of instructional explanations is to convey knowledge, which is apparently interpreted as conveying known facts.

> *"A reasonable learning outcome could only be: We still do not know many important connections, maybe someone among you wants to investigate something like that in the future" (G_C)*

Another distinctive feature of the material is, that teachers rarely refer to the adequacy of the explanation in relation to the real phenomenon.

Didacticians, in contrast to teachers, refer positively to implemented non-knowledge. The reasons give indication that didacticians do not conceptualise non-knowledge as a lack of facts, but as a learning element of 'nature of science'.

> *"Good: Uncertainty included in explanations instead of positivist true/false dichotomy" (Ö$_C$)*

Didacticians also regard complex explanations much less frequently to be vague in an inappropriate way. Their views on these points do not coincide with those of students and teachers.

They also refer positively to the adequacy of the explanation in relation to the real phenomenon when rating complex explanations. Such reasons were not found among teachers.

> *"Good: Multi-causality instead of OGOD (one gene, one disease), including the gene-environment interaction dynamic" (G$_C$)*
>
> *"Pointing out the difficulty of explanations when many factors are involved is important --> promotes a realistic understanding" (Ö$_C$)*

Student teachers use similar criteria as didacticians, but they evaluate those criteria differently. Unlike didacticians they refer to non-knowledge in a neutral or descriptive way. But there are also indications, that the 'nature of science' (NOS) aspect is considered.

> *"It is clear that there is still much to be investigated in science and that we do not know everything yet" (G$_C$)*

Like didacticians, student teachers refer rarely to vagueness. And they refer positively to the adequacy of the explanation in relation to the real phenomenon—but only when rating complex explanations.

> *"The complex pathogenesis of MS is made clear" (G$_C$)*

Based on the results, it would even be plausible for students to rate the complex variants better than the linear pendants.

11.4 Discussion

The results indicate a preference for linear explanations, particularly among students, as well as among student teachers and teachers. In contrast to this trend, didacticians tend to prefer complex explanations. A possible conditional factor for these tendencies could be that complex explanations occur less frequently in biology lessons (Grotzer, 2012; Jacobson & Wilensky, 2006; Potochnik, 2013). For students, student teachers and teachers, these explanations may therefore be less familiar or may not conform to expectations. Another aspect that could partly influence the preference for linear explanations is CL (Sweller, 2011). Grotzer et al. (2017) showed that CL is higher for complex phenomena. However, since the video vignettes were of similar length, the complex explanations contained more factors and interactions, but the linear explanations contained more details about the focussed cause-effect-relationship (e.g. the mechanism). Visualisations of complex phenomena can provide a higher extraneous CL, so that the CL isn't increased directly by the complexity of the phenomena, but by the learning material.

Perceptions of complexity, non-knowledge and uncertain knowledge were indicated by the participants' answers in the open text field. For students, student teachers and teachers, complex explanations tend to be vague, which is rated negatively. While students criticise the lack of a clear result in the sense of a take-home message, the reasons given by student teachers and teachers focus on the fact that in complex explanations only possible factors are addressed. On the one hand, these statements are related to findings by Jacobson (2000) who describes that when thinking in and about complex systems, it is common to assume that the effects are completely predictable while overlooking chance and probabilistic relationships (Jacobson, 2000). On the other hand, the teaching and learning culture is describe as fact-orientated and focussed on tests (Dittmer, 2010), so the findings might also be interpreted as a result of this culture.

Furthermore, the participants do not respond to the references of non-knowledge and uncertain knowledge in explanations to such an extent that uncertainty or even a loss of trust in science can be expected, as is sometimes assumed (Kampourakis & McCain, 2020; Varwig, 2020). One potential explanation is that the explanatory objects in question are not directly engaged in a process of decision-making, so students are not confronted with uncertainty (Gustafson & Rice, 2020; Lübke & Heuckmann, 2024).

A noteworthy finding is that students do not incorporate non-knowledge and uncertain knowledge in their reasoning at all. This may suggest that students are capable of handling scientific non-knowledge and uncertainty. This aligns with the findings of Gustafson and Rice (2020), who found that uncertainty was primarily caused by controversial opinions rather than other types of knowledge gaps. But as the presented study is a cross-sectional design with students in the 10th grade, it remains unclear to what extent younger students can also deal with non-knowledge and uncertain knowledge in instructional explanations.

Differences in the responses regarding the inclusion of non-knowledge and uncertain knowledge exist between student teachers, teachers and didacticians: On the one hand, student teachers and didacticians refer to ideas about the 'nature of science' when giving reasons, which leads to a more positive assessment of the inclusion of non-knowledge and uncertain knowledge. On the other hand, the reasons given by teachers reflect the fact that addressing gaps in scientific knowledge is not part of the expectations of good instructional explanations—non-knowledge and uncertain knowledge is not perceived as content. Since textbook knowledge is typically conveyed through instructional explanations, this finding appears to be consistent with previous research (Krüger et al., 2013; Latour, 1987). However, from a didactical perspective—particularly concerning the 'nature of science'—this is problematic: If teachers do not address non-knowledge and uncertain knowledge, students develop a distorted view of biology as a scientific discipline. During the process of didactical reduction, teachers must adapt scientific knowledge for students (Kattmann, 2008). The question, however, is how to ensure that non-knowledge and uncertain knowledge are not omitted—not due to learning psychology considerations, but simply because they are absent from teachers' preparatory materials. It is therefore essential to incorporate these aspects into textbooks and educational materials, as well as to integrate them into the concept of didactical reduction. Teachers can then use their professional expertise to determine whether the level of complexity, along with the existing non-knowledge and uncertainty, aligns with their teaching objectives and students' abilities.

From a methodological point of view, the following can be concluded: The triangulation of perspectives can be considered fruitful due to the different ratings and reasonings—the methodological approach contributes to a deeper understanding of the research subject (Fielding & Fielding, 1986).

11.5 Limitations

The following limitations of the study should be highlighted: The answers in the open text field are rather short and only include the most evident criteria, due to a time limit for the online questionnaire. As only a few of the numerous criteria available are mentioned, the frequencies of each criterion per instructional explanation are rather low—especially concerning the data of the evaluative qualitative content analysis. Therefore, the possibilities for reconstructing reasons related to complexity, non-knowledge, and uncertain knowledge for the rating remain constrained and the conclusions are therefore tendencies that need to be explored in further research.

11.6 Implications

Based on the assumption that non-knowledge and uncertain knowledge are relevant characteristics of scientific knowledge that should be taught to students, the findings have implications for conceptual approaches in biology didactics, for biology teacher education and for classroom practices: Biology didacticians should implement the aspects of non-knowledge and uncertain knowledge in the concept of 'nature of science'—the aspect of tentativeness of scientific knowledge is not entirely sufficient for a broad understanding of contemporary challenges and the public discourses surrounding them. With regard to the training of biology teachers, the aspects of non-knowledge and uncertain knowledge need to be integrated in teacher training programs to ensure that non-knowledge is not perceived as a contradiction to the communication of knowledge through instructional explanations.

Finally, the integration of complex instructional explanations and consequently complex phenomena into biology lessons is necessary. It is essential that not only prototypes and principles are focused, but also real-world phenomena with their idiosyncratic characteristics. However, it is important to consider that comprehending complex phenomena can be challenging for novices at the subject-specific level. Therefore, examples that do not immediately require decision-making and contain ethical complexity are essential for learners to become familiar with the multifaceted aspects of complexity.

11.7 Conclusion and Further Research

This study indicates a preference for linear instructional explanations among students, student teachers, and teachers, whereas didacticians are more receptive to complex explanations incorporating non-knowledge and uncertain knowledge. The findings also indicate that teachers do not perceive the integration of non-knowledge and uncertain knowledge as learning objectives in biology classes. Additionally, the data indicates that including non-knowledge and uncertainty in explanations does not create uncertainty among students.

Further research should also investigate the cognitive processes involved in dealing with complex instructional explanations. Investigating the relationship between cognitive load, prior knowledge, and the perception of complexity could provide insights into how explanations should be structured to facilitate understanding. Moreover, longitudinal studies could accompany teacher training programs that address the implementation of non-knowledge and uncertain knowledge. These studies should investigate how teachers' attitudes and beliefs about these concepts evolve over time, identifying factors that contribute to positive shifts in perception and how these changes influence classroom practices.

Given that instructional explanations also play a role in shaping students' understanding of the subject matter and an understanding of complexity is one of the essential abilities in dealing with current ecological and societal challenges, it is crucial to consider the inclusion of complexity, non-knowledge, and uncertain knowledge in biology classes.

Acknowledgements This project was part of the "Qualitätsoffensive Lehrerbildung", a joint initiative of the Federal Government and the *Länder* which aims to improve the quality of teacher training. The program was funded by the Federal Ministry of Education and Research (ref. 01JA1512). The authors are responsible for the content of this publication.

References

Bar-Yam, Y. (1997). *Dynamics of complex systems*. Westview Press.

Braaten, M., & Windschitl, M. (2011). Working toward a stronger conceptualization of scientific explanation for science education. *Science Education, 95*(4), 639–669.

Christensen, C., & Fensham, P. J. (2012). Risk, uncertainty and complexity in science education. In B. Fraser, K. Tobin, & C. McRobbie (Eds.), *Second international handbook of science education* (pp. 751–769). Springer.

Dittmer, A. (2010). *Nachdenken über Biologie. Über den Bildungswert der Wissenschaftsphilosophie in der akademischen Biologielehrerbildung*. Springer Fachmedien.

Dittmer, A., & Ehras, C. (2024). Nature of bioscience. Komplexität und Ethik als Merkmale eines biologiedidaktischen NOS-Ansatzes. In B. Heinisch, D. Mahler, & D. Krüger (Eds.), *Biologiedidaktische Nature of Science-Forschung: Zukunftsweisende Praxis* (pp. 7–18). Springer.

Ehras, C. (2024). *Perspektiven auf gute instruktionale Erklärungen im Biologieunterricht – Komplexität als didaktische Herausforderung* (unv. Diss.). Universität Regensburg: Universitätsbibliothek.

Fielding, N., & Fielding, J. (1986). *Linking data*. SAGE.

Grotzer, T. A. (2012). *Learning causality in a complex world. Understandings of consequence*. Rowman & Littlefield Education.

Grotzer, T. A., Solis, S. L., Tutwiler, M. S., & Cuzzolino, M. P. (2017). A study of students' reasoning about probabilistic causality. Implications for understanding complex systems and for instructional design. *Instructional Science, 45*(1), 25–52.

Gustafson, A., & Rice, R. E. (2020). A review of the effects of uncertainty in public science communication. *Public Understanding of Science, 29*(6), 614–633.

Härtig, H., Kauertz, A., & Fischer, H. E. (2012). Das Schulbuch im Physikunterricht. Nutzung von Schulbüchern zur Unterrichtsvorbereitung in Physik. *MNU, 65*(4), 197–200.

Jacobson, M. J. (2000). Problem solving about complex systems: Differences between experts and novices. In B. Fishman & S. O'Connor-Divelbiss (Eds.), *Fourth international conference of the learning sciences* (pp. 14–21). Erlbaum.

Jacobson, M. J., & Wilensky, U. (2006). Complex systems in education: Scientific and educational importance and implications for the learning sciences. *The Journal of the Learning Sciences, 15*(1), 11–34.

Kampourakis, K., & McCain, K. (2020). *Uncertainty. How it makes science advance*. Oxford University Press.

Kattmann, U. (2008). Biologie als Wissenschaft im Unterrichtsfach. In H. Gropengießer & U. Kattmann (Eds.), *Fachdidaktik Biologie* (pp. 39–63). Aulis.

Krauss, S., Bruckmaier, G., Lindl, A., Hilbert, S., Binder, K., Steib, N., & Blum, W. (2020). Competence as a continuum in the COACTIV-study: The "cascade model." *ZDM, 52*(2), 311–327.

Krüger, J., Ruhrig, J., & Höttecke, D. (2013). Lehrerperspektiven auf unsichere Evidenz II: Ergebnisse einer Gruppendiskussionsstudie. In S. Bernholt (Ed.), *Inquiry-based learning - Forschendes Lernen. Gesellschaft für Didaktik der Chemie und Physik, Jahrestagung in Hannover 2012* (pp. 728–730). IPN Kiel.

Kuckartz, U. (2018). *Qualitative Inhaltsanalyse. Methoden, Praxis, Computerunterstützung*. Beltz Juventa.

Kuhlmann, M. (2007). Theorien komplexer Systeme. Nicht-fundamental und doch unverzichtbar? In A. Bartels & M. Stöckler (Eds.), *Wissenschaftstheorie. Ein Studienbuch* (pp. 307–356). Mentis.

Larreamendy-Joerns, J., & Muñoz, T. (2010). Learning, identity, and instructional explanations. In M. K. Stein & L. Kucan (Eds.), *Instructional explanations in the disciplines* (pp. 23–40). Springer.

Latour, B. (1987). *Science in action*. Harvard University Press.

Lederman, N. G. (2007). Nature of science: Past, present and future. In S. K. Abell & N. G. Lederman (Eds.), *Handbook of research on science education* (pp. 831–880). Lawrence Erlbaum Associates.

Lübke, B., & Heuckmann, B. (2024). Umgang mit Ungewissheit als Charakteristikum von Nature of Science. Eine Begriffsbestimmung und Konzeptionalisierung für die Integration in Lehr-Lernkonzepte. In B. Reinisch, D. Krüger, & D. Mahler (Eds.), *Biologiedidaktische Nature of Science-Forschung: Zukunftsweisende Praxis* (pp. 59–70). Springer.

Mainzer, K. (2008). *Komplexität*. Wilhelm Fink Verlag GmbH & Co.

McComas, W. F., & Olson, J. K. (1998). The nature of science in international science education standards documents. In W. F. McComas (Ed.), *The nature of science in science education* (pp. 41–52). Kluwer Academic Publishers.

Mehren, M., Mehren, R., Ohl, U., & Resenberger, C. (2015). Die doppelte Komplexität geographischer Themen. Eine lohnenswerte Herausforderung für Schüler und Lehrer. *Geographie und Schule, 37*(216), 4–11.

Mitchell, S. D. (2008). *Komplexitäten. Warum wir erst anfangen, die Welt zu verstehen*. Suhrkamp.

Potochnik, A. (2013). Biological explanation. In K. Kampourakis (Ed.), *The philosophy of biology. A companion for educators* (pp. 49–65). Springer.

Schilcher, A., Krauss, S., Kirchhoff, P., Lindl, A., Hilbert, S., Asen-Molz, K. et al. (2021). FALKE: Experiences from transdisciplinary educational research by fourteen disciplines. *Frontiers in Education, 5*(579982), 112.

Seidel, T., & Thiel, F. (2017). Standards und Trends der videobasierten Lehr-Lernforschung. *Zeitschrift Für Erziehungswissenschaft, 20*(1), 1–21.

Sweller, J. (2011). Cognitive load theory. In J. P. Mestre & B. H. Ross (Eds.), *The psychology of learning and motivation: Cognition in education* (pp. 37–76). Elsevier Academic Press. https://doi.org/10.1016/B978-0-12-387691-1.00002-8

Varwig, C. (2020). Kommunizieren oder verschweigen - Wie geht man mit wissenschaftlicher Unsicherheit um? In J. Schnurr & A. Mäder (Eds.), *Wissenschaft und Gesellschaft: Ein vertrauensvoller Dialog. Positionen und Perspektiven der Wissenschaftskommunikation heute* (pp. 209–214). Springer.

Wehling, P. (2008). Wissen und seine Schattenseite: Die wachsende Bedeutung des Nichtwissens in (vermeintlichen) Wissensgesellschaften. In T. Brüsemeister & K.-D. Eubel (Eds.), *Evaluation, Wissen und Nichtwissen* (pp. 17–34). VS Verlag für Sozialwissenschaften.

Chapter 12
Investigating Relationships Between Prospective Biology Teachers' Knowledge of Models and Their Knowledge of the Modeling Process

Kim Eleni Lobner, Tom Bielik, and Moritz Krell

12.1 Introduction

In the natural sciences, various methods are used to investigate and explain natural phenomena (Kind & Osborne, 2017). In addition to observation and experimentation, modeling plays an essential role in the scientific endeavor and models fulfil a dual function in biology (Krell et al., 2013; Svoboda & Passmore, 2013). On the one hand, models are used as media to represent, share, and communicate existing knowledge about phenomena. On the other hand, they serve as research tools for the exploration and prediction of phenomena (Chiu & Lin, 2019; Schwarz et al., 2009).

Modeling competence is defined "as a system of the knowledge, skills, and abilities necessary to engage in the process of developing and using models for reasoning in science, including motivational dispositions to apply these capabilities for problem-solving in specific situations" (Göhner et al., 2022, p. 1357) and is perceived as one of the main goals of science education (Chiu & Lin, 2022). Teachers and prospective teachers in biology should gain a high level of modeling

K. E. Lobner (✉) · M. Krell
Biology Education, IPN – Leibniz Institute for Science and Mathematics Education, Kiel, Germany
e-mail: lobner@leibniz-ipn.de

M. Krell
e-mail: krell@leibniz-ipn.de

T. Bielik
Radboud University, Institute for Science Education, Nijmegen, The Netherlands
e-mail: tom.bielik@ru.nl

O. Morin et al. (eds.), *Challenges of a Changing World in Biology Education*, Contributions from Biology Education Research,
https://doi.org/10.1007/978-3-032-05346-6_12

competence to be able to use models adequately in their lessons and to impart appropriate knowledge about models and modeling to their students (Krell & Krüger, 2016).

Literature suggests that modeling competence encompasses modeling metaknowledge (consisting of knowledge about models [MKM] and knowledge of the modeling process [MKP]), modeling practices, and the modeling product (Chiu & Lin, 2019; Nicolaou & Constantinou, 2014), as well as related motivational and volitional dispositions (Ammoneit et al., 2024). However, studies have shown that teachers often have limited modeling metaknowledge and that models are predominantly used as representations at school (e. g. Cheng et al., 2021; Göhner et al., 2022; Nielsen & Nielsen, 2021). The modeling metaknowledge is assumed to be essential and helpful for guiding and supporting students' modeling activities as well as for evaluating and reflecting on modeling products (Chiu & Lin, 2019; Nielsen & Nielsen, 2021; Schwarz et al., 2009). Therefore, fostering the development of MKM and MKP seems necessary to achieve the prescribed educational goals. However, models are seldom used as research tools in biology education (Werner et al., 2019) and science teachers rarely implement 'metamodeling talk' in their lessons (Gray & Rogan-Klyve, 2018). Therefore, it is necessary to gain more insights into teachers' MKM and MKP and the relationship between them.

12.2 Theoretical Background

12.2.1 Knowledge About Models (MKM)

According to Schwarz et al. (2009), MKM includes understanding on "how models are used, why they are used, and what their strengths and limitations are" (p. 634). To further differentiate and operationalize MKM, the competence framework by Krüger and Upmeier zu Belzen (2021) can be used. In this framework, MKM is divided into the five aspects of *nature of models, multiple models, purpose of models, testing models* and *changing models* (Table 12.2) with three levels each, ranging from a naïve to a comprehensive understanding. At level 1, models are perceived as copies of biological phenomena and are utilized to describe such phenomena. At level 2, models are regarded as idealized representations of phenomena, serving explanatory purposes that are modified when new insights about the phenomenon become available. Finally, at level 3, models are conceptualized as hypothetical entities and theoretical constructs of phenomena, which are used to derive and test hypotheses about the phenomenon (Krüger & Upmeier zu Belzen, 2021).

12.2.2 Knowledge About the Modeling Process (MKP)

According to Justi and van Driel (2005), the MKP includes "steps to be followed in the modelling process and factors on which the modeling process depends" (p. 555). Following Engelschalt et al. (2023) and referring to the diagram of the modeling process suggested by Krell et al. (2019), the modeling process consists of the following steps and components. The process begins with the perception and exploration of an unknown phenomenon. After the exploration of the phenomenon, an initial model is developed (*develop evidence-based model*). This model is continuously tested for consistency (*test model with data*) and used for (further) exploration of the phenomenon by testing and deriving hypotheses (*predict with the model*). The resulting findings about the phenomenon can lead to a change in the model (*revise the model*) and re-initiate the modeling process. The modeling process is therefore an iterative and cyclic process, in which the steps and components can be repeated and do not necessarily follow a strict order (Krell et al., 2019; Passmore et al., 2014). When modeling, the modeler navigates between the model world and the experiential world, engaging in constructing a *model of something* and applying the *model for something* (Giere et al., 2006; Krell et al., 2019).

12.2.3 Relationship Between MKM and MKP

Nicolaou and Constantinou (2014) suggest modeling metaknowledge encompassing the two subdimensions of MKM and MKP. Similarly, Engelschalt et al. (2023), argue that MKM and MKP focus on different aspects and, therefore, must be regarded as two separate knowledge constructs. In their review, Nicolaou and Constantinou (2014) have identified the need to develop approaches for assessing these constructs validly. As a response to that need, Engelschalt et al. (2023) developed a diagram-task to assess MKP independently of MKM. In an initial validation study of this diagram-task, the authors found a significant positive correlation between MKM and MKP. Similar findings on the relationship between MKM and MKP, though obtained with different testing instruments and in different contexts, can also be found in the studies of Koch et al. (2015) and Sins et al. (2009).

12.3 Aim and Research Question

The aim of this study is to replicate the findings from Engelschalt et al. (2023) about the relationship of prospective biology teachers' MKM and MKP. In the spirit of good scientific practice, it is necessary to conduct replication studies in order to generalise findings (Yong, 2012). Therefore, in this study, both MKM and MKP are assessed, using the same instruments as Engelschalt et al. (2023).

Hence, the research question of this study is: *What is the relationship between prospective biology teachers' MKM and MKP?*

12.4 Research Design and Methodology

12.4.1 Study Design

Our study is conducted in a mixed-method quasi-experimental design, using established quantitative and qualitative tools to assess MKM and MKP (Engelschalt et al., 2023). The sample consists of $n = 148$ prospective biology teachers from the Bachelor- and Master program at Kiel University, Germany. To this end, the first author attended the lectures and seminars of the winter and summer semesters 2023/2024 and asked the students to voluntarily and anonymously participate in the study.

First, to assess their MKM, the participants answered an online questionnaire consisting of five open-ended questions, each addressing one aspect of MKM (Krell & Krüger, 2016). In order to assess their MKP, participants were then asked to develop a diagram of the modeling process (Engelschalt et al., 2023) using the online tool CoMapEd (Mühling, 2014). Of the 148 completed questionnaires, 19 questionnaires had to be excluded because the diagram task was not completed and 4 questionnaires did not contain any answers to the open-ended questions, leaving 125 complete data sets for analysis (Table 12.1).

Table 12.1 Overview of the participants organized by gender and field of study

	Bachelor	Master	Others
Male	24	9	2
Female	60	29	/
Divers	1	/	/

Table 12.2 Distribution of response levels regarding the five aspects of knowledge about models from the open-ended questionnaire ($n = 125$)

	n responses level I	*n* responses level II	*n* responses level III	NA[a]
Aspect				
Nature of models	18	87	8	12
Multiple models	8	69	40	8
Purpose of models	51	51	23	/
Testing models	9	48	23	45
Changing models	13	102	10	/
Total	99	357	104	65

[a] Missing values due to questions not answered or the answer could not be rated

12.4.2 Data Analysis

The data was analyzed following qualitative content analysis; that is, using existing coding schemes and checking for interrater agreement (Schreier, 2012).

The responses regarding MKM were analyzed by two raters (Cohen's κ = 0.63–0.73), using the coding scheme by Krell and Krüger (2016). The responses are each assigned to a level of the competence framework by Krüger and Upmeier zu Belzen (2021), whereby the highest level is always assigned, even if parts of the response can be assigned to a lower level. In addition, for each participant, the average level of all five aspects was calculated in order to make statements about the overall MKM (MKM level).

Two raters analyzed the diagrams using a coding scheme adapted from Engelschalt et al. (2023), which included a component and a structure score ($\kappa = 0.80$ component score, $\kappa = 0.87$ structure score). The component score (level 1–3) describes whether the components *develop evidence-based model, revise the model, test the model with data,* and *predict with the model* can be found in the diagrams (level 1: one or two components; level 2: three components; level 3: all four components). The structure score (level 1–3) describes the extent to which a cyclic and iterative modeling process is depicted in the diagram (level 1: linear modeling process without model revision; level 2: cyclic process without the collection of new data or with only one event of model revision; level 3: cyclic process with model revision that includes further data collection steps; see example in Fig. 12.1). At the end, the component and structure score of each participant were added together and then the average was calculated (MKP level) in order to make statements about the overall MKP.

To define the relationship between MKM and MKP, correlation analyses were conducted using SPSS (Field, 2013) and effect sizes were interpreted based on Cohen (1988).

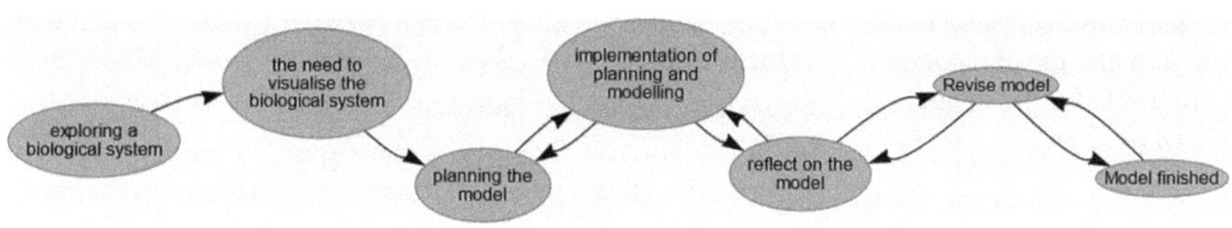

Fig. 12.1 Example of a diagram of the modeling process, receiving component score level 1 (including the components *develop evidence-based models* and *revise the model*) and structure score level 2 (cyclic structure with model revision), resulting in a total MKP level 1.5

12.5 Results

12.5.1 Knowledge About Models (MKM)

The findings show that for MKM, most responses can be assigned to levels 1 and 2, with level 3 being underrepresented (Table 12.2). It is noticeable that most responses ($n = 51$) were categorized in level 1 in the aspect *Purpose of models*, while, in contrast, most responses ($n = 40$) were assigned to level 3 in the aspect *multiple models*. Furthermore, it is striking that many of the responses ($n = 45$) for the aspect *testing models* could not be assigned to any level.

12.5.2 Knowledge About Modeling Process

The analysis of the diagrams shows that most of the diagrams could be assigned to level 1 or 2 for both the structure and the component score, while level 3 is underrepresented in the sample (Table 12.3).

Additionally, the examined diagrams show that the MKP components *develop evidence-based model* and *revise the model* were identified most frequently, whereas the component *predict with the model* were only included in 25 of the diagrams and were, therefore, underrepresented (Table 12.4).

Correlation analyses (Spearman) revealed a large positive and significant correlation ($r = 0.656$, $p < 0.001$) between the component score and the structure score (Table 12.5).

Table 12.3 Classification of the diagrams according to the structure and the component score levels for the diagram task ($n = 125$)

	Level I	Level II	Level III
Structure score	34	64	27
Component score	68	39	18
Total	102	103	46

Table 12.4 Number of named components within the diagrams for the diagram task ($n = 125$)

Components	Develop evidence-based model	Predict with the model	Test the model with data	Revise the model
	93	25	70	91

Table 12.5 Overview of the correlations between MKM and MKP

			1	2	3	4	5	6	7	8	9	10	11	12	13
1	**MKM level**	*r*	–												
		n	125												
2	Nature of models	*r*	0.458**	–											
		n	113	113											
3	Alternative models	*r*	0.492**	0.137	–										
		n	117	106	117										
4	Purpose of models	*r*	0.590**	0.103	−0.073	–									
		n	125	113	117	125									
5	Testing models	*r*	0.639**	0.141	0.028	0.244*	–								
		n	80	74	75	80	80								
6	Changing models	*r*	0.481**	0.095	0.257**	0.044	0.235*	-							
		n	125	113	117	125	80	125							
7	**MKP level**	*r*	0.356**	0.118	0.057	0.340**	0.184	0.080	–						
		n	125	113	117	125	80	125	125						
8	Structure score	*r*	0.300**	0.135	0.049	0.268**	0.101	0.052	0.911**	–					
		n	125	113	117	125	80	125	125	125					
9	Component score	*r*	0.337**	0.073	0.057	0.325**	0.233*	0.078	0.902**	0.656**	–				
		n	125	113	117	125	80	125	125	125	125				
10	Develop evidence-base model	*r*	0.139	0.015	0.054	0.065	0.086	0.053	0.368**	0.245**	0.425**	–			
		n	125	113	117	125	80	125	125	125	125	125			
11	Revise model	*r*	0.314**	0.115	0.150	0.280**	−0.059	0.133	0.791**	0.843**	0.538**	0.259**	–		
		n	125	113	117	125	80	125	125	125	125	125	125		

(continued)

Table 12.5 (continued)

			1	2	3	4	5	6	7	8	9	10	11	12	13
12	Test model with data	*r*	0.180*	−0.058	−0.006	0.221*	0.359**	−0.127	0.638**	0.504**	0.673**	0.145	0.364**	–	
		n	125	113	117	125	80	125	125	125	125	125	125	125	
13	Predict with model	*r*	0.272**	0.138	0.010	0.270**	0.141	0.119	0.615**	0.412**	0.703**	0.156	0.306**	0.282**	–
		n	125	113	117	125	80	125	125	125	125	125	125	125	125

r: Spearman correlation coefficient
n: Number of participants
**Significance level $p < 0.01$
*Significance level $p < 0.05$

12.5.3 Relationship Between Knowledge About Models and Knowledge About the Modeling Process

A correlation analysis (Spearman) revealed a medium positive and significant correlation ($r = 0.356, p = 0.001$) between the MKM level and the MKP level (Table 12.5). Additionally, there are significant correlations between the MKM level and the components *test the model with data* ($r = 0.180, p = 0.045$; i.e., small correlation), *predict with the model* ($r = 0.272, p = 0.002$; i.e., medium correlation), and *revise the model* ($r = 0.314, p < 0.001$; i.e., medium correlation).Additionally, significant correlations were identified between the aspects *purpose of models* ($r = 0.340, p < 0.001$) and the MKP level. Furthermore, a significant medium correlation was found between the MKM aspect *purpose of models* and the MKP component *test the model with data* ($r = 0.221, p = 0.013$), the MKM aspect *purpose of models* and the MKP component *predict with the model* ($r = 0.270, p = 0.002$), the MKM aspect *purpose of models* and the component *revise the model* ($r = 0.280, p = 0.002$) as well as between the MKM aspect *test with models* and the MKP component *test model with data* ($r = 0.359, p = 0.004$). There was no significant correlation between the aspect *changing models* and the component *revise the model.*

12.6 Discussion

The findings from the correlation analyses (Spearman) regarding the relationship between MKM and MKP also showed a significant positive correlation in this study and therefore align with the results found by Engelschalt et al. (2023). Participants with a higher level of MKM appeared to be able to draw more elaborate diagrams, which was reflected in a higher MKP level.

Furthermore, a positive correlation was found between the components *revise the model, test the model with data* and *predict with the model* and the overall level of MKM. Higher scores in MKM are associated with an understanding of models that is based on the use of models to predict and explore phenomena and where the modification and testing of models is also based on hypotheses falsified by the model (Engelschalt et al., 2023; Krüger & Upmeier zu Belzen, 2021). Participants with a high level of knowledge in this area therefore used the corresponding components more often in their diagrams for the modeling process.

Participants achieving high level in the aspects *purpose of models* also had a higher level of MKP.

Knowledge about using models as tools for predicting (aspect *purpose of models*, level 3) seemingly enable participants to incorporate the component *predict with the model* into their modeling diagram and to perceive the modeling process as iterative. The diagrams that received a high component score were therefore always diagrams that contained the *predict with the model* component, which was found in the fewest diagrams overall. This is consistent with the findings of Engelschalt et al. (2023).

The positive correlation between the MKM aspects *purpose of models* and the MKP components *predict with models* supports this conclusion.

Furthermore, the moderate correlation also indicates potential for development in this area. For instance, some participants had knowledge about using models to predict phenomena (aspect *purpose of models*, level 3), but did not integrate this into their modeling process (component *predict with model*). It seems significant for the development of an elaborate understanding of models, as well as for the enhancement of sophisticated knowledge about the modeling process, to focus on, address, and reflect on these aspect and component during work with models and throughout the modeling process. The focus on these elements seems to have the potential to address and promote both knowledge constructs (MKM and MKP) equally (Engelschalt et al., 2023). In addition to focusing on the component *predict with the model* or the aspect *purpose of models*, greater consideration is also required for the aspect *testing models* (e. g. Justi & Gilbert, 2003; Krell & Krüger, 2016; Lee & Kim, 2014). On the one hand, a positive correlation was found between the aspect *testing models* and the component *test model with data*. Participants with a high level of understanding of the aspect *testing models* used the component *test model with data* more frequently in their diagram. On the other hand, however, it is also noticeable that many responses could not be coded for the aspect *testing models*. Similar findings have already been reported by Krell and Krüger (2016), who attributed this to a more didactic rather than scientific view of models among participants. The analysis of the uncoded responses in our sample supports this interpretation. It therefore seems all the more sensible and necessary to address and reflect on this aspect.

Furthermore, the results of the study confirmed that the MKM of the participants was insufficient, which is in line with the results of previous studies (e. g. Cheng et al., 2021; Göhner et al., 2022; Göhner & Krell, 2022; Krell & Krüger, 2016; Nielsen & Nielsen, 2021). Particularly, the aspect *purpose of models* is underrepresented, indicating that participants predominantly perceive models as tools for representation rather than tools for predicting and exploring phenomena (Engelschalt et al., 2023). Regarding MKP, it also became evident that the component *predict with the model* (similar to the MKM aspect of the *purpose of models*) is underrepresented in the participants' diagrams. Similar findings are also reported by Engelschalt et al. (2023) and Vo et al. (2015).

12.7 Conclusion

One aim of the study was to use the tool (diagram-task to measure MKP) developed by Engelschalt et al. (2023) and to replicate the results and existing findings of the study on the relationship between the MKM and MKP of prospective biology teachers. As with Engelschalt et al. (2023), a significant positive correlation between MKM and MKP was found. This means that prospective teachers with a higher understanding of models were also more likely to be able to draw more elaborate diagrams of the modeling process. Similar to Engelschalt et al. (2023), this study also identified a positive correlation between the aspect *purpose of models* and the component *predict with the model*, as well as between the aspect *testing models* and the component *test model with data*. Thus, these two elements (*purpose of models/predict with models* and *testing models/test model with data*) were identified as having the potential to promote both MKM and MKP equally. However, it seems necessary to address and reflect on the elements in both areas of knowledge equally. This is because it could be shown that participants, even if they understood models as tools for predicting phenomena, did not necessarily integrate the component in their modeling process, and thus were not able to equally use the potential of their knowledge about models in their diagram of the modeling process.

While this study added to a growing body of evidence about the two dimensions of modeling metaknowledge (i.e., MKM and MKP), further research is needed to develop and establish instruments to coherently and validly assess the other dimensions of modeling competence (Nicolaou & Constantinou, 2014). For example, instruments to separately and coherently assess the modeling practices and the quality of the modeling product have just recently been suggested (Göhner et al., 2022) but require further attention by the science education research community.

Acknowledgements Funded by the Deutsche Forschungsgemeinschaft (DFG, German Research Foundation); project number 510978671.

References

Ammoneit, R., Göhner, M., Bielik, T., & Krell, M. (2024). Why most definitions of modeling competence in science education fall short. *Science Education, 108*(2), 443–466.

Cheng, M.-F., Wu, T.-Y., & Lin, S.-F. (2021). Investigating the relationship between views of scientific models and modeling practice. *Research in Science Education, 51*(1), 307–323.

Chiu, M. H., & Lin, J. W. (2019). Modeling competence in science education. *Disciplinary and Interdisciplinary Science Education Research, 1*, 1–11.

Chiu, M. H., & Lin, J. W. (2022). Research on modeling competence in science education from 1991 to 2020 with cultural and global implications. In M. M. Atwater (Ed.), *International handbook of research on multicultural science education* (pp. 961–1000). Springer.

Cohen, J. (1988). *Statistical power analysis for the behavioral sciences* (2nd ed.). L. Erlbaum.

Engelschalt, P., Bielik, T., Krell, M., Krüger, D., & Upmeier zu Belzen, A. (2023). Investigating pre-service science teachers' metaknowledge about the modelling process and its relation to metaknowledge about models. *International Journal of Science Education, 46*(7), 691–714.

Field, A. (2013). *Discovering statistics using IBM SPSS statistics* (4th ed.). SAGE Publications.

Giere, R., Bickle, J., & Mauldin, R. (2006). *Understanding scientific reasoning.* Thomson/Wadsworth.

Göhner, M., Bielik, T., & Krell, M. (2022). Investigating the dimensions of the modeling competence among pre-service science teachers. *Journal of Research in Science Teaching, 59*(8), 1354–1387.

Göhner, M., & Krell, M. (2022). Preservice science teachers' strategies in scientific reasoning. *Research in Science Education, 52*, 395–414.

Gray, R., & Rogan-Klyve, A. (2018). Talking modelling. *International Journal of Science Education, 40*(11), 1345–1366.

Justi, R., & Gilbert, J. K. (2003). Teacher's views on the nature of models. *International Journal of Science Education, 25*(11), 1369–1386.
Justi, R., & Van Driel, J. (2005). The development of science teachers' knowledge on models and modelling. *International Journal of Science Education, 27*(5), 549–573.
Kind, P., & Osborne, J. (2017). Styles of scientific reasoning. *Science Education, 101*(1), 8–31.
Koch, S., Krell, M., & Krüger, D. (2015). Förderung von Modellkompetenz durch den Einsatzeiner Blackbox. *Erkenntnisweg Biologiedidaktik, 14*, 93–108.
Krell, M., & Krüger, D. (2016). Testing models. *Journal of Biological Education, 50*(2), 160–173.
Krell, M., Upmeier zu Belzen, A., & Krüger, D. (2013). Students' understanding of the purpose of models in different biological contexts. *International Journal of Biology Education, 3*(1a), 1–34.
Krell, M., Walzer, C., Hergert, S., & Krüger, D. (2019). Development and application of a category system to describe pre-service science teachers' activities in the process of scientific modelling. *Research in Science Education, 49*, 1319–1345.
Krüger, D., & Upmeier zu Belzen, A. (2021). Kompetenzmodell der Modellierkompetenz. *Zeitschrift für Didaktik der Naturwissenschaften, 27*, 127–137.
Lee, S., & Kim, H.-B. (2014). Exploring secondary students' epistemological features depending on the evaluation levels of the group model on blood circulation. *Science & Education, 23*, 1075–1099.
Mühling, A. (2014). *CoMapEd*. https://www.comaped.de/
Nicolaou, C., & Constantinou, C. (2014). Assessment of the modeling competence. *Educational Research Review, 13*, 52–73.
Nielsen, S., & Nielsen, J. (2021). A competence-oriented approach to models and modelling in lower secondary science education. *Research in Science Education, 51*, 565–593.
Passmore, C., Gouvea, J., & Giere, R. (2014). Models in science and in learning science. In M. Matthews (Ed.), *International handbook of research in history, philosophy and science teaching* (pp. 1171–1202). Springer.
Schreier, M. (2012). *Qualitative content analysis in practice*. Sage.
Schwarz, C., Reiser, B., Davis, E., Kenyon, L., Achér, A., Fortus, D., Shwartz, Y., Hug, B., & Krajcik, J. (2009). Developing a learning progression for scientific modeling. *Journal of Research in Science Teaching, 46*(6), 632–654.
Sins, P., Savelsbergh, E., van Joolingen, W., & van Hout-Wolters, B. (2009). The relation between students' epistemological understanding of computer models and their cognitive processing on a modelling task. *International Journal of Science Education, 31*(9), 1205–1229.
Svoboda, J., & Passmore, C. (2013). The strategies of modeling in biology education. *Science & Education, 22*, 119–142.
Vo, T., Forbes, C. T., Zangori, L., & Schwarz, C. V. (2015). Fostering third-grade students' use of scientific models with the water cycle. *International Journal of Science Education, 37*(15), 2411–2432.
Werner, S., Förtsch, C., Boone, W., Kotzebue, L., & Neuhaus, B. J. (2019). Investigating how German biology teachers use three-dimensional physical models in classroom instruction. *Research in Science Education, 49*, 437–463.
Yong, E. (2012). Bad copy. *Nature, 485*, 298–300.

Chapter 13
Biology Undergraduates' Internship Combining Lab Training and Making Learning Videos

Tetiana Krushynska and Iryna Kompanets

13.1 Introduction

13.1.1 Educational Videos to Support the Teaching and Learning of Biology

Films for learning have always been popular in biology education due to their possibilities for clear visualization. They reflect the Universal Design for Learning approach (Mcguire et al., 2006), according to which offering content in various formats helps to accommodate diverse learning needs. These aspects became crucial during the Covid-19 pandemic when only online learning was possible. During this period, many educational videos were produced, and instructors gained valuable experience of integrating them with relevant technologies (Elma et al., 2024; Stadlinger et al., 2021). Scientific videos can attract students' attention, fostering curiosity and motivation (Nasution & Adlini, 2022). They contribute to understanding of experimental principles, the ability to apply techniques, and engagement. Both teachers and students appreciate video-based learning in biological disciplines (Neofotistos et al., 2024; Saifi & Sharabi, 2023). Viewing teaching videos before linked laboratory sessions assists learners with little prior technical experience in processing core concepts, and it improves the quality of hands-on practical training (Lacey & Wall, 2021).

T. Krushynska (✉)
Finnish Institute for Educational Research, University of Jyväskylä, Jyväskylä, Finland
e-mail: tetiana.y.krushynska@jyu.fi

I. Kompanets
Ruhr-University Bochum, Bochum, Germany
e-mail: Iryna.Kompanets@ruhr-uni-bochum.de

O. Morin et al. (eds.), *Challenges of a Changing World in Biology Education*, Contributions from Biology Education Research,
https://doi.org/10.1007/978-3-032-05346-6_13

Videos are excellent for practical studies in molecular biology, biochemistry, microbiology, and similar disciplines where laboratory equipment is expensive, lab methods are complicated and time-consuming, reagents and tested materials may be hazardous, and bioethical issues become prominent (Hamilton et al., 2023; Miezah & Whajah, 2023). Video-recorded experiments can expose students to state-of-the-art experimental techniques in lecture-based classes or help them better prepare for the experiments that they conduct in lab-based courses (Huang et al., 2015). The video demonstrations enabled students to enter the in-person lab sessions with a solid understanding of the processes and desired outcomes, reducing the strain on facilities and staff (Onyeaka et al., 2023). This is particularly necessary for low-income or war-affected countries, where universities often lack the resources to offer students laboratory training in biological disciplines.

Learning videos vary in nature, from enormously diverse YouTube content to thematic video blogs. However, such videos often have insufficient technical quality and are too long to be incorporated into educational activities. Moreover, their content may not be aligned with curricula, and videos may feature terminology and equipment different from that used by students (Patrick & Zhu, 2022). Hiring professional filmmakers to create customized educational videos is beyond the budget of most universities. Fortunately, current technologies offer non-professional video tools allowing university teachers to produce their own instructional films tailored to lab schedules and learning outcomes. Moreover, students can become involved in this creative process.

13.1.2 Benefits of Students' Participation in Creating Visual Educational Content

The involvement of students in making visual learning tools is not a new practice. It can allow visually engaging representation of scientific data and concepts, including realistic anatomical models and microscopic slides, or schematic material, such as infographics and charts. The creative task of sharing knowledge via self-made videos makes it more engaging (Martin-Garcia et al., 2020) and enhances the understanding and communication of complex learning topics. Arzaga (2023) compared the academic performance of students who created videos on cytology topics with that of students who watched downloaded YouTube videos. The results of the comparison favored the students' creations. In a similar vein, Young (2020) found that student-produced videos, in which students played the roles of molecules, organelles, or cell types, encouraged scientific discussion, promoted teamwork skills and self-awareness, and facilitated student integration in international settings. Video-capture appears to foster creativity, collaboration, and a learning environment while enhancing students' skills in problem-solving, presentation, and the capacity to independently discover principles, concepts, and facts (Lam & Yunus, 2023). When students create videos oriented toward teaching a hypothetical audience, they demonstrate better topic comprehension than if they compiled simple video reports of their laboratory practice (Zarate et al., 2024).

The empirical studies above are aligned with Skill Acquisition Theory. When the students' behavior is not aimed merely at working out some learning content but presenting some content as an educational video, it can motivate deeper learning, turning "knowledge what" into "knowledge how" or, in more theoretical terms, turning declarative knowledge into procedural knowledge (Dekeyser, 2007), i.e., knowledge incorporating higher levels of skill acquisition.

13.1.3 Video-Making as a Way of Sharing Knowledge and Experience of Academic Mobility

Academic mobility is of particular value for students from low-income countries, or countries suffering from conflict. It allows students to study for a certain period in the laboratories of leading world universities. International exchange programs are always oriented to the personal growth of participants, with benefits also for their home universities (Herazo et al., 2023). The goal of an international internship is not only for students to gain new knowledge and experiences but also to share these with peers on returning to their home country. In the pro-social and societal dimensions of mobility, students recognize themselves as carriers of knowledge and culture (Papatsiba, 2005). Creative forms of communication, such as personal stories, are effective for sharing the understanding of cultural contexts and complicated theoretical concepts, and they also enhance critical thinking (Balslev et al., 2020). However, intense competition among students, driven by a desire to outperform classmates, can sometimes hinder the willingness to share knowledge (Majid & Wey, 2011). Some students wish to keep their knowledge and intercultural skills to themselves with the aim of achieving value in the labor market (Juhász et al., 2024). Knowledge transfer, organized as a creative film-making activity, could mitigate this unwelcome trend.

Videos can realistically present knowledge and experience, especially given that the democratization of media has worked towards the availability of equipment and the freedom to create content (Lalon, 2023). Students display a positive attitude towards knowledge-sharing and an appreciation of its significance in facilitating peer-to-peer learning in cases where the technological aspect is the significant determinant of knowledge-sharing behavior (Jesuthasan et al., 2022). Even if student-made videos are not at the standard of professionally produced ones, they can be more adjusted to students' comprehension.

The knowledge-sharing objective requires students to grasp course material at a deeper level, since they must improve their communicative skills to convey it adequately to the audience. Working with educational content in the new conditions of video-making facilitates changes in students' mental operations. According to the Conscious Competence Learning Model (Cannon et al., 2010), this situation moves students from conscious competence (in which learning tasks are perceived as requiring deliberate effort), or even from unconscious competence (in which tasks are effortless), towards conscious incompetence. In this latter case, tasks perceived as problems make learners aware of deficiencies in their own knowledge and skills and encourage learners to overcome these deficiencies.

Pedagogical theories and recent educational research have indicated the benefits of using videos in education and involving students in creating learning videos, especially now when there are no technical obstacles to amateur film-making. One is then led to the possibility of directly integrating the creating of videos as part of natural science classes. Our research explored the compatibility of students' video-making with their learning activities in biological laboratories, posing the following research questions:

- How do students perceive the task of making educational films about their practice in biological laboratories, within the challenging conditions of international mobility?
- How does video-making, integrated with lab classes, influence students' academic performance?
- What are the didactic prospects for student-made videos?

13.2 Research Design and Methodology

13.2.1 Organizational Framework

The research was conducted as a case study on the materials of an international students' exchange program. Ukrainian students were invited to Ruhr-University Bochum (RUB), Germany, for a two-month training in modern laboratories. The placement was set up within the framework of the project "Digital Laboratories for Students at Risk (DigiLabStars)" initiated by the DAAD. The requirements for participation included a high score in academic performance and a good command of English. Proficiency in film-making was not required.

The students mastered lab techniques while simultaneously recording and editing videos of lab procedures for their classmates who remained in Ukraine, and who therefore had no access to real lab classes due to online learning under wartime conditions. Twenty students from Ukrainian biology, chemistry, and physics faculties visited RUB. The present study, however, focused solely on the outcomes of the DigiLabStars project related to biology education.

13.2.2 *Participants, Settings, and Activities*

Ten biology students (two at bachelor's level and eight at master's level) from Kyiv National University participated in the case study. All were female due to travel restrictions on men during wartime in Ukraine. Students were organized into small groups of 2–3 people. Each group had training in one of the laboratories of Molecular Embryology, Developmental Biology, Molecular Cell Biology, and Human Genetics. During their internship, the students mastered several research methods and filmed laboratory procedures. Some students already possessed basic video-editing skills. The staff at the Centre for Teaching and Learning conducted a workshop on video editing software, including Shotcut and DaVinci Resolve. Subtitles and animations were incorporated in the videos to enhance understanding and improve learner engagement. Student reports and videos were carefully reviewed by teachers from RUB who supervised the internship.

13.3 Data Collection and Analysis

Data were collected on students' attitudes to combining videomaking with performing other learning tasks, their impressions of the training program, and its possible influence on their further studies. The data were obtained via an anonymous questionnaire based on the Motivated Strategies for Learning Questionnaire (Duncan et al., 2015). Our questionnaire contained both Likert-scale questions and open-response questions related to students studying value, goal orientation, self-efficacy, and expectancy components. The answers for the Likert-scale items were treated as ordered-categorical data. Due to the small sample size, the standard functions of Excel were sufficient for organizing and processing these data. Content analysis was applied to all the answers to the open-response questions.

The students' performances were assessed via a qualitative analysis of several documents. These consisted of internship reports and academic records at the home university. Semi-structured supervisors' feedback provided data on student achievements in scientific and collaborative aspects. Expert evaluation of the student-made videos (which formed the material output of their internship) supplied data on the applicability of this educational content for practical teaching and learning.

13.4 Findings and Discussion

13.4.1 Students' Reflection on the Internship Experience

Answering open-response questions, the students expressed their reasons to participate in the DigiLabStars project, which were categorized as follows: "Professional development, new knowledge, and skills" (6 respondents), "International experience" (5 respondents), and "Personal challenge" (5 respondents). In students' responses, the latter was formulated in terms of curiosity, trying to "do something new." In this context, "doing something new" could be assumed to refer, specifically, to the creation of videos, given that "international experience"—which was also new for almost all the students—was mentioned separately.

The students gave an overall rating of the internship on a 5-point scale (from 'very poor' to 'excellent'). The average rating was 3.78 ± 0.63, with a predominant grade of 4 (good). Most students agreed that the training was successful and that they had achieved their goals. The difficulty level and pace of the internship program were rated as "just right" by 7 out of 10 participants. Regarding the learning methods, 9 out of 10 of the students agreed (somewhat or strongly) that the training had an engaging design, was well prepared and supported with helpful handouts. Hence, the students did not indicate any problem with combining biology studies with the creation of educational films.

Students were particularly appreciative of the friendly educational environment, which was seen as facilitating creativity (cf. Kalogeratos et al., 2023). Overall, there was strong agreement (by 8 out of 10 students) that the professional commitment of the supervisors and the respectfulness shown toward the students, in terms of considering students' contributions and suggestions, promoted their learning and video-making, and left the students satisfied with the mobility program.

Students' perceptions of the internship program were assessed in terms of the interest, importance, and utility (i.e., the task value), revealed that among the key takeaways that students acquired at the training, in this evaluation, lab skills were placed first, and biological knowledge was placed second. Confidence in these two components is mandatory for successful presentation of students' lab experiences to their peers. Skills in presenting scientific information and social/ intercultural skills (both of which are significant for biological lab practice and creating videos) shared third and fourth positions. Students believed that the skills they had acquired would be more valuable for their future professional careers (7 responses) than for completing course or diploma projects or for pursuing further higher education (3 responses each). This finding is in line with the study by Castro and Barbosa (2021) indicating that a prime motivation for students' participation in exchange programs is the promotion of their subsequent careers.

Students appreciated the acquisition of scientific techniques. However, for 50% of them, the research methods they studied were not directly applicable to their current study subject (Fig. 13.1). This does not mean that the students considered the

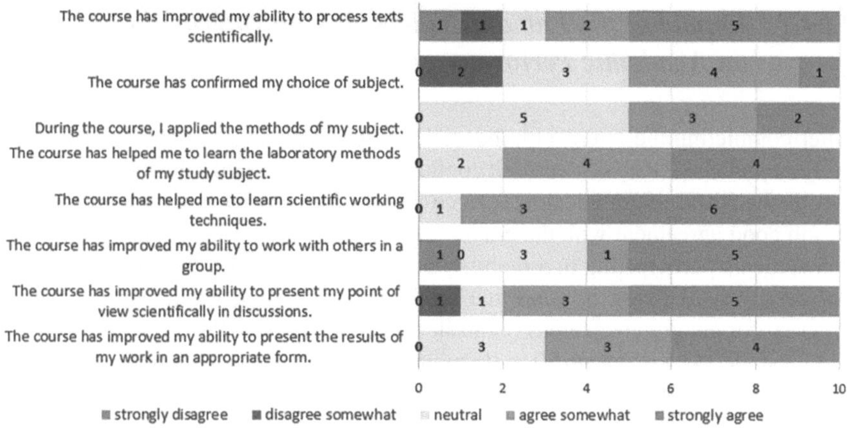

Fig. 13.1 Students' self-reflections on the impact of the internship

research methods to be useless for them; rather, the response seemed to be bound up with critical rethinking on the field of their further specialization.

It was notable that the students did not focus on a specific utility regarding film-making skills, considering them as learning means rather than learning outcomes. About one-third of the participants believed that the video-making had been useful for memorization, while an equal number felt it had contributed to a deeper understanding of lab methods. Only one student identified these skills as important for their future teaching career. Additionally, three out of ten students rated the video-making activities as unhelpful for studying biology. This finding is in line with Goulet-Pelletier et al. (2023), who highlight the complex relationship between creative abilities (such as those used in video-making) and two forms of academic motivation, i.e., the motivation to learn and the motivation to explore. The skepticism of some students regarding the usefulness of the video-making experience appeared partly to contradict their emotional reflections on creating videos. Students might not be aware that a positive emotional experience of film-making (including feelings of success and pride) had helped them to acquire biological knowledge and lab skills.

One function of emotions is to inform the individual of progress in goal pursuit and permit behavior calibration that influences motivation (Gendolla, 2017). Emotions are causal variables of motivation, along with needs, desires, and necessities. They are integrated with thinking and can generate prompt impulses, behaviors, and reactions. However, emotional motivation is not a sufficient reason to act for work performance, given that it persists for only a limited time, i.e., until the emotional intensity decreases (Popa & Salanta, 2013). Hence, students were excited that they were able to create learning videos. Typical reactions included "Making videos is great!" and "It opens a new insight in biology". However, this did not necessarily motivate them to further improve their video-making skills.

13.4.2 *Impact of the Lab and Video-Making Internship on Academic Performance*

After completing the internship, the students self-assessed their ability to perform professional tasks. The students felt confident in understanding and applying laboratory methods and communicating science (Fig. 13.2). The observed high self-efficacy was in good agreement with the grades given by the supervisors. It can be expected that students participating in a mobility program will improve their scores to some degree, always bearing in mind that levels are not homogeneous across mobility programs or geographical areas (González-Baixauli et al., 2018).

Our students scored from 90 to 100 points (with 100 as the maximum grade), with an average score of 94.73 ± 2.56. In the semi-structured assessments, the supervisors mentioned that students performed well in both individual and group work, were highly motivated, and possessed the necessary biological and communicative skills. The teachers emphasized their enthusiasm and creativity, independence, reliability, and desire for perfection. It can reasonably be assumed that the satisfaction with the internship as expressed by the visiting students and the hosting supervisors originated to some degree from their creative collaboration in producing learning films. This would be in line with the findings of Mohyaldinn et al. (2020), who examined student-faculty partnerships in which students were involved in developing instructional videos.

Video-creation can be expected to have a positive effect on academic performance, giving support to memorizing and understanding laboratory procedures and their scientific background. When students re-examine their recordings, edit videos, and demonstrate them to others along with explanations, they can reach the procedural level in the knowledge dimension, and at least the level of analysis in the dimension of cognitive processes according to Bloom's taxonomy of learning outcomes (Anderson et al., 2001).

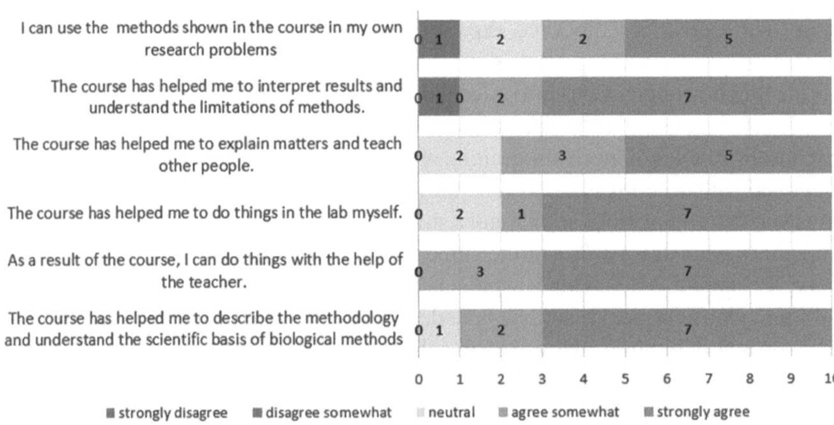

Fig. 13.2 The students' confidence in the laboratory methods

Unlike Bacolod (2022), who found student-generated videos improved test scores, our study showed a weak impact of video-making on academic performance. It should be noted that there were strict selection criteria for the participants in the international mobility program, with one of these being excellent academic performance in home universities. Hence, there was little scope for a visible increase in students' scores. At the same time, it is worth noting that there was no actual decrease in academic performance under the conditions of the challenging multitasking program followed in the internship. The stability of the scores indicates that in any case, the video-making did not hinder the development of biological competencies. These findings suggest that the use of video recording alongside other creative activities in biology lab classes can be integrated without adversely affecting the completion of learning tasks.

13.4.3 Learning Videos as an Output from Academic Mobility

The students created 16 Live Action tutoring-style videos. Individual videos had a duration of 1–24 min, but the usual duration was 4–8 min. The experts, i.e., university teachers from biological and medical faculties, evaluated the students' videos (Table 13.1). They evaluated all the videos as being suitable for educational use, and as relevant to studying modern immunological, genetic, and biochemical methods. The videos could be used in their original form, or after further enhancement, involving, for example, conversion of them into interactive formats using applications such as H5P integrated with Moodle.

The technical quality of the student-made videos was relatively high, even if the students had no experience in this area. Errors in the scientific aspect were related to lab safety issues, or to the omission of some steps of the lab procedures. The pedagogical quality was also evaluated, with the assessment that in all the videos the materials and lab procedures were shown correctly and comprehensively. However, the students did not pay enough attention to the structure of their videos. Only two videos included the most crucial elements of educational materials, i.e., the introduction and the learning objectives, and all the participants ignored the summary and references.

The internship program did not include pedagogical training, on the grounds that in Ukraine, graduates from university biology faculties usually get double qualifications as biologists and biology teachers. All of them study teaching methods in biology as

Table 13.1 Evaluation of the student-made learning videos

Characteristics of the videos	Average grade (min. = 1, max. = 5) and standard deviation
Scientific accuracy	4.5 ± 0.61
Pedagogical delivery	3.9 ± 1.0
Technical quality	4.25 ± 0.66

a mandatory discipline. Hence, it was expected that our participants would possess some pedagogical competencies and require no special training in pedagogics related to film-making. This observed drawback highlights the importance of instructional support for academic mobility, with proper guidance given for all planned activities.

13.5 Conclusions

Video-making was a crucial component of the internship, and students executed it successfully. Capturing and editing instructional videos was a new task for most of them. Students perceived it as challenging and engaging, thus demonstrating a strong intrinsic goal orientation towards the internship objectives in general. However, they perceived video-making as an auxiliary activity, belonging to a separate domain from the mastery of biological research methods. Students found the activity helpful for memorizing and understanding lab procedures and improving their communication skills, but did not regard it as essential for their future professions. So, the task value could be regarded as moderate. Students' expectancy (i.e., the belief that efforts will lead to outcomes) related to acquiring biological competencies prevailed over developing video-making skills, especially in long-term perspectives. Nevertheless, within the favorable educational environment of the internship, the observed expectancy and task value appeared sufficient to motivate students to produce learning videos that positively influenced their self-efficacy.

The participants in the internship, which combined lab practice and video-making, demonstrated excellent academic performance in their home university and in the mobility program, with no significant differences in the scores they achieved. The limited number of participants and the specificity of their selection made it difficult to assess the kinds of learning gains that could be associated with video-making. A quasi-experimental study with pre-test and post-test measurement of students' scores could shed further light on this question. In any case, it appeared that the creating of educational videos did not impede students in terms of learning biology and did not decrease their academic performance.

Overall, it can be suggested that making videos in the lab classes motivated students to study biology more deeply, and further, that the films they produced had distinct value as visual educational content. The case study also indicated that students could present their lab practice results through self-recorded videos on par with traditional written reports, hand-drawn illustrations, photographs of micro- and macro-objects, and so on. Integrating video recordings with lab reports posed no technical challenges, as students could easily capture the relevant material via their smartphones or tablets without requiring extra time. The quality of these student-produced videos was satisfactory for educational purposes. Further editing of the video recordings to make complete instructional videos would have been technically feasible for the students, but additional pedagogical assistance would then have been beneficial.

It can be claimed that integrating video-making within the schedules of mobility-program students from low-income and war-affected countries has great potential utility. Videos created by students within mobility programs can allow their peers at home to familiarize themselves vicariously with scientific and intercultural experiences relevant to future careers.

Acknowledgements The authors express their sincere gratitude to colleagues from Ruhr-University Bochum, as follows: Prof. Kornelia Freitag—Vice-rector of Academic and International Affairs; Dr. Peter Salden—Director of the Centre for Teaching and Learning; Prof. Beate Brand-Saberi—Head of the Department of Anatomy and Molecular Embryology; Prof. Stephan Wise—Head of the Cell Morphology and Molecular Neurobiology Department; Prof. Huu Phuc Nguyen—Head of the Department of Human Genetics; eLearning experts Ralf Otto and Sabine Romer. Special thanks are also due to Project Officer Dr. Stephanie Heimgartner.

The authors would like to thank Prof. Taina Saarinen and Dr. Jani Ursin from the Finnish Institute for Educational Research for the all-around support, and Donald Adamson for the helpful comments and proof-reading the manuscript.

References

Anderson, L. W., Krathwohl, D. R., Airasian, P. W., Cruikshank, K. A., Mayer, R. E., Pintrich, P. R., Raths, J., & Wittrock, M. C. (2001). *A taxonomy for learning, teaching, and assessing: A revision of bloom's taxonomy of educational objectives*. Longman.

Arzaga, J. S. (2023). Student-created versus YouTube videos: Which type is more effective in enhancing biology learning in a virtual classroom? *European Journal of Contemporary Education and E-Learning, 1*(3), 73–81. https://doi.org/10.59324/ejceel.2023.1(3).07

Bacolod, D. (2022). Student-generated videos using green screen technology in a biology class. *International Journal of Information and Education Technology, 12*(4), 339–345. https://doi.org/10.18178/ijiet.2022.12.4.1624

Balslev, H., Andersson, V., & Schaltz T. (2020). Sharing experiences and the co-creation of knowledge through personal stories – tools for critical thinking: Student perspectives. *Innovative Practice in Higher Education, 4*(1). https://vbn.aau.dk/ws/portalfiles/portal/399338422/194_774_1_PB.pdf

Cannon, H., Feinstein, A., & Feinstein, D. (2010). Managing complexity: Applying the conscious-competence model to experiential learning. *Developments in Business Simulation and Experiential Learning, 37*, 172–182. https://www.researchgate.net/publication/261949851

Castro, C. F., & Barbosa, M. R. (2021). International student mobility: An empirical study based on FEUP mechanical engineering data. *Education Sciences, 11*(2), 36. https://doi.org/10.3390/EDUCSCI11020036

Dekeyser, R. (2007). Skill acquisition theory. In B. VanPatten & J. Williams (Eds.), *Theories in second language acquisition: An introduction* (pp. 97–113). Lawrence Erlbaum Associates.

Duncan, T., Pintrich, P., Smith, D., & Mckeachie, W. (2015). *Motivated Strategies for Learning Questionnaire (MSLQ) manual*. https://doi.org/10.13140/RG.2.1.2547.6968

Elma, E., Saputra, H. N., & Ilcham, I. (2024). Development of video learning materials on animal and human muscular organs. *Progres Pendidikan, 5*(1), 51–57. https://doi.org/10.29303/prospek.v5i1.456

Gendolla, G. H. E. (2017). Comment: Do emotions influence action? – Of course, they are hypo-phenomena of motivation. *Emotion Review, 9*(4), 348–350. https://doi.org/10.1177/1754073916673211

González-Baixauli, C., Montañes-Brunet, E., & Perez-Vazquez, P. (2018). Effects of mobility programmes on university students' academic performance. In *4th International Conference on Higher Education Advances* (pp. 553–562). Valencia, Spain. https://doi.org/10.4995/HEAD18.2018.8034

Goulet-Pelletier, J.-C., Gaudreau, P., & Cousineau, D. (2023). Do students motivated to learn have better creative abilities? *Learning and Individual Differences. 106*(3). https://doi.org/10.1016/j.lindif.2023.102327

Hamilton, K. L., Kuo, Y.-C., Horneffer, P., Stein, T. P., & Goldberg, G. S. (2023). Video didactic preparation augments problem-based learning for first year medical students. *Journal of Medical Education and Curricular Development, 10*. https://doi.org/10.1177/23821205231177862

Herazo, M., Restrepo, R. M., & Álvarez-Ruíz, L. (2023). Perception about the mobility experience of students in universities: A descriptive study. *Journal of Higher Education Theory and Practice, 23*(13), 155–164. https://doi.org/10.33423/jhetp.v23i13.6324

Huang, Z. J., Elmer, J. J., Comolli, N. K., & Kelly, W. J. (2015). Preparation of biology review and virtual experiment/training videos to enhance learning in biochemical engineering courses. In *2015 ASEE Annual Conference & Exposition*. Seattle, Washington. https://doi.org/10.18260/p.24582. https://peer.asee.org/24582

Jesuthasan, S., Umakanth, N., & Dineshkumar, S. (2022). Role of knowledge sharing and learning process of students. *Journal of Business Studies, 8*(1), 98–116. https://doi.org/10.4038/jbs.v8i1.76

Juhász, T., Csongrádi, G., & Tóth, A. (2024). Students' willingness to share knowledge in higher education in Hungary. *Society and Economy, 46*(1), 79–98. https://doi.org/10.1556/204.2023.00022

Kalogeratos, G., Anastasopoulou, E., Tsagri, A., Tseremegklis, C., & Asimakopoulou, S. (2023). Enhancing creativity in the school environment. A narrative examination. *Technium Education and Humanities, 6*, 84–97. https://doi.org/10.47577/teh.v6i.10223

Lacey, K., & Wall, G. J. (2021). Video-based learning to enhance teaching of practical microbiology. *Fems Microbiology Letters, 368*(2). https://doi.org/10.1093/FEMSLE/FNAA203

Lalon, M. G. (2023). A conceptual model for introducing videos for creative skill-based education. In A. Chakrabarti & V. Singh (Eds.), *Design in the era of industry* (Vol. 4.0, No. 3, pp. 1009–1022). Springer. https://doi.org/10.1007/978-981-99-0428-0_82

Lam, J., & Yunus, M. (2023). Student-produced video for learning: A systematic review. *Journal of Language Teaching and Research, 14*(2), 386–395. https://doi.org/10.17507/jltr.1402.14

Majid, S., & Wey, S. M. (2011). Knowledge sharing behavior of graduate students. In E. J. Murray (Ed.), *Global aspects and cultural perspectives on knowledge management* (pp. 113–125). IGI Global. https://doi.org/10.4018/978-1-60960-555-1.CH008

Martín-García, R., López-Martín, C., & Arguedas-Sanz, R. (2020). Collaborative learning communities for sustainable employment through visual tools. *Sustainability, 12*(6), 2569. https://doi.org/10.3390/SU12062569

Mcguire, J. M., Scott, S. S., & Shaw, S. F. (2006). Universal design and its applications in educational environments. *Remedial and Special Education, 27*(3), 166–175. https://doi.org/10.1177/07419325060270030501

Miezah, W., & Whajah, G. (2023). A survey on the effects of using videos in biology lessons on student performance. *International Journal of Scientific Research in Science and Technology, 10*(4), 118–132. https://doi.org/10.32628/IJSRST5231047

Mohyaldinn, M., Kaur, A., Alhassan, H., Hasan, M., Yousif, A., & Ayoub, M. (2020). Students' perception toward the involvement in developing instructional videos using PowToon tool. *A Journal of Engineering, 4*(1), 64–72. https://doi.org/10.61762/pajevol4iss1art7129

Nasution, A. N., & Adlini, M. (2022). Development of scientific approach-based learning videos to improve student learning outcomes on cell biology. *Journal Pendidikan. MIPA, 23*(3), 844–853. https://doi.org/10.23960/jpmipa/v23i3.pp844-853

Neofotistos, V., Mavropoulou, E., Stergiou, C., Aslanidou, S., & Oikonomou, A. (2024). Exploring the role of educational videos in teacher training: Usability, satisfaction, and pedagogical intentions. *Education Quarterly Reviews, 7*(4), 71–82. https://doi.org/10.31014/aior.1993.07.04.252

Onyeaka, H., Passaretti, P., Miri, T., Hart, A., Favero, C., Anumudu, C. K., & Robbins, P. (2023). Pre-lab video demonstrations to enhance students' laboratory experience in a first-year chemical engineering class. *Biochemistry and Molecular Biology Education, 51*(1), 29–38. https://doi.org/10.1002/bmb.21688

Papatsiba, V. (2005). Student mobility in Europe: An academic, cultural and mental journey? Some conceptual reflections and empirical findings. *International Perspectives on Higher Education Research, 3*, 29–65. https://doi.org/10.1016/S1479-3628(05)03003-0

Patrick, J. E., & Zhu J. Z. (2022). Video making tips for laboratory instructors. *CourseSource, 9*. https://doi.org/10.24918/cs.2022.45

Popa, M., & Salanta, I. I. (2013). The emotions' role in the motivation process. *Proceedings of the Managerial Challenges of the Contemporary Society*, *6*, 42–47. Cluj-Napoca, Romania. https://www.researchgate.net/publication/330651477

Saifi, A. G., & Sharabi, S. (2023). Using educational video to enhance creative thinking among eight grade students. In S. Affouneh, S. Salha, A. Tlili, & S. Abu-Eisheh (Eds.), *Education in the post-COVID-19 era—opportunities and challenges. ICLTDW 2022. Lecture notes in educational technology* (pp.137–160). Springer Nature. https://doi.org/10.1007/978-981-99-7293-7_9

Stadlinger, B., Jepsen, S., Chapple, I. L. C., Sanz, M., & Terheyden, H. (2021). Technology-enhanced learning: A role for video animation. *British Dental Journal, 230*(2), 93–96. https://doi.org/10.1038/S41415-020-2588-1

Young, P. (2020). Student-produced video of role-plays on topics in cell biology and biochemistry: A novel undergraduate group work exercise. *Frontiers in Education, 5*, 68–74. https://doi.org/10.3389/FEDUC.2020.00115

Zarate, D., Sheahan, K., & Li, J. (2024). Effective learning through a remote lab kit in an undergraduate invertebrate biology course. *The Journal of College Science Teaching, 53*(2), 111–120. https://doi.org/10.1080/0047231x.2024.2316391

Chapter 14
Defining the Genus *Homo*: A New Curricular Challenge. What if the Chimpanzee Belonged to the Genus *Homo*?

Corinne Fortin

14.1 Introduction

In recent years, research in fields such as paleoanthropology, molecular biology, genetics, and primatology has led scientists to re-examine the definition of the genus *Homo*. In the 2000s, a debate emerged regarding whether chimpanzees should be included in the genus *Homo* due to their genetic proximity to humans (Page & Goodman, 2001; Wildman et al., 2003). Subsequent studies have highlighted significant genetic differences between chimpanzees and humans, supporting the distinction between the two genera. Although the prevailing view is that chimpanzees should remain in the genus *Pan*, the debate remains open, as evidenced by recent work from some researchers (Roffman et al., 2019) who propose reclassifying chimpanzees into the genus *Homo* based on their genetic proximity and certain cognitive, behavioral, and socio-cultural traits shared with humans.

Given that "genus" is not a natural entity but a product of taxonomic rules based on different epistemological approaches, the question of including chimpanzees in the genus *Homo* extends beyond the scientific community. Indeed, such a change could have socio-cultural and educational implications, as the genus *Homo* would no longer be reserved exclusively for *Homo sapiens* (Silva et al. 2018).

What differences concerning the genus *Homo* are discussed in the scientific community, and have these discussions penetrated the field of education? This question can be explored by examining the gaps between scientific research and curricula. This paper investigates the differing scientific characterization of the genus *Homo* and

C. Fortin (✉)
Laboratoire de Didactique André Revuz (LDAR), Université Paris Cité (UPC), Inspé Université Paris Est Créteil (UPEC), Paris, France
e-mail: corinne.fortin@u-pec.fr

O. Morin et al. (eds.), *Challenges of a Changing World in Biology Education*, Contributions from Biology Education Research,
https://doi.org/10.1007/978-3-032-05346-6_14

examines how epistemological and socio-cultural choices influence the knowledge that is taught and its implications for science education. We emphasize the importance of scientific controversy as a key to understanding how science is conducted, viewing it as "the great way to enter into the science that is being done. If we consider controversies not as fights, but as moments where we start by not knowing, and where we discuss, it is the royal road by which to enter into scientific activity" (Latour, 2006, p. 118).

14.2 The Genus *Homo* Under Debate in the Scientific Community

Since Linnaeus's *Systema Naturae* (1758), binomial nomenclature has assigned each organism a genus name and a species name. For humans, this is the genus *Homo* and the species *sapiens*. Other fossil species with similar anatomy are typically also placed in the genus *Homo*.

Approximately fifteen fossil species are currently classified within the genus *Homo*, while none are classified as the genus *Pan*. This discrepancy has raised questions in the scientific community: is this due to a lack of fossils or is it a classification issue? Some researchers argue that the genetic and behavioral similarities between chimpanzees and humans render the division into two genera artificial. They propose renaming chimpanzees as *Homo pan troglodytes* and bonobos as *Homo pan paniscus*. Such changes would lead to significant taxonomic revisions, replacing genera like *Australopithecus* or *Paranthropus* with names such as *Homo afarensis* and *Homo robustus*. This would entail a comprehensive overhaul of the Hominidae classification, a prospect that some researchers support while others oppose.

The genus, as a taxonomic unit, is a historical construct shaped by various scientific perspectives within the scientific community. The debate encompasses arguments both for and against retaining certain fossils in the genus *Homo* or including chimpanzees within *Homo*. For instance, in the 1960s, fossils discovered in Olduvai Gorge led to the designation of new species such as *Homo habilis*, despite its small brain size of 600 cm^3. Leakey et al. (1964) argued that these fossils exhibited upright posture, bipedalism, and dexterity, justifying their inclusion in *Homo*. However, some paleoanthropologists think that the similarities with *Australopithecus* fossils suggest they should belong to *Australopithecus* rather than *Homo*.

Currently, there are four primary classifications, depending on how the genus is defined:

- Biological classification: According to Ernst Mayr, "a genus is a group of one or more species with a common ancestry. These species differ significantly from other groups and are separated by clear morphological gaps" (Mayr, 1950, p. 110). In this view, a genus is a monophyletic group, meaning it includes all descendants of a common ancestor. However, these species can occupy different ecological zones and may be monophyletic or paraphyletic. This classification produces phylograms.

- Cladistic classification: Willi Hennig (1966) proposed a cladistic approach that focuses on shared derived characters (synapomorphies) and groups species according to their exclusive common ancestor. In this case, paraphyly is not allowed. In this system, the genus is defined as a clade of species that share a common ancestor, based on character transformations from ancestral (plesiomorphic) to derived (apomorphic) states. This approach produces cladograms, wherein each genus is a necessary category containing a monophyletic group.
- Mixed classification: Collard and Wood (2015) define a genus as a monophyletic group occupying a single adaptive zone, including species that share similar adaptive strategies with other taxa. According to Wood, for species such as *H. rudolfensis*, *H. habilis*, *H. ergaster*, *H. erectus*, *H. heidelbergensis*, and *H. neanderthalensis* to belong to the genus *Homo*, they must be closer to *H. sapiens* than to *Australopithecus*. However, phylogenetic analyses suggest that *H. rudolfensis* and *H. habilis* may belong in a clade with *Australopithecus* and *Paranthropus*, rather than *Homo*. This could lead to a renaming: *A. habilis* and *A. rudolfensis*.
- Phenetic classification: This approach emphasizes the relationship between similarity and proximity. However, similarity can result from homoplasy (convergent evolution or reversion). This method can produce paraphyletic groups with shared ancestral characters (symplesiomorphies) or monophyletic groups with shared derived characters (synapomorphies). Watson et al. (2001) demonstrated that the genetic distance between chimpanzees and humans is comparable to the distances between species within the same mammalian genus. Based on this, they argued that the genus *Homo* should include chimpanzees.

Thus, the definition of the genus is not self-evident but depends on different epistemological approaches according to various fields of expertise, such as genetics, paleontology, taxonomy, etc. The debate on the definition of the genus *Homo* appears as competing views between different research programs (Dascal, 2008).

It is important to note that our intention here is not to advocate for one epistemological approach over another, but rather to highlight the educational issues in the curricular choices made in teaching the genus *Homo*. How do curricula address the different definitions of the genus *Homo*?

14.3 Theorical Framework: Curricular Didactic Approach

Our study is based on curriculum theory (Young, 2014) and focuses on the curriculum teaching framework (Martinand, 2016) to explain the translation from academic level of science to the curriculum. Curriculum guidelines are seen as a method of teaching. They also guide how knowledge is applied in the classroom. Forquin (2008, p. 88) defines the curriculum as "*what is supposed to be taught and learned*," organized in a specific sequence for a studies cycle. The curriculum is more than just a syllabus; it

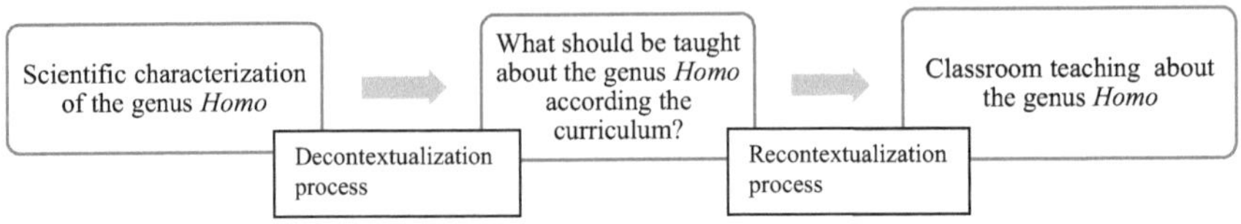

Fig. 14.1 Diagram of didactic transposition

reflects the educational policy rooted in a socio-cultural project. To study curricula, we consider two components of curriculum instruction.

At first, the didactic transposition process refers to "the transformations an object or a body of knowledge undergoes from the moment it is produced, put into use, selected, and designed to be taught until it is actually taught in a given educational institution" (Chevallard & Bosch, 2020). Two stages can be distinguished in didactic transposition: decontextualization and recontextualization:

- The "decontextualization process" corresponds to the extraction of scientific knowledge from its initial epistemological context (contextualization) to be transformed into knowledge to be taught within an institutional educational system (decontextualization).
- The "recontextualization process" refers to the transformation of the knowledge to be taught within a specific pedagogical context, particularly in teacher-student interactions in the classroom.

Below is a diagram that summarizes the two stages of the didactic transposition for the genus *Homo* teaching (Fig. 14.1).

Second, the curricular coherence approach (Lebeaume, 2004) seeks the coherence of the knowledge to be taught by linking three related categories:

- The "educational purpose," which refers to the goals of teaching about the genus *Homo*, linked to tasks and social practices.
- The "tasks" that students are expected to perform according to the curriculum. These tasks require students to use or develop knowledge and involve cognitive process connected to the educational purpose and social practices. These tasks are shaped by the scientific conceptions of the genus *Homo*.
- The "social practices" which are the real-world activities linked to the tasks and the educational purpose. These practices, as defined by Martinand (2001), involve the transformation of a natural or human reality and are used as a reference for teaching. In short, reference practice is every social activity (research, medicine, engineering, daily life, etc.) that is used as a teaching guide.

The curricular coherence approach can be illustrated by the following diagram (Fig. 14.2).

The circulation of scientific knowledge about the genus *Homo* is thus the result of both a didactic transposition and a curricular coherence approach. This circulation is not a simple reduction of scientific knowledge to teachable content. It is not merely a "top-down" transfer from science to curricula. Instead, it is a process of

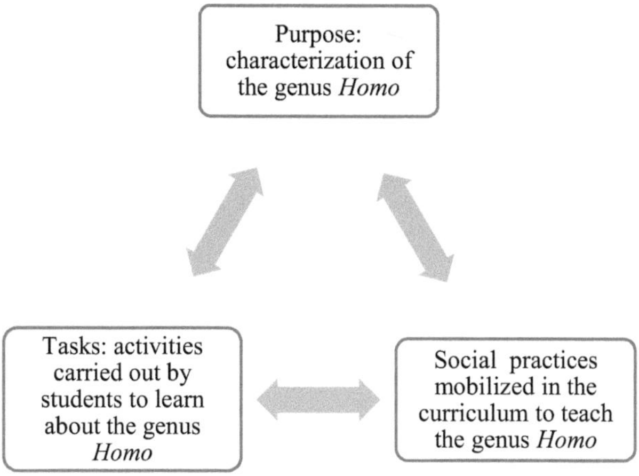

Fig. 14.2 Curricular coherence between purpose, tasks, practices of reference to characterize the genus *Homo*

translating the "message" (Callon & Latour, 1991) about the characterization of the genus *Homo* from one field to another. To understand this "translation process," we see the curriculum as a tool for selecting scientific data about the genus *Homo*.

14.4 Research Questions

Selecting one or more definitions of the genus *Homo* is not a neutral act; it reflects a pedagogical choice regarding what the curricula aim to teach about the genus *Homo*.

In this study, we are interested in:

On the one hand, the gap between the field of scientific research and the pre-service teacher training curriculum at the master's level;

On the other hand, the gap between this curriculum and the secondary education curriculum.

The goal of this study is to determine the characterization of the genus *Homo* and whether the inclusion of chimpanzees in the genus *Homo* is considered during of the didactic transposition.

To achieve this goal, this study aims to answer the following research questions:

- What scientific knowledge about the genus *Homo* is incorporated into the curriculum of university pre-service teacher training?
- What knowledge about the genus *Homo* is transferred from the pre-service teacher training curriculum to the secondary education curriculum?

We hypothesize that these gaps reflect a reconfiguration of the genus *Homo*, resulting from a selection of scientific knowledge, social practices, and tasks proposed to students (Fig. 14.3).

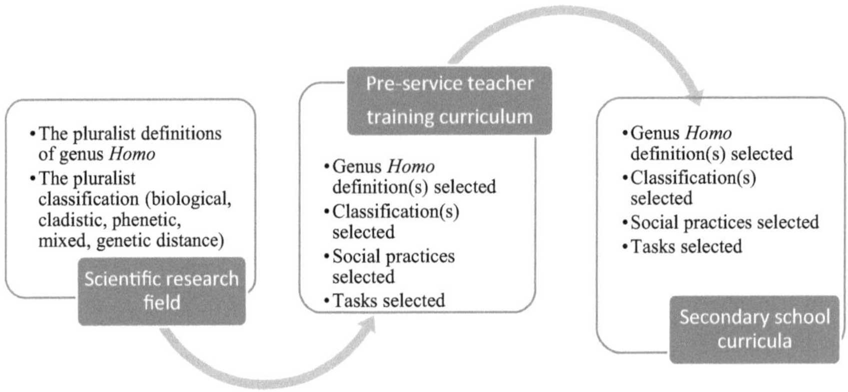

Fig. 14.3 Modalities of the didactic choices of the genus *Homo* between scientific research and curricular fields

14.5 Methodology

14.5.1 Curriculum Instruction

To address these two research questions, we relied on the official curricula documents published by the French Ministry of National Education between 2018 and 2022 (Ministère de L'Éducation Nationale, 2018, 2019, 2020a, 2020b, 2022).

- On the one hand, these curricula documents concern the training of pre-service teachers who are preparing for the competitive examination at the master's level called Agrégation sciences de la vie et de la Terre (Life & Earth Sciences,). These documents determine what future teachers are expected to know about the genus *Homo* (Ministère de l'Éducation Nationale, 2020a, 2022)
- On the other hand, these curricula documents concern secondary education (Ministère de l'Éducation Nationale, 2018, 2019, 2020b). They indicate what should be taught about the genus *Homo* at this level, allowing to understand what students are expected to learn.

These official curricula documents enable us to identify which characterization of the genus *Homo* are discussed and whether the inclusion of chimpanzees is addressed in both pre-service teacher training and secondary education.

14.5.2 Curricular Analysis

To examine the didactic transposition of scientific knowledge at both the master's and secondary school level, we conducted a comparative qualitative content analysis

(Bardin, 2013) of the curricula using an analysis grid (Table 14.1) that includes three categories and with their criteria:

- **Scientific Components**: Presence or absence of scientific controversies on the genus *Homo* limits; genus *Homo* characteristics; relationship between humans and chimpanzees.
- **Social Practices mobilized for the study of the Genus** *Homo*: The contribution of scientific debate on the genus *Homo* limits as a necessary scientific activity; the choice to favor certain types of data in classification, reflecting competition between different characterization of the genus *Homo*.
- **Tasks assigned to future teachers at the master's level or to secondary school students**: Activities designed to develop their skills to characterize the genus *Homo* (e.g., comparing, arguing, classifying, etc.).

14.6 Results

14.6.1 Genus Homo*: From Scientific Research to Master's Curriculum*

To address question n°1: *"What scientific knowledge about the genus Homo is incorporated into university pre-service teacher training curricula (master's level)?"*, we focus on the significance of the gap between the field of scientific research and the pre-service teacher training curriculum.

The analysis of the curricula documents indicates that, between the field of scientific research and the master's level curriculum, the genus *Homo* is limited by the selection of its paleontological data and the common origin of humans and chimpanzees.

Figure 14.4 illustrates the "decontextualization process" as the extraction of scientific knowledge from the field of research and its transformation into knowledge to be taught within the university pre-service teacher training curriculum.

It highlights that the gap between the scientific field and the curriculum is reflected in the lack of consideration of both molecular and paleontological approaches to the characterization of the genus *Homo*.

More specifically, our analyses of the curricula revealed the following aspects:

- Scientific components: The scientific community's disagreement on the classification of the genus *Homo* is not mentioned in the curriculum. The genus *Homo* is primarily presented in a paleontological context, based on anatomical data (skull volume, bipedalism, mandibles, etc.) as well as tool manufacture and cultural activities. The phylogenetic relationships between *Homo sapiens* and other modern primates are established based on morpho-anatomical and genetic data.

Table 14.1 Extract of comparative analysis grid of scientific components, social practices and tasks in the field of scientific research, Master's level in pre-service teachers and secondary school curricula (+ present, −absent)

Criteria	Scientific research field	Citations from master' s level curriculum		Citations from secondary education curriculum	
Controversies about the genus *Homo limits*	Inclusion of the chimpanzee in the genus *Homo*	Not in the curricular documents	−	"'*The precise construction of the phylogenetic tree of the* genus *Homo is controversial in detail.*'"	+
	Inclusion of *habilis* & *rudolfensis* in the genus *Australopithecus*				
	Homo species (*H. sapiens* + fossil *Homo* species) Phylogeny of the human lineage	*The current human species & its closest fossil relatives*"	+	"*Human evolution*" "*The current human species & its closest fossil relatives*"	+
	H. paniscus & H. *troglogytes*, *H.gorilla*	Not in the curricular documents			
Paleontological characteristics of the genus *Homo*	Reduction of prognathism	"*Fossils, witnesses of evolution*" "*Origin and evolution of hominids*"	+	"*Modern humans and some fossils*" & "*Human relatedness to the human lineage*" It is specified: "*The genus Homo groups together the current human species and fossil species characterized in particular by the development of cranial capacity. Several human species of the genus Homo have coexisted on Earth*" "*Scientific arguments derived from the comparative analysis of fossils enable us to reconstruct the history of our origins. The study of fossils from 3 to 7 million years ago reveals innovations characteristic of the human lineage (extended bipedalism, shape of the mandible).*"	+
	Parabolic mandible		+		+
	Centrally positioned foramen magnum		+		+
	Increase in brain volume		+		+
	Bipedalism		+		+

(continued)

Table 14.1 (continued)

Criteria	Scientific research field	Citations from master' s level curriculum		Citations from secondary education curriculum	
	Tools manufacture		+		+
	Not in the curricular documents			**"***Paleoanthropology constructs a scientific story of our origins based on the fossil record. Phylogeny studies the relationships between modern species and hominid fossils"*	+
Paleogenetic characteristics of the genus *Homo*	Paleogenetic analysis of *H. neanderthalensis* & *H. denisovensis*	Not in the curricular documents		***"Scientific arguments (paleogenetic) to clarify the relationship between H. sapiens and other Homo species (Neanderthal and Denisova)"*** *"Phylogeny based on paleontological and paleogenetic data"*	+
Relationship between humans and chimpanzees	Genetic similarities	*"Place of the human species in the classification"* *"the kinship relations of the human species within the Primates"*	+	***"The place of humans in classification"*** ***"Human evolutionary relationships within primates"*** ***"The present-day human species (H. sapiens) belongs to the primate group and is more specifically related to great apes, with which it shares morphological, anatomical, and genetic similarities."*** ***"It shares its most recent common ancestor with the chimpanzee."*** *"The production of complex tools and a variety of cultural practices are associated with the genus Homo but not exclusively."*	+
	Chromosomal similarities		+		+
	Partial/permanent bipedalism		+		+
	Tools use and manufacture		+		+
	Common ancestor		+		+

(continued)

Table 14.1 (continued)

Criteria	Scientific research field	Citations from master' s level curriculum		Citations from secondary education curriculum	
Social practices	Choice of data for phylogenetic trees	*"Current systematics methods (cladistics and phenetics)"*	+	*"Phylogeny makes it possible to study the evolutionary relationships between extant and fossil hominid species."*	+
	The scientific contribution of the debate on the limits of the genus *Homo*	Not in the curricular documents			
Tasks	-	*"Compare anatomical traits and genetic similarities among primates (degree of relatedness and construction of a phylogenetic tree)"*			+
		Not in the curricular documents		*"Arguing the construction of the phylogenetic tree of the genus Homo"*	+
				"Identify the characters specific to the human lineage"	+
				"Analyze scientific evidence to refine the relationship of H. sapiens with other Homo species (Neanderthal and Denisova)"	+

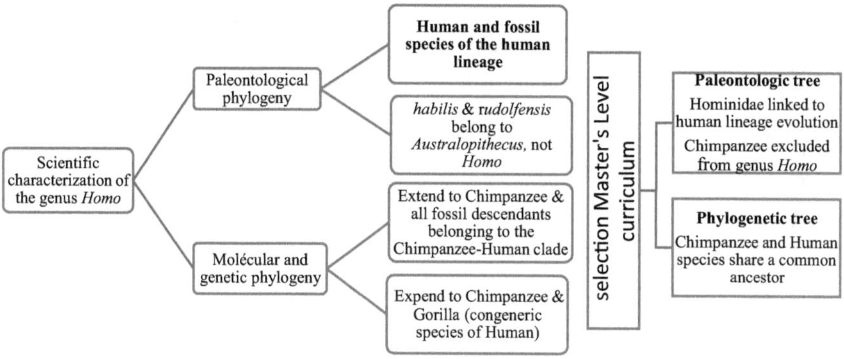

Fig. 14.4 Curricular selection from scientific characterization of the genus *Homo* to master's level exams

- Social practices: The boundaries of the genus *Homo* (e.g. the inclusion of chimpanzees in the genus *Homo*, the inclusion of *H. habilis* and *H. rudolfensis* in the genus Australopithecus or the controversies about the genetic distance between chimpanzees and humans) are not questioned or discussed in the curriculum. Nor is the choice of data used to construct a phylogeny a matter of discussion.
- Tasks: The focus is on comparing and arguing based on fossil data to place the species within the genus *Homo* in the phylogeny of the human lineage

14.6.2 Genus Homo*: From the Master's Level to Secondary Education Curricula*

To address question n°2: *What knowledge about the genus Homo is transferred from the university pre-service teacher training curriculum to the secondary school curriculum?* We focused on the significance of the gap between pre-service teacher training and secondary education curricula.

The implementation of the analysis grid of curricula documents (Table 14.1) highlights that the presentation of the genus *Homo* at the master's level is similar to secondary school curricula. The secondary school curriculum primarily emphasizes the paleontological aspects of the genus *Homo* and the shared ancestry of humans and chimpanzees. However, it also introduces a paleogenetic dimension concerning Neanderthal and Denisova.

Below, Fig. 14.5 shows the commonalities (paleontological and phylogenetic trees) and the difference at the paleogenetic level.

More specifically, our analyses of the curricula revealed the following aspects:

- **Scientific components:** Characterizing the genus *Homo* on the basis of paleontology and paleogenetics excludes other criteria, such as tool use, manufacture, and social organization among great apes.

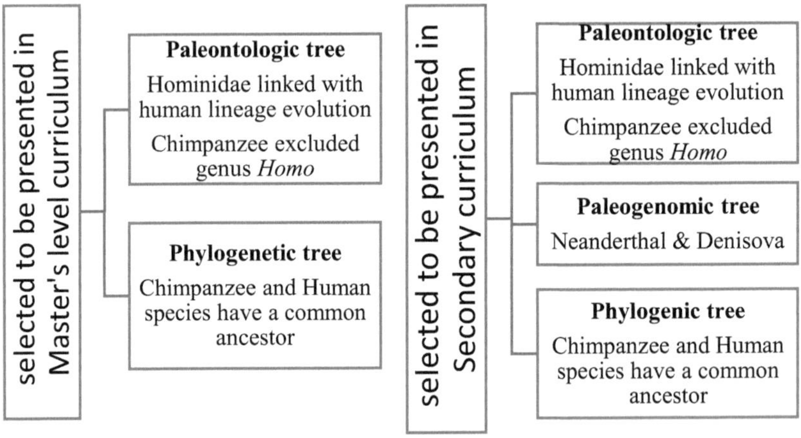

Fig. 14.5 Selections from master's level in pre-service teachers to Secondary Curricula

- **Social practices**: Similar to the master's level curriculum, the secondary curriculum does not address debate about the genus *Homo* limits or the selection of data used to construct phylogenies.
- **Tasks**: several tasks are mentioned explicitly in the curriculum. Students should be able to argue, identify, and analyze the relationships between different *Homo* species within the human lineage.

Notably, and somewhat paradoxically, the secondary education curriculum is more comprehensive than the master's level curriculum, as it acknowledges that "the precise construction of the phylogenetic tree of the genus *Homo* is controversial in detail."

14.7 Discussion

The French curricula for pre-service teacher education and for secondary education share the same coherence in terms of scientific goals, social practices, and tasks (Fig. 14.6).

The common choice of these two curricula to characterize the genus *Homo* on the basis of paleontology alone leads future teachers or students to engage in tasks aimed at defining the genus *Homo* exclusively through paleontological classifications. This choice is logical, as it provides students and future teachers with access to paleontological data. The curricula tend to focus on the process of hominization, even if the term itself is not used.

Chimpanzees are described in a subtractive way, focusing on what humans possess and chimpanzees do not possess. For example, chimpanzees are depicted with plesiomorphic characters (reduced skull volume, eccentric foramen, large

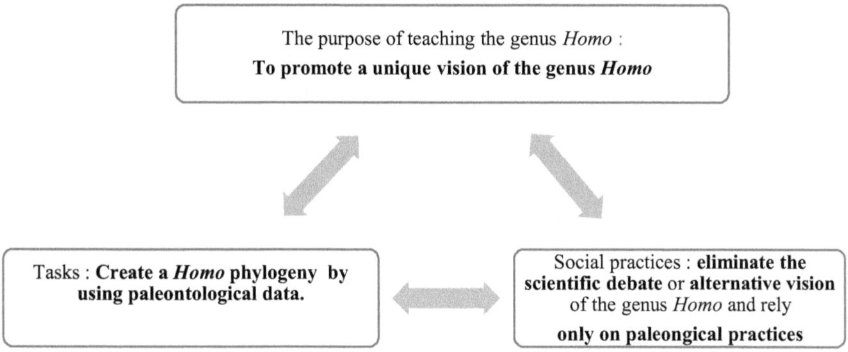

Fig. 14.6 Curricular coherence pre-service teachers and secondary school curricula

canines, opposable toes, etc.), as if they possessed no apomorphic characters. In contrast, humans are depicted exclusively with apomorphic characters (developed skull volume, centered foramen, small canines, etc.). Curricula present chimpanzees as an out-group to highlight the link between fossil *Homo* species and *H. sapiens.* This approach does not focus on the relationship between the *Pan* and *Homo* genera, reinforcing the idea of an unbridgeable gap between the two.

This curricular approach may limit the understanding of how the genus *Homo* is constructed by the scientific community. As a result, future teachers may not be able to explain to their students that the genus *Homo* is a scientific concept used to describe a set of shared characteristics (paleontological, genetic, behavioral) among several species. In a context where the position of the human species in evolutionary history is not always culturally accepted, it is surprising that, at the master's level, the critical dimension of questioning the limits of the genus *Homo* is not considered a key training issue. There is, therefore, a real risk of essentializing the genus *Homo* and, by extension, all other genera. Encouraging students to consider the possibility that the chimpanzees might belong to the genus *Homo* could help prevent an anthropocentric vision of the genus *Homo.*

Our study is focused on the French situation, but it would be interesting to know what is the situation in other countries and if and how other countries embrace other curricular approaches to the genus *Homo.* Future studies performing such comparative analysis may identify possible patterns and allow to understand how these contribute to students' anthropocentric vision of the genus *Homo.*

14.8 Conclusion

The curricular approach to the genus *Homo* in France does not question the separation between the genus *Homo* and the genus *Pan.* However, this could be an opportunity to raise awareness among future teachers of a pluralist approach, helping to avoid any form of essentialization of the genus *Homo* and opening the minds of future teachers and students to alternative epistemological perspectives. As Goodall (1998) suggested "Now we must redefine tool, redefine Man, or accept chimpanzees as humans" (p. 2084).

For both university teacher training and secondary education, it is important to move beyond the idea that scientific data (paleontological or molecular) speak for themselves. Students need to understand the link between empirical data and the theoretical framework of evolution (Fortin, 2015; Fortin & Gobert, 2024). We suggest that curricula should also incorporate different epistemological points of view, encouraging a critical approach that considers our relationship with other great apes.

Rethinking our relationship with great apes means understanding that our phylogenetic proximity is not just morpho-anatomical, genetic, or molecular. It is also behavioral and, more broadly, cultural (Whitehead et al., 2019). We, now, know that we also share a common psychic activity based on self-awareness and the consciousness of others (Kaneko & Tomonaga, 2011). The evolution of great primates is now understood as a co-evolution of genes and culture, involving the development of abstract cognitive reasoning (Rey et al., 2018).

However, acknowledging these similarities does not mean denying the differences between humans and non-human primates, nor does it imply naturalizing or biologizing culture. There is no need to oppose great non-human primates to humans or to establish a hierarchy between them. They should be recognized both in their singularity and in their diversity. In both humans and chimpanzees, there are similarities in behavior and social life, as well as differences resulting from a shared evolutionary history and their respective specific evolution.

Finally, this study on the genus *Homo* limits reflects work in science education on the role of scientific controversies as valuable opportunities (Niaz, 2016) in understanding the nature of science (NOS). Moreover, curricula should treat the teaching of the limits of genus *Homo* as a controversial socioscientific issue (Zeidler, 2014) because this controversy offers a way to re-examine, both scientifically and culturally, the identity of the human species within the evolutionary history of primates.

References

Bardin, L. (2013). Chapitre IV. L'analyse propositionnelle du discours. *Quadrige*, 243–254.

Callon, M., & Latour, B. (1991). *La science telle qu'elle se fait. Anthologie de la sociologie des sciences de langue anglaise.* La Découverte, Paris.

Chevallard, Y., & Bosch, M. (2020). Didactic transposition in mathematics education. *Encyclopedia of Mathematics Education*, 214–218.

Collard, M., & Wood, B. (2015). Defining the Genus *Homo*. In W. Henke & I. Tattersall (Eds.), *Handbook of paleoanthropology* (pp. 2107–2144). Springer.

Dascal, M. (2008). Dichotomies and types of debate. *Controversy and Confrontation: Relating Controversy Analysis with Argumentation Theory*, 27–49.

Fortin, C. (2015). Evolutionary theory in secondary schools: Some teaching issues. In T. Heams, P. Huneman, G. Lecointre, & M. Silberstein (Eds.), *Handbook of evolutionary thinking in the sciences* (pp. 897–910). Springer.
Fortin, C., & Gobert, J. (2024). *Evolution of life: Teaching, learning and training*. New approaches in Current Reasearch in Didactics of Evolution. ISTE-Wiley.
Forquin, J.-C. (2008). *Sociologie du curriculum*. PUR, Rennes.
Goodall, J. (1998). Learning from the chimpanzees: A message humans can understand. *Science, 282*, 2184–2185.
Hennig, W. (1966). *Phylogenetic systematics*. University of Illinois Press.
Kaneko, T., & Tomonaga, M. (2011). The perception of self-agency in chimpanzees (Pan troglodytes). *Proceedings of the Royal Society B: Biological Sciences, 278*(1725), 3694–3702.
Latour, B. (2006). Entretien avec Bruno Latour, Fossier et Gardella Tracés. *Revue De Sciences Humaines, 10*, 113–129.
Leakey, L. S. B., Tobias, P. V., & Napier, J. R. (1964). A new species of the genus *Homo* from Olduvai Gorge. *Nature, 202*, 7–9.
Lebeaume, J. (2004). Designing technology education at the junior high school level: Propositions from the French school curriculum. *The Journal of Technology Studies, 30*(3), 2–9.
Martinand, J. L. (2001). Pratiques de référence et problématique de la référence curriculaire. In A. Terisse (Ed). *Didactique des disciplines. Les références au savoir*. De Boeck Université (pp. 17–24).
Martinand, J. L. (2016). La question de la référence en didactique du curriculum. *Investigações Em Ensino De Ciências, 8*(2), 125–130.
Mayr, E. (1950). Taxonomic categories in fossil hominids. *Cold Spring Harbor Symposia on Quantitative Biology, 15*, 109–118.
Ministère de L'Éducation Nationale. (2018). *Bulletin officiel de l'Éducation nationale n° 30*.
Ministère de L'Éducation Nationale. (2019). *Bulletin officiel de l'Éducation nationale n° 8*.
Ministère de L'Éducation Nationale. (2020a). *Concours externe de l'agrégation du second degré. Section sciences de la vie – sciences de la Terre et de l'Univers. Programme de la Session 2020*.
Ministère de L'Éducation Nationale. (2020b). *Bulletin officiel de l'Éducation nationale n° 31*.
Ministère de L'Éducation Nationale. (2022). *Concours externe de l'agrégation du second degré. Section sciences de la vie – sciences de la Terre et de l'Univers. Programme de la Session 2022*.
Niaz, M. (2016). Nature of science in science education: An integrated view. *Chemistry Education and Contributions from History and Philosophy of Science*, 37–89.
Page, S. L., & Goodman, M. (2001). Catarrhine phylogeny: Noncoding DNA evidence for a diphyletic origin of the mangabeys and for a human-chimpanzee clade. *Molecular Phylogenetics and Evolution, 18*(1), 14–25.
Rey, A., Minier, L., Malassis, R., Bogeart, L., & Fagot, J. (2018). Regularity extraction across species: Associative learning mechanisms shared by humans and non-human primates. *Topics in Cognitive Science, 11*(3), 573–586.
Roffman, I., Mizrachi, Y., Chiarelli, B., Savage-Rumbaugh, S., Rubert-Pugh, E., Fontaine, L., Panchevre, M., Ronen, A., & Nevo, E. (2019). Pan survival in the 21st century: Chimpanzee cultural preservation, rehabilitation, and emancipation manifesto. *Human Evolution, 34*(1–2), 1–19.
Silva, H. M., Penaloza, G., Tomasco, H., & Carvalho, G. S. (2018). Chimpanzee included in the genus Homo. How biology teachers from three Latin American countries conceive it. *Journal of Biological Education, 53*(5), 506–515.
Watson, E. E., Penny, D., & Easteal, S. (2001). *Homo* genus: A review of the classification of humans and the great apes. In *Humanity from African naissance to coming millennia: colloquia in human biology and palaeoanthropology* (pp. 1000–1012). Firenze University Press.
Whitehead, H., Laland, K. N., Rendell, L., Thorogood, R., & Whiten, A. (2019). The reach of gene–culture coevolution in animals. *Nature Communications, 10*(1), 2405.
Wildman, D. E., Uddin, M., Liu, G., Grossman, L. I., & Goodman, M. (2003). Implications of natural selection in shaping 99.4% nonsynonymous DNA identity between humans and chimpanzees:

Enlarging genus Homo. *Proceedings of the National Academy of Sciences of the United States of America*, USA, *100*(12), 7181–7188.
Young, M. (2014). What is a curriculum and what can it do? *The Curriculum Journal, 25*(1), 1–7.
Zeidler, D. L. (2014). Socioscientific issues as a curriculum emphasis: Theory, research, and practice. In N. G. Lederman & S. K. Abell (Eds.), *Handbook of research on science education, volume II* (pp. 697–726). Routledge.

Part III
Students' and Teachers' Conceptions

Chapter 15
Beyond Pen-and-Paper: Challenges and Potentials in Capturing Students' Conceptions by Using the Digital "Draw"-A-Biologist Test

Bianca Reinisch, Tom Bielik, Moritz Krell, and Daniela Mahler

15.1 Introduction

Stereotypical conceptions of scientists and their work appear to influence students' interest in science (Scholes & Stahl, 2020). Also, a positive view of scientists and their work makes students more likely to choose a career in science after leaving school (Finson, 2002). At the same time, Stamer et al. (2020) point to the potential negative impact of unrealistically positive conceptions, noting that these may contribute to "students dropping out of university" when their expectations are not met (p. 170). Therefore, the aim of science education in schools should be, among other things, to foster authentic ideas about scientists and their work (Reinisch & Krell, 2023; Reinisch et al., 2017).

Conceptions of students about scientists and their work have been researched for many decades (Chang et al., 2020; Miller et al., 2018). One of the most widely used instruments is the Draw-A-Scientist Test (DAST). With the DAST, students'

B. Reinisch (✉)
University of Potsdam, Potsdam, Germany
e-mail: bianca.reinisch@uni-potsdam.de

T. Bielik
Radboud University, Nijmegen, Netherlands
e-mail: tom.bielik@ru.nl

M. Krell
IPN - Leibniz Institute for Science and Mathematics Education, Kiel, Germany
e-mail: krell@leibniz-ipn.de

D. Mahler
Freie Universität Berlin, Berlin, Germany
e-mail: daniela.mahler@fu-berlin.de

O. Morin et al. (eds.), *Challenges of a Changing World in Biology Education*,
Contributions from Biology Education Research,
https://doi.org/10.1007/978-3-032-05346-6_15

conceptions without writing and possibly even reading skills can be assessed by asking them to express their conceptions in drawings (Chang et al., 2020). Numerous studies confirm the widespread image of a (mostly male) scientist wearing a lab coat and goggles while conducting (dangerous) experiments in a laboratory, usually alone (see Miller et al., 2018).

For a long time (and still), the DAST checklist (DAST-C; Finson et al., 1995) has been used to analyse DAST drawings (see Chang et al., 2020; for a more detailed historical report on the DAST, see Reinisch et al., 2017). The DAST-C contains the aforementioned stereotypical elements that were previously identified in different studies. The more of these elements can be recognised in a drawing, the more stereotypical a scientist's conception is. More recent studies have made adjustments to the DAST prompt. Farland-Smith (2012), for example, modified the prompt so that the respondents are asked to explicitly reflect on three aspects of their drawings: the appearance, the location, and the activity of scientists. Accordingly, the category system for analysing the drawings has become more extensive but also more meaningful than the DAST-C.

Studies show that assessment using the DAST (or modified versions) and the corresponding analysis of drawings may be associated with validity issues (e.g., Brochey-Taylor & Taylor, 2024; Losh et al., 2008; Reinisch et al., 2017). A key factor influencing the pictures appears to be the specific prompt given to test persons. For example, it plays a role in whether one or more scientists are asked for (Reinisch et al., 2017). But even with a carefully considered prompt, it is difficult to collect and analyse conceptions with drawings in this area. Methodological studies show that drawing skills, for example, influence the outcome (Losh et al., 2008). Reinisch et al. (2017) demonstrated that around one-third of the pre-service teachers they surveyed were unable to depict their conceptions through drawings. Also, studies in which participants were asked to produce multiple drawings show that female scientists appear more frequently in subsequent drawings (Losh et al., 2008).

Lamminpää and Vesterinen (2020) were able to show in the so-called Draw-A-Science Comic that their surveyed students seemed to think of scientists in specific disciplines, most frequently chemistry, biology, astronomy, and robotics when creating the comics. The students assigned different attributes and activities to the scientists they drew, such as dangerous activities carried out by chemists. Thus, it is worth considering whether assessing conceptions about scientists and their work in a specific scientific discipline would be valuable. Such an approach can be justified by the fact that different disciplines involve specific procedures and methods. For instance, biologists must consider ethical norms related to living beings as research objects, acknowledging their rights and dignity (Resnik, 2007). This 'special ethical position of biology' (Köchy, 2008, p. 203, [own translation]) contrasts with other fields, such as chemistry or physics, where concerns like environmental impact or hazardous materials are more prominent. Also, in biology, the prospect of scientific work with living organisms in their habitats may be more attractive for students to pursue a scientific career. It is conceivable that some students might not imagine the laboratory as a place where they would like to spend their working day.

The problems described with the DAST highlight the need for alternative testing instruments to validly capture and evaluate students' concepts (Brochey-Taylor & Taylor, 2024; Chang et al., 2020; Reinisch & Krell, 2023; Reinisch et al., 2017). In sum, three essential aspects can be deduced: (1) The prompt must be formulated in a way that certain conceptions are not (unintentionally) predetermined (e.g., the number of scientists). (2) Students should not be asked generally about their conceptions regarding scientists and their work; instead, the assessment should be more specifically tied to a scientific discipline. (3) Drawing skills influence the representation of one's conceptions. Therefore, a method should be chosen that does not depend on the drawing skills of the participants. One way to address the last issue could be to use technology-based tools (Chang et al., 2020).

15.2 Research Aim and Questions

This study aims to investigate whether a digital version of the DAST, which does not require drawing skills, can be validly used to interpret students' visualised conceptions. In psychological and educational research, validation is understood as a process involving the creation and evaluation of arguments that provide a solid scientific basis for the interpretation and use of test scores (argument-based approach to validation; Kane, 2013). In this study, we use an existing web-based software (Pixton) to develop an alternative instrument for capturing students' conceptions of biologists and their work. The first research question (RQ1) is: *In what ways does evidence support the interpretation of student pictures as indicators of their conceptions about biologists and their work?* Methodical literature describes different sources of evidence for valid test score interpretation, including evidence based on response processes, test content, internal structure, relations to other variables, and consequences of testing (AERA et al., 2014). In this study, we focus on evidence based on response processes, which is "the fit between the construct and the detailed nature of the performance or response actually engaged in by test takers" (AERA et al., 2014, p. 15).

While the DAST requires drawing skills, the use of digital tools requires digital media competencies. Students today are considered to be "fluent in acquiring and learning all sorts of new technology" (Mete et al., 2017, p. 69). Nevertheless, the DAST was designed especially for younger children to avoid reading and writing skills (Losh et al., 2008), as these are extraneous to what the DAST aims to measure ('construct-irrelevance'; AERA et al., 2014). Thus, an alternative digital instrument should be useable not only for older but also for younger students as well. It is questionable from which age on Pixton can be used for the intended purpose without students encountering technical difficulties.

In addition to the potential technical challenges, Pixton is an English-language software that is not available in the native language of the participating students. Although English is widely taught as a second language in Germany, beginning in elementary school, limited English proficiency could introduce another source of construct-irrelevance, potentially hindering students' effective use of the tool. Thus, our research question (RQ2) is: *To what extent do students encounter construct-irrelevant difficulties when using a digital tool (i.e., Pixton), such as* (a) *technical difficulties and* (b) *language-related difficulties?*

15.3 Methods

15.3.1 The Software "Pixton"

Pixton is a web-based comic creation application. Students can create comics that can be viewed by their teachers. The application is available in English, French and Spanish. When creating a comic, there are many options for personalisation. At the time of the study, the tabs were arranged horizontally: Background, Characters, Focus, Words, Faces and Actions (Fig. 15.1).

At the beginning, users are in the Background tab by default. Here, you have the option of selecting from various preset backgrounds. It is possible to filter for various keywords using a search bar. Various characters can be added under the Characters tab. It is possible to either design a character individually or to select one from the predefined ones and adapt it if necessary (e.g., body shape, age, skin and hair colour, outfit, hairstyle). This is followed by the Focus tab. Here, you can choose between different sections within the selected background and vary the position of the character. A title for the image can be created under the Words category. Characters can also be given speech or thought bubbles. Various facial expressions can be selected under the Faces tab, which can be filtered according to different emotions or activities, for example. Similarly, the Actions tab allows you to assign

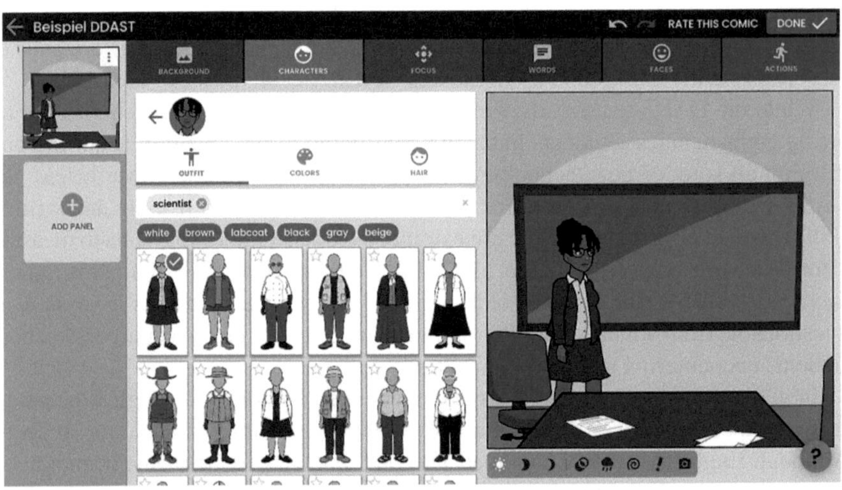

Fig. 15.1 Screenshot of Pixton under the 'Characters' tab

one of many different poses to the character (e.g., sitting, lying, standing). In addition, various objects (e.g. pens, signs, notepads) can be selected, which can then be placed in a hand.

15.3.2 *Sample and Study Design*

A total of 31 students from three different schools in two federal states of Germany voluntarily agreed to participate in the study. They were recruited through a convenience sampling approach.

Elementary school: $n_{\text{grade6}} = 10$ (M_{age}: 11.50, SD_{age}: 0.53)
Comprehensive school: $n_{\text{grade7}} = 11$ (M_{age}: 12.55, SD_{age}: 0.52)
Grammar school: $n_{\text{grade11}} = 10$ (M_{age}: 16.73, SD_{age}: 0.79)

In individual sessions, the test persons' activities were audio and video recorded during the study using the screencast software Camtasia. The students were given instructions on thinking aloud before the study began (Sandmann, 2014). Students were then given the task of using Pixton to create a picture of biologists going about their daily work (modified DAST; Farland-Smith, 2012; Reinisch et al., 2017; Fig. 15.2). The survey process was standardised using a previously piloted and adapted guideline. After creating the first picture, the students could voluntarily create another one. The processing time was limited to 90 min but lasted usually between 30 and 60 min. The interview sessions, which took place in a separate room at the students' schools, were conducted by one of three graduate students who acted as test administrators and were responsible for conducting the interviews.

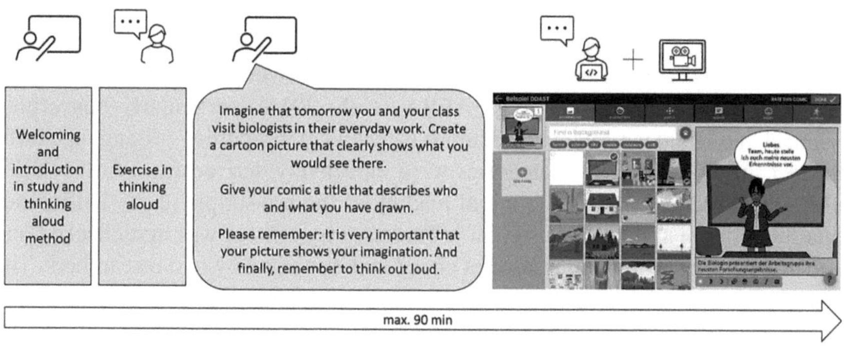

Fig. 15.2 The procedure of the study. *Source* Picture on the right: Screenshot created with Pixton (https://www.pixton.com)

15.3.3 Data Processing and Analysis

The data was processed by transcribing the verbal data using a transcription guideline and embedding all the data (Pixton pictures, transcripts from verbal data, screencast video) in the MAXQDA analysis software.

In the first step of the analysis, the pictures were coded concerning the appearance of the biologists, the locations depicted, and the activities of the biologists using an existing DAST category system (Reinisch et al., 2017). After the first independent coding of five pictures by two researchers, an 'almost perfect' Cohen's κ of .91, .80, .64 was found (Landis & Koch, 1977, p. 165). Hence, one researcher analysed the rest of the pictures. In total, 42 pictures were created by the students, of which five were not appropriate for analysis (e.g., because they depicted a school class in a lab). It is important to note that some students, on their initiative, created a series of pictures showing the same biologist(s) engaged in a sequence of activities. Since these pictures are not independent, each series is considered as a single picture for the analysis.

We then analysed the extent to which the data from the think-aloud protocols ($N = 29$) matched the coding of the 42 students' pictures. For this purpose, we formed six validation categories into which the students' think-aloud statements were categorised, one of which was added inductively in the course of coding (i.e., 'Category for image coding not available'; Table 15.1). One category is marked as 'unclear' as the absence of statements does not necessarily indicate missing validity, as students might have perceived the element but chose not to comment on it. This represents an open issue for the interpretation of test validity, rather than direct evidence against it. Three think-aloud protocols were coded by two researchers, which resulted in an 'almost perfect' Cohen's κ of 0.80 (Landis & Koch, 1977, p. 165). Differently coded statements were discussed and resolved. Both researchers then coded half of the remaining think-aloud protocols.

For RQ 2a and 2b, students' difficulties while creating the pictures were coded using predefined categories based on theoretical frameworks addressing construct-irrelevant variance in testing (e.g., AERA et al., 2014), particularly concerning technical and language-related difficulties ('Technical difficulties', 'Language difficulties'). Additionally, new categories were inductively derived from further difficulties that emerged during the initial reading of the transcripts (e.g., 'Difficulties regarding administration consistency'). The category system was first checked and further developed using one transcript coded and discussed by two researchers. The subsequent coding of three transcripts resulted in a 'substantial' Cohen's κ of 0.64 (Landis & Koch, 1977, p. 165). One researcher then carried out the final coding. In sum, 29 transcripts were analysed. Again, two think-aloud protocols had to be excluded because the students did not fulfil the task according to the prompt.

Table 15.1 Validation categories representing evidence for and against test validity

Validation category	Description	Example from the Appearance categories
Evidence for test validity		
Picture = verbal data	Students' statement matches coding for a picture element	Picture coding: *Female scientist* Verbal data: "I think scientists are often women [clicks character]." (student, grade 7)
Evidence against test validity		
Element in picture coding missing	An element that is considered relevant by the student in the verbal data is not present in the coding of the picture	Picture coding: The biologist in the picture is standing behind a table so that the skirt *(Daily wear)* is no longer recognizable Verbal data: "Maybe with a skirt or something [clicks outfit]." (student, grade 7)
Element in picture coding not intended	A picture element that was coded in the Pixton picture was not intended	Picture coding: *Further protective cloth* Verbal data: "Then [clicks on an outfit] here, maybe gloves not quite so important, yes, why not." (student, grade 11)
Element in picture coding differs	A picture element was coded differently than students' statements suggest	Picture coding: *Young / middle aged* Verbal data: "Let's take a fat old man." (student 11)
Category for picture coding not available	Category for picture coding not available in DAST category system	Picture coding: no category for scientists' emotions available Verbal data: "Now I choose the face and they often look happy, I think." (student, grade 7)
Unclear		
No statement about element in picture coding	There are no statements in the students' statement on the coded Pixton picture element	Picture coding: *Lab coat, Glasses, further protective cloth, Young / middle aged* Verbal data: No statement from student (grade 7) about these elements

Note The interview data were originally in German and have been translated

15.4 Findings

15.4.1 *Comparison of Coded Picture Elements and Students' Thought Processes (RQ 1)*

For an initial overview, Table 15.2 shows the coded picture elements according to the category system used in this study (Reinisch et al., 2017).

The comparison of the picture codings and the verbal data (RQ 1) shows that most of the categories identified in the pictures correspond to the verbal data (Fig. 15.3). However, the DAST subcategories of the category 'Appearance' fall short. The students most frequently describe picture elements that cannot be categorised due to the absence of suitable categories. For example, the students repeatedly described the emotions of the biologists depicted (Table 15.3).

Table 15.2 DAST categories with the number of codings in 37 Pixton pictures

Category	*n*		Category	*n*
Appearance			*Location*	
Lab coat	22		Inside	28
Glasses	21		Outside	3
Further protective cloth	15		Inside and outside	2
Daily wear	17		Research instruments	22
Hat	2		Signs of knowledge	15
1 scientist	13		Signs of technology	13
> 1 scientist	20		Signs of danger	3
Male	25			
Female	21		*Activity*	
Not determinable sex	3		Inquiry	26
Young / middle aged	31		Scientific exchange	20
Old	2		Teaching	2
Not determinable age	2		Unrealistic	8
(Half-)bold hair	6		Enterprising	2
Beard	1			

Note There is a maximum of one coding per picture and category (depending on whether the element is present at least once)

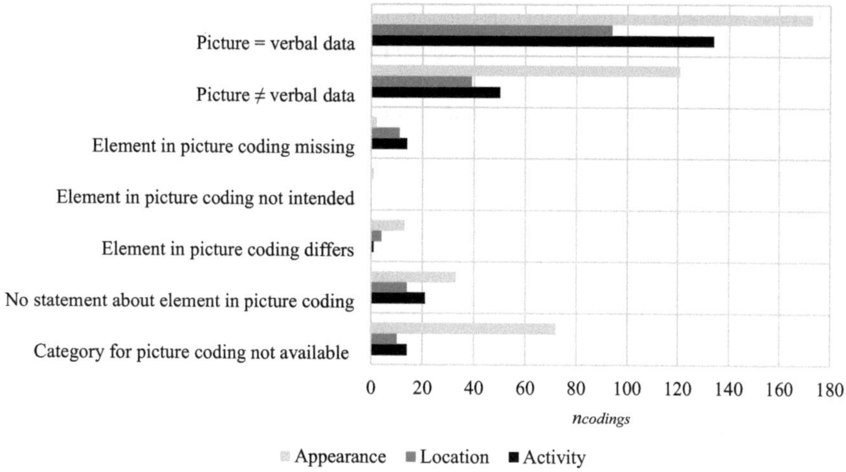

Fig. 15.3 Number of codings per validation category and DAST category (Appearance, Location, Activity) in $N = 29$ transcripts. *Note* The category 'Picture ≠ verbal data' summarises the number of codings below

Table 15.3 Possible new categories for 'Appearance'

Suggested category	Example quote	$n_{students}$ (n_{codes})
Emotion	"But she looks a bit scared [P. clicks on filter] [P. clicks on a face]." (student, grade 11)	15 (30)
Hair style / colour	"I think they always have to wear ponytails, too." (grade, 7)	10 (17)
Feature of personality	"Maybe look a bit smart because he is a scientist." (student, grade 7)	3 (5)
Appearance not important	"Because theoretically it doesn't really matter, because every scientist looks different." (student, grade 7)	2 (4)

Note Only categories named by more than one student are suggested; the interview data were originally in German and have been translated

15.4.2 Difficulties While Creating the Pictures (RQ 2)

The most frequently coded text passages are difficulties anticipated by the test administrators ('Difficulties regarding administration consistency'). Contrary to the instruction to let the students develop the pictures independently, the test administrators gave precautionary instructions, especially in the lower grades (Table 15.4). Often, the students had difficulties at first but were able to solve them independently ('Independent resolution of a difficulty').

Almost all students had technical difficulties. Out of 236 codes, 99 related to the non-intuitive design of the user interface and 61 related to technical difficulties not

Table 15.4 Number of codings in 29 transcripts regarding the difficulties in creating the picture

Description	Example quote	gr 6 (10)	gr 7 (8)	gr 11 (11)	∑
CATEGORY: DIFFICULTIES REGARDING ADMINISTRATION CONSISTENCY					
Function in Pixton is explained by the test administration without prior request by student	"If you still want words … you can make speech bubbles there [shows] at Words [student clicks Words]." (test administration)	158 (10) [15.8]	96 (8) [12.0]	54 (11) [4.9]	308 (29) [10.6]
CATEGORY: INDEPENDENT RESOLUTION OF A DIFFICULTY					
A difficulty arises, which student resolves independently	"Can I have the (…) [clicks outfit] [s. clicks back] How? (…) Ah, here. [s. clicks add character]" (student, gr 7)	48 (9) [5,3]	17 (7) [2,4]	41 (10) [4.1]	106 (26) [4.0]
CATEGORY: TECHNICAL DIFFICULTIES					
Difficulties with Pixton, as student is restricted regarding various changes or the implementation of own conception	"Can I also write something on the blackboard [which is shown in the picture]? Or can you … No." (student, gr 6)	36 (10) [3.6]	7 (3) [2.3]	56 (9) [6.2]	99 (22) [4.5]
There is no restriction of the searched function in Pixton, but the interface is unclear (i.e. the function is not understood or found by the student)	"Can you also add speech bubbles at the end?" Test administration: "If you want, you can click on Words, exactly, and then you can add speech bubbles." (student, gr 6)	42 (10) [4.2]	12 (4) [3.0]	22 (9) [2.4]	76 (23) [3.3]
Technical difficulties that do not affect Pixton but whose source is clear (e.g. mouse handling, long loading time) that irritates students	"Hm, that's annoying that it doesn't work so well with the scrolling." (student, gr 11)	37 (9) [4.11]	12 (6) [2.0]	12 (6) [2.0]	61 (21) [2.9]

(continued)

Table 15.4 (continued)

Description	Example quote	gr 6 (10)	gr 7 (8)	gr 11 (11)	Σ
CATEGORY: LANGUAGE DIFFICULTIES					
Student looks up the translation of a word or asks the test administration	"What is 'lab coat' [Laborkittel] in English?" (student, gr 6)	17 (8)	7 (4)	7 (5)	31 (17)
CATEGORY: DIFFICULTY WITH UNKNOWN SOURCE					
Difficulty occurs whose source is not clear	"What shall I do now?" (student, gr 7)	3 (1)	2 (1)	1 (1)	6 (3)

Note gr = grade; (x) = number of students; [X] = average number of codings per student; the interview data were originally in German and have been translated

related to Pixton (Table 15.4). Language difficulties, however, were rare and could be solved by translating a single word.

15.5 Discussion

15.5.1 Evidence for the Interpretation of Pixton Pictures as Indicators of Conceptions About Biologists and Their Work

The findings for RQ1 suggest that Pixton pictures can largely be interpreted as valid indicators of students' conceptions about biologists and their work, as the majority of picture elements across the categories of appearance, location, and activity aligned with students' verbal statements. However, it is problematic that the students did not verbally address some of the coded picture elements. This absence of verbal references might suggest that students did not ascribe significant importance to these elements, which could be random artefacts of the picture creation process. For instance, elements such as the age of the scientists, which were often highlighted in previous DAST studies (see Miller et al., 2018), were present in the pictures but not explicitly mentioned by the students. This raises questions about whether these elements genuinely reflect students' conceptions or if they are merely byproducts of the visual tool.

At the same time, the frequent mentioning of emotions depicted in the biologists opens a promising avenue for future research. Emotions were not part of the traditional DAST coding system, yet students often referenced them in their verbal descriptions (Table 15.3). This could indicate that emotions play a more prominent role in how students perceive biologists than previously thought (Lamminpää &

Vesterinen, 2020). As Christidou et al. (2021) suggest, these emotional depictions may reflect students' positive or negative attitudes toward scientific professions and could even serve as an indicator of their interest in these careers. Thus, while some traditional elements like age seem less relevant to students, the introduction of new categories, such as emotions, could provide deeper insights into their perceptions of scientists and may warrant inclusion in future adaptations of corresponding category systems.

While our study primarily focused on response processes as a source of validity evidence, other sources could also contribute to a more robust interpretation of students' conceptions based on Pixton pictures. For example, relations to other variables (AERA et al., 2014), such as students' prior experiences with science, could provide further insights into the extent to which their pictures reflect their conceptual understandings. Future research could explore these additional sources of validity evidence to develop a more comprehensive and sound scoring system for interpreting student-generated pictures.

15.5.2 *Technical and Language Difficulties in Using Pixton for Capturing Students' Conceptions*

The investigation into the technical and language difficulties students encountered when using Pixton (RQ 2) revealed several insights into the usability and accessibility of this digital tool for capturing students' conceptions.

Many students experienced technical difficulties, particularly regarding Pixton's user interface, which at times proved unintuitive (Table 15.4). Younger students (i.e., grade 6) were more likely to require assistance in navigating the software, with frequent requests for help from the test administrators. This suggests that while students today are often regarded as "digital natives" (Mete et al., 2017), this does not necessarily mean they are automatically adept at using all digital tools, especially new or unfamiliar ones. A preparatory exercise or tutorial session could help students become familiar with Pixton's features before beginning the task. The need for help decreased as the student's grade level increased, indicating that older students were generally more comfortable using the software independently. This grade-specific trend implies that future studies should take age or grade level into account when planning how much guidance students may require while using digital tools. By focusing on grade-level specific interventions, future research could ensure that younger students are adequately supported, while older students may benefit from more complex tasks that challenge their existing digital literacy.

Another explanation for the technical difficulties encountered could be related to the software itself, as the user interface showed some design issues that impacted ease of use. It must be noted that Pixton constantly provides updates, which could lead to improvements in the user interface. However, these updates would need to be reassessed to determine whether they effectively address the usability challenges.

This is a common issue with many digital tools, as updates may introduce new features or changes that can either improve or complicate the user experience, necessitating ongoing evaluation.

Language barriers were less prominent than initially expected, even though Pixton is only available in English (and several other languages). Most students across all grade levels were able to navigate the language-related challenges either on their own or with minimal assistance. Occasional translation issues, such as needing to translate terms like "lab coat", were easily resolved and did not appear to significantly hinder students' progress. However, younger students (particularly those in grade 6) were more likely to face these minor language-related obstacles. Providing a basic glossary or integrating translation tools for specific terms could help mitigate any remaining language difficulties, especially for younger students or those with lower English proficiency.

15.5.3 Further Emerging Difficulties in Using Pixton for Capturing Students' Conceptions

Beyond the identified technical and language challenges, additional difficulties emerged during the study that warrant discussion (Table 15.4). One notable issue was the high frequency of administrative interventions. Test administrators often provided guidance, especially to younger students, even though the task was designed to be completed independently. This frequent intervention may have unintentionally influenced the students' creative processes, potentially limiting the authenticity of their conceptions. The necessity for guidance, particularly in the lower grades, suggests that younger students may require more structured support in navigating open-ended tasks, even in digital environments. Future studies might consider a more controlled approach to minimising external assistance, such as offering clear, step-by-step instructions or video tutorials in advance of the task.

Additionally, a small number of students encountered cognitive or conceptual difficulties unrelated to the software or language. Some struggled with understanding the task itself or how to represent their conceptions through the digital medium. These issues point to potential gaps in students' conceptual understanding of the work of biologists, as well as their ability to translate abstract conceptions into visual representations. This challenge highlights the need for further scaffolding, such as pre-task discussions or examples that clarify the expectations and goals of the task, ensuring that students can better articulate their conceptions.

Moreover, the predefined nature of Pixton's backgrounds, characters, and available actions may have constrained students' ability to fully express their conceptions of biologists and their work. While the broad range of options provided flexibility, the inability to entirely create custom elements might have influenced the authenticity and depth of students' conceptions. Future studies should consider the impact of such software-imposed limitations on students' conceptual expressions and explore alternative digital tools that allow for more open-ended visualisations.

15.6 Conclusions

The collected evidence from this study demonstrates the potential of digital drawing tools to validly capture students' conceptions of biologists and their work. By reducing the dependence on traditional drawing skills, such tools offer an accessible alternative to traditional methods like the DAST (Reinisch et al., 2017), especially for students who may struggle with freehand drawing (Losh et al., 2008). The alignment between verbal data and visual representations suggests that students can effectively communicate their conceptions through digital media, providing researchers with a valuable means of assessing conceptual understandings.

However, this study also highlights areas for further refinement, particularly in terms of expanding the category system to better capture the emotional and personal characteristics of depicted scientists, which were frequently mentioned by students but not fully addressed by existing categories.

While most students navigated the digital tool without significant language barriers, technical challenges related to user interfaces need to be addressed to ensure that such difficulties do not introduce construct-irrelevant factors into the assessment process. Future research should focus on refining both the digital tools used for these tasks and the analytical frameworks employed to interpret the visual data generated.

Despite the discussed challenges, this study supports the broader argument that digital drawing tools hold significant promise as valid instruments in educational research (Chang et al., 2020), providing innovative ways to capture and interpret students' conceptions of scientists and their work across different contexts.

As students nowadays are often considered digital natives, they tend to engage intuitively with digital tools and expect technology to be integrated into their learning environments. Research suggests that incorporating digital media into education can enhance student motivation and engagement, as digital tools align with their everyday experiences and learning preferences (see Mete et al., 2017). Compared to traditional freehand drawing, digital platforms like Pixton provide structured yet customisable options that can reduce frustration related to artistic skill while still allowing for creative expression (Chang et al., 2020). By engaging in digital visual representations, students may become more aware of their implicit stereotypes and conceptions, fostering a more nuanced understanding of scientific professions. Future studies could explore how such tools might be used to actively support inquiry-based learning, promote self-reflection, and encourage a deeper engagement with scientific careers.

References

AERA (American Educational Research Association), American Psychological Association, & National Council on Measurement in Education. (2014). *Standards for educational and psychological testing*. AERA.

Brochey-Taylor, J., & Taylor, J. A. (2024). Synthesizing validity and reliability evidence for the Draw-A-Scientist Test. *Educational Research and Reviews, 19*(3), 44–52.

Chang, H., et al. (2020). A systematic review of trends and findings in research employing drawing assessment in science education. *Studies in Science Education, 56*, 77–110.

Christidou, V., et al. (2021). Drawing a scientist. *Research in Science & Technological Education.* https://doi.org/10.1080/02635143.2021.1998770

Farland-Smith, D. (2012). Development and field test of the modified draw-a-scientist test and the draw-a-scientist rubric. *School Science and Mathematics, 112*, 109–116.

Finson, K. D. (2002). Drawing a scientist: What we do and do not know after fifty years of drawings. *School Science and Mathematics, 102*, 335–345.

Finson, K. D., Beaver, J. B., & Cramond, B. L. (1995). Development and field test of a checklist for the draw-a-scientist test. *School Science and Mathematics, 95*, 195–205.

Kane, M. T. (2013). Validating the interpretations and uses of test scores. *Journal of Educational Measurement, 50*(1), 1–73.

Köchy, K. (2008). *Biophilosophie zur Einführung*. Junius.

Lamminpää, J., & Vesterinen, V. M. (2020). Draw-A-Science-Comic: Alternative prompts and the presence of danger. *LUMAT: International Journal on Math, Science and Technology Education, 8*(1), 319–339.

Landis, J. R., & Koch, G. G. (1977). The measurement of observer agreement for categorical data. *Biometrics, 33*(1), 159–174. https://doi.org/10.2307/2529310

Losh, S. C., Wilke, R., & Pop, M. (2008). Some methodological issues with "Draw a Scientist Tests" among young children. *International Journal of Science Education, 30*(6), 773–792.

Mete, R., Riegel, C., Kozen, A., & Polka, W. (2017). What every contemporary teacher educator needs to know about technology...but, didn't know who to ask! *Critical Issues in Teacher Education, 26*, 41–55.

Miller, D., et al. (2018). The development of children's gender-science stereotypes. *Child Development, 89*, 1943–1955.

Reinisch, B., & Krell, M. (2023). Assessing pre-service teachers' views of scientists, their activities, and locations: The VoSAL Instrument. *Research in Science Education, 53*(1), 139–153.

Reinisch, B., Krell, M., Hergert, S., Gogolin, S., & Krüger, D. (2017). Methodical challenges concerning the Draw-A-Scientist Test: A critical view about the assessment and evaluation of learners' conceptions of scientists. *International Journal of Science Education, 39*(14), 1952–1975.

Resnik, D. B. (2007). *The price of truth: How money affects the norms of science.* Oxford University Press.

Sandmann, A. (2014). Lautes Denken—die Analyse von Denk-, Lern- und Problemlöseprozessen [Thinking aloud—the analysis of thinking, learning, and problem-solving processes]. In D. Krüger, I. Parchmann & H. Schrecker (Eds.), *Methoden in der naturwissenschafts-didaktischen Forschung* (pp. 179–188). Springer. https://doi.org/10.1007/978-3-642-37827-0_15

Scholes, L., & Stahl, G. (2020). 'I'm good at science but I don't want to be a scientist': Australian primary school student stereotypes of science and scientists. *International Journal of Inclusive Education, 26*(9), 927–942.

Stamer, I., Pönicke, H., Tirre, F., Laherto, A., Höffler, T., Schwarzer, S., & Parchmann, I. (2020). Development & validation of scientific video vignettes to promote perception of authentic science in student laboratories. *Research in Science & Technological Education, 38*(2), 168–184.

Chapter 16
Unlocking Minds: Revealing Students' Preconceptions of Biotechnology

Michaela Horniaková

Modern biotechnology stands as one of the most rapidly evolving fields, with profound implications for society. This study underscores the critical importance of understanding students' thinking in the realm of modern biotechnology. The research aims to find out the preconceptions and misconceptions held by students. Misconceptions, cognitive structures that deviate from accepted scientific knowledge, are becoming more prevalent, particularly in an era where media significantly influences younger generations. The research involved semi-structured interviews with 40 students in the Czech Republic, including 20 pupils from the final year of lower secondary school (9th grade, ages 12–14) and 20 students from the final year of upper secondary school (12th grade, ages 17–19) covering topics such as biotechnology, genetic engineering, genetically modified organisms, and cloning. The findings indicate omnipresent misconceptions, such as associating biotechnology with machinery and misunderstanding genetic engineering as genealogy. Moreover, there is a notable gap in understanding genetic engineering. The study suggests educational approaches for addressing students' preconceptions and reconstructing misconceptions in science education. It offers a foundation for further research and the development of strategies to enhance the teaching of biotechnology and improve understanding among students and the broader public.

M. Horniaková (✉)
Department of Biology, Faculty of Education, Palacký University Olomouc, Olomouc, Czech Republic
e-mail: michaela.horniakova01@upol.cz

O. Morin et al. (eds.), *Challenges of a Changing World in Biology Education*, Contributions from Biology Education Research,
https://doi.org/10.1007/978-3-032-05346-6_16

16.1 Theoretical Background

Preconceptions are learner's subjective understandings of the content of a given domain, formed before instruction (pre-instructional conceptions) about the phenomenon (Duit & Treagust, 2003; Slavík et al., 2017). Misconceptions are preconceptions that are incompatible with scientific explanations of phenomena; they are also referred to as alternative conceptions (Duit & Treagust, 2003). A characteristic feature of misconceptions is that they lead to incorrect explanations of phenomena, interpretations or solutions to problematic situations (Horniaková, 2024). The theoretical framework guiding this study integrates two key educational theories: The Theory of Conceptual Change (TCC; Duit & Treagust, 2003; Posner et al., 1982; Strike & Posner, 1982) and The Model of Educational Reconstruction (MER; Duit et al., 2012; Kattmann et al., 1997). These frameworks emphasize the dynamic relationship between learners' pre-existing conceptions (preconceptions) and new scientific knowledge, with a focus on how misconceptions can be effectively addressed through educational interventions. The Model of Educational Reconstruction, in particular, is designed to bridge scientific content with students' preconceptions while also structuring the learning environment in ways that facilitate deeper understanding. This involves aligning scientific principles with students' mental models and reshaping the curriculum to foster conceptual change. In the modern information age, where data is constantly accessible from numerous sources, the accuracy and reliability of information are paramount, especially in fields like biotechnology, which often spark debate and controversy. Biotechnology, as defined by the OECD (2013), refers to "the application of science and technology to living organisms, as well as parts, products, and models thereof, to alter living or non-living materials for the production of knowledge, goods, and services" (p. 156). This definition highlights the broad scope of biotechnology, making it essential to construct accurate educational strategies that take into account both scientific complexity and students' existing ideas about the subject.

Modern biotechnology, as distinguished from traditional biotechnology practices such as breeding and fermentation (which have been used for millennia), emerged between the 1960s and 1980s (Clark & Pazdernik, 2012). Unlike traditional methods, modern biotechnology rely heavily on genetic manipulation to bring about change, with techniques like genetic engineering and cloning being prime examples. Drobník (2008) identifies the recombinant DNA method as the cornerstone of this modern field, a view echoed by Clark and Pazdernik (2012), who emphasize that DNA manipulation, particularly for creating recombinant DNA, forms the foundation of modern biotechnology. Despite its scientific advancements, students often harbour misconceptions about biotechnology that differ significantly from the accepted scientific understanding (Dawson & Schibeci, 2003). Biotechnology, especially DNA technologies, is often perceived as highly risky, akin to other technologies involving chemical substances, nuclear energy, radioactive waste, and electromagnetic fields (Welser, 1991). Additional misconceptions include beliefs such as hygiene preventing genetic disorders (Casanoves et al., 2015 in Horniaková,

2024) and viewing in vitro fertilization as a form of genetic engineering (Dawson & Soames, 2006 in Horniaková, 2024). Despite extensive studies disproving significant health risks associated with genetically modified (GM) foods, public misconceptions about GM products persist as a major concern (Jones et al., 2000; López & Carrau, 2002; Prokop et al., 2007). Addressing and correcting students' misconceptions is crucial for achieving meaningful learning outcomes In an era where information flows rapidly from diverse sources, it is essential to foster critical thinking. Accurate scientific concepts serve as the foundation for constructive discussions, which are crucial for navigating societal debates on these topics.

Dawson and Schibeci (2003, p. 66) pose a vital question: "If science curricula are supposed to prepare students to become citizens, can we continue to pay insufficient attention to this area of science?" This question was asked over 22 years ago and unfortunately, it still stands. There is still a need to raise awareness of the lack of coverage of the subject in the school curriculum. It really highlights the importance of understanding what students actually know or do not know about biotechnology. Furthermore, grasping accurate concepts about any phenomenon is a fundamental skill for making informed decisions in everyday life (Harms, 2002).

16.2 Aims of the Study, Research Design and Methodology

The research aims to find out the preconceptions and mostly the misconceptions held by students about modern biotechnology, with an emphasis on misconceptions so far not described in scientific literature. The main research instrument used in the study is a semi-structured interview with individual respondents (Posner & Gertzog, 1982). The research instrument was chosen in order to find out as much quality information as possible for the pilot research. Questions for the interview were designed based on the content analysis (see Mayring, 2014) of Framework Education Program for Basic Education in the Czech republic (FEP BE; Jeřábek & Tupý, Eds., 2007) and school textbooks (20 textbooks and 7 workbooks) for lower and upper secondary schools, with a focus on the field of biotechnology. Questions for the interview were validated by 4 experts in the fields of biotechnology, genetics, and didactics of biology. Altogether, the interviews consisted of 28 questions. The questions were divided into four concepts: biotechnology, genetic engineering, genetically modified organisms (GMO), and cloning. Interviews were administered by the author. The interview were gathered from 20 pupils from the final year of lower secondary school (9th grade, 12–14 years old) and 20 students from the final year of upper secondary school (12th grade, 17–19 years old) in the Czech Republic. All the interviews were recorded on a dictaphone and transcribed. Data were analyzed in MAXQDA 2022, a data analysis software. The gathered data were analyzed both qualitatively and quantitatively. For the qualitative analysis, the Grounded theory (Strauss & Corbin, 1990); inductive coding was used. The quantitative analysis was based on the absolute and relative abundance of a specific category identified through qualitative analysis. This was followed by a Student's t-test, applied to the results of the quantitative analysis. The Student's t-test was used as a statistical method to compare the knowledge of lower secondary school students and upper secondary school students. This test determines whether there are significant differences in the measurements between the two groups.

16.3 Findings

The results of the research were categorized according to the categories of questions asked in the interviews: biotechnology, genetic engineering, genetically modified food (GMO) and cloning. The data in all categories were analysed by independent Students' t-test. There was found no significant difference in the results between the lower and upper secondary school students ($t = -0.743$; $p = 0.595$). Nevertheless, the students have a considerable amount of misconceptions or lack of knowledge about the phenomenon.

16.3.1 Biotechnology

The results of the research in the field of biotechnology (in general) showed that both lower and upper secondary students had difficulties in defining biotechnology. Students found the term 'biotechnology' most difficult.

Table 16.1 shows that students mainly used the combination of the words 'bio' and 'technology' to explain the meaning of biotechnology: "*Technology somehow related to biology, but I don't know more*" (student 1.10.). However, this also led to frequent misconceptions. Table 16.2 shows that for every subcategory, which includes students' statements about the concept of biotechnology itself, its uses or products, there were statements that the word 'technology' reminded both lower secondary school and upper secondary school students robotics or machinery as an outcome of biotechnology, rather than its inherent connection to living organisms. Pupils from lower secondary school could only imagine that biotechnology would be taught in biology or chemistry (Table 16.1), but were unable to give any further explanation. Both groups of respondents frequently (Table 16.1) associated biotechnology with medicine: "*So, for example, some vaccinations, vaccines, drugs ...*" (student 2.2) or agriculture "*So certainly medicine, agriculture ...*" (student 2.2).

Misconceptions were also common in the subcategory 'Usage'. Upper secondary students (3/15%) and also lower secondary students (2/10%) mentioned use in the military, in space industry or for the police. However, they mentioned them as misconceptions when they mentioned fingerprints or various high-tech used in space: "*For example, in space, if we want to inhabit a planet, some technology would probably be needed*" (student 2.1).

Table 16.1 Absolute and relative frequencies of the 3 most frequent codes (correct answers) about biotechnology of students from the final year of lower secondary school and final year of upper secondary school

Category	Codes	Final year of lower secondary school (%)	Final year of upper secondary school (%)
Biotechnology	Etymology	8/40	7/35
	Biology/Chemistry	7/35	1/5
	Medicine	5/25	7/35

Table 16.2 Examples of students' misconceptions about biotechnology from subcategories from both groups of respondents

Category	Subcategory	Examples of answers
Biotechnology	Concept	"*....that human work is replaced by machines*" (student 2.7)
	Usage	"*... some kind of device with a scalpel that would cut more precisely than human hands ...*" (student 2.9)
	Product	"*...the only thing I can think of, like I don't know if you're familiar with R.U.R.* [Czech science fiction drama by Karel Čapek], *there are these robots, that it's something between a robot and a human, that's the only thing I can think of. Something between man and technology ...*" (student 2.12)

16.3.2 Genetic Engineering

To students, genetic engineering most frequently evoked things about medicine, as the most frequent code among the correct concepts in Table 16.3 was 'medicine'. Also, a commonly mentioned example of a use besides medicine was, for example, agriculture: "*So some genetic engineering, for example the development of some pharmaceutical things, or for example some fertilizer for agriculture*" (student 2.2). The second most frequently mentioned code was 'genetics' (Table 16.3). Pupils commented, based on the etymology of the word 'genetic' in the term that it would have something to do with genetics: "*So, for example, research, in the field of genetics*" (student 2.7). However, they were often unable to provide the details.

Table 16.3 Absolute and relative frequencies of the 3 most frequent codes (answers) about genetic engineering of students from the final year of lower secondary school and final year of upper secondary school

Category	Codes	Final year of lower secondary school (%)	Final year of upper secondary school (%)
Genetic engineering	Medicine	8/40	9/45
	Genetics	12/60	4/20
	DNA manipulation	2/10	13/60

Table 16.4 Examples of students' misconceptions about genetic engineering from subcategories from both groups of respondents

Category	Subcategory	Examples of answers
Genetic engineering	Concept	"*...like the genetics of how it progresses in that family tree or something like that.*" (student 1.11)
	Usage/Making	"*I guess... it is artificial insemination.*" (student 2.15)
	Product	"*...that some of those fingerprint scanners are some of those genetic products.*" (student 1.14)
	Biotechnology and genetic engineering	"*Biotechnology would be like focused more on plants and animals and the genetic engineering on human beings.*" (student 2.1)

Although lower secondary school pupils recognized genetic engineering's involvement in DNA manipulation (Table 16.3), 80% of them lacked a comprehensive understanding of this concept. Furthermore, concerning genetic engineering, 35% of lower secondary school pupils incorrectly believed it to be linked to 'genealogy', see Table 16.4 (concept), or: "*Hmm, a device with which to examine some heredity*" (student 2.3).

A frequent problem for pupils in both groups was the misconception that artificial insemination was considered to be an example of genetic engineering (Table 16.4—Usage/Making). Similar to the biotechnology category, the genetic engineering category had (4/20%) examples such as "machines, mechanics, apparatus, prosthetic limbs, fingerprint machines" (Table 16.4). Pupils from both groups lacked a conceptual understanding of the whole phenomenon. This can be seen, for example, in the answers to the question from the interview: "*How is genetic engineering related to biotechnology?*" (Table 16.4—Biotechnology and Genetic engineering). Biotechnology there has been seen, for example, as machines that are used to break into genes (as a part of genetic engineering): "*Yes, so for example to get into that gene*" (student 1.9). For these subcategories (Biotechnology and Genetic engineering), misconceptions were common among both groups of respondents (5/25%—lower secondary; 12/60%—upper secondary school) or pupils were unable to answer (15/75%—lower secondary school; 7/35%—upper secondary school).

16.3.3 Genetically Modified Organisms (GMO)

Genetically modified organisms are a rather more realistic concept for pupils and students than biotechnology or genetic engineering. Pupils (Table 16.5) know that a modified organism should have a particular change that benefits either society, the environment or the organism itself: "*These are organisms that originally had a different genetic structure and we have altered their functions for a specific purpose*" (student 2.2.). They know that genetic manipulation must have taken place: "*I would say it's something where somebody changed the gene of that organism*" (student

2.12.). Students from upper secondary school are more aware of this fact (14/70%; Table 16.5). The code 'side-effects for humans' means that pupils are thinking that GMOs are not natural and may raise fears about potential risks. However, this implies reluctance rather than resistance or rejection.

Despite the fact that GMOs are easier for pupils to imagine, many of them have difficulty understanding the concept itself or its production: "*Some kind of like linked genes from the parents*" (student 1.3) or "*For example, some dogs... so those genetics, that they somehow copulate.*" (student 1.4). Pupils, also due to their lack of knowledge of genetics as such, imagine the production of GMO by sexual intercourse of two individuals (Table 16.6—Concept, Making).

The problem also arises when pupils have to explain the environmental benefits (Table 16.6—+Environment). The pupils understand that GMOs may have environmental benefits, but they think, for example, that they will arise from some chemical

Table 16.5 Absolute and relative frequencies of the 3 most frequent codes (answers) about genetically modified organisms (GMO) of students from the final year of lower secondary school and final year of upper secondary school

Category	Codes	Final year of lower secondary school (%)	Final year of upper secondary school (%)
GMO	Advantages	9/45	9/45
	DNA manipulation	7/35	14/70
	Side-effects for humans	11/55	5/25

Table 16.6 Examples of students' misconceptions about genetically modified organisms (GMO) from subcategories from both groups of respondents

Category	Subcategory	Examples of answers
GMO	Concept	"*...he got some of those genes from his mother and some from his father...So he's modified by those parents.*" (student 2.18)
	Making	"*Sexual intercourse between two species*". (student 2.7)
	Products	"*... it is possible only with plants*" (student 2.15)
	+Society	"*That we would lack enough wise people to do certain jobs and we could take advantage of that and expand the population with those smart ones.*" (student 2.2)
	−Society	"*Like incest...Then, are those individuals like with some kind of disability (...)*" (student 2.7)
	+Environment	"*It may not be so good for the organism, it's chemically modified...*" (student 1.9)
	−Environment	"*I'm sure it will get better, I doubt there are negatives.*" (student 2.18)
	Breeding and GMO	"*Breeding...to me it evokes animals for example dogs and modification is with plants. An animal can´t be modify right away, like those genes humans can't modify.*" (student 2.20)

spraying: "*For example, if there is a bad crop, it will be possible to spray those potatoes or that crop and it will be done*" (Student 1.4). The research also revealed the problem that pupils cannot distinguish between cross-breeding/breeding of organisms and genetic modification (Table 16.6—Breeding and GMO). Some pupils even crossbreed humans with other animal species: "*With those blood types and human genes ... and there's been a lot of attempts to crossbreed humans with animals*" (student 2.7). It can also be seen that pupils think that only plants can be modified (Table 16.6—Products, Breeding and GMO).

16.3.4 Cloning

The results in the 'cloning' category indicate that students understand what a clone is (Table 16.7—Two identical organisms/copy). However, the greatest difficulty was for pupils to explain or describe how such a clone is created. For pupils from the lower secondary school, 8 (40%) knew nothing at all about the creation of clones, 6 (30%) knew that DNA is incorporated into cloning and 4 (20%) of pupils had misconceptions (Table 16.8—Making, Products). When it comes to product or example of cloned organism, pupils (Table 16.8—Don´t know any example) from the lower secondary school could not give any example of a cloned organism, 5 pupils (25%) remembered Dolly the sheep. In contrast, only one student from an upper secondary school remembered Dolly because it will be a question in the final exam.

Based on the results of the research, the category 'cloning' appears to be science fiction for pupils in some areas. One of the reasons for this is the media, which is keen to use this field of science. In the research, pupils' answers referring to the science fiction series of Star Wars films appeared: "*I've seen Star Wars and there's a lot there! ... Well like the creation of an army, that but that's from those Star Wars and just with me this cloning evokes that*" (student 2.20).

The students also thought that clones are created by sexual reproduction and did not mean a clone in the form of identical twins at an early stage (Table 16.8—Ethics/Law, Products). We can see in Table 16.8 (Ethics/Law) the reasoning of the student who could not imagine how the clone would be made if a mouse and a giraffe were to be paired (sexually).

Table 16.7 Absolute and relative frequencies of the 3 most frequent codes (answers) about cloning of students from the final year of lower secondary school and final year of upper secondary school

Category	Codes	Final year of lower secondary school (%)	Final year of upper secondary school (%)
Cloning	Two identical organisms/copy	19/95	15/75
	Created by DNA	6/30	10/50
	Don´t know any example	12/60	8/40

Table 16.8 Examples of students' misconceptions about cloning from subcategories from both groups of respondents

Category	Subcategory	Examples of answers
Cloning	Concept	"*Imitation, so it depends how good a clone it is,..*" (student 2.1)
	Usage	"*Honestly, I don't see somehow those reasons... there are no advantages.*" (student 1.2)
	Ethics/Law	"*So, I guess we can clone without restriction. If we look at animals. I guess somewhere it won't work if we have a giraffe and a mouse* (sexual intercourse) ..." (student 2.13)
	Making	"*That we put it in something and it comes out twice.*" (student 2.3)
	Products	"*...a female of one species and a male of the other species and they would have a child together.*" (student 2.13)

16.4 Implications for Using Preconceptions in the Field of Modern Biotechnology in School Science and Biology Education

In the 1980s, researchers began to explore how students' preexisting concepts, or misconceptions, impact their understanding of new subject matter. This growing interest led to the development of the Theory of Conceptual Change, which investigates the interaction between students' prior knowledge and new ideas (Chi et al., 1994; Duit & Treagust, 2003). TCC, as described by Posner et al. (1982), Strike and Posner (1982) and Duit and Treagust (2003), explores how individuals modify core beliefs when confronted with conflicting information. Looking at the results of the research, however, we can conclude that the respondents have too little information about the phenomena in question to be able to address the sometimes quite controversial issues. This finding is supported by a number of existing research studies. For example, in the Wisch et al. (2018) study, when discussing recombinant DNA, students often included general definitions such as 'modified DNA' or 'replicated/copied DNA'. While technically accurate, these answers do not demonstrate a deep understanding of recombinant DNA, as a foundation for genetic engineering. Also, it can be stated that most students hold misconceptions about the concept of genetic engineering in general. The same result was reached by the study of Suryanti et al. (2018). Research by Suryanti et al. (2018) indicates that university students hold the most misconceptions about the general concept of genetic engineering. Students' inability to relate the concept to real life is the main problem they see. Research shows that having a correct understanding of one concept, such as biotechnology, does not necessarily translate to an understanding of related concepts like genetic engineering or cloning, which are often seen as unrelated. According to a review study by Horniaková (2024), research on students' misconceptions about biotechnology has been corroborated by multiple studies (e.g., Šorgo & Ambrožič-Dolinšek, 2010; Veličković et al., 2015, pp. 7–9). These misconceptions often stem from associations with fields like transportation (mechanics) or radioactivity, leading students to perceive biotechnology as a potential risk ("Perception of personal health risks"). The authors suggest that these misconceptions arise not only from students' insufficient factual knowledge but also from a lack of understanding of biotechnology's practical applications. The results of the studies (e.g. Bal et al., 2007; Prokop et al., 2007; Kooffreh et al., 2021) state that enhancing awareness and education about biotechnology and its applications across various fields is critical, particularly among high school students, to foster improved knowledge and understanding. Research recommends that some of the key scientific trends that regularly appear in the media, including modern biotechnology, should also be included in the curriculum (e.g., Alanazi, 2021; Dawson & Schibeci, 2003; Paš et al., 2019; Prokop et al., 2007).

Teachers need to give students the right stimulus to think, use the appropriate method, get them to evaluate their thinking and prepare them to be willing to reconstruct their concept and offer them the right concepts to think about. For conceptual change to occur, students must first experience dissatisfaction with their current understanding, triggering a conflict between their existing beliefs and new ideas (Tekkaya, 2002). For conceptual change to take place, the new idea must be intelligible, plausible, and fruitful (Duit & Treagust, 2003). These conditions increase the likelihood of change, making it essential to include foundational knowledge about such topics in the school curriculum. However, new concepts may also conflict with one another (Duit & Treagust, 2003).

Toulmin (1972, as mentioned in Posner et al., 1982; Duit & Treagust, 2003) refers to this dynamic as 'conceptual ecology', where students' beliefs influence how they integrate new ideas. They evaluate the intelligibility, plausibility, and usefulness of new concepts, deciding whether to replace or adapt their existing beliefs.

Also the emotional dimension plays a significant role in whether students accept new information. Research by Šorgo & Ambrožič-Dolinšek (2010) suggests that attitudes toward GMOs are shaped not just by scientific knowledge but also by emotional responses. In the case of topics like GMOs, acceptance of information is deeply intertwined with emotional factors. Therefore, simply presenting new facts is insufficient; educators must also actively engage students in critical discussions that guide them from informal to formal reasoning based on scientific evidence (Šorgo & Ambrožič-Dolinšek, 2010). Teachers play a key role in helping students recognize the relevance of the content, particularly in controversial fields like modern biotechnology. Research shows that explicitly teaching biotechnology increases acceptance, understanding, and reduces uncertainty (Maes et al., 2018). Identifying and addressing students' preconceptions and misconceptions is crucial for effective biotechnology education. Instruction should create opportunities for cognitive conflict, prompting conceptual change. This approach, tailored to students' preexisting ideas, is particularly important in science education (Suryanti et al., 2018; Witzig et al., 2013). Moreover, teaching strategies must confront the 'defences' students use to resist new concepts (Posner et al., 1982). The more dissatisfaction students feel with their current understanding, the more likely they are to adopt new ideas (Duit & Treagust, 2003; Posner et al., 1982).

One successful method for promoting meaningful learning and reconstruct misconceptions is the Model of Educational Reconstruction (MER; Duit et al., 2012;

Kattmann et al., 1997). Based on The Theory of Conceptual Change, this model provides a framework for fostering conceptual reconstruction in students. It emphasizes designing instructional materials and activities that encourage this change, viewing students' preconceptions as valuable starting points for learning. The model focuses on three key areas: (a) empirical study of students' pre-scientific conceptions, (b) clarification of scientific subject matter, and (c) design and evaluation of learning environments (Duit et al., 2012; Kattmann et al., 1997). As a result of the research, Wisch et al. (2018) recommend making a list of misconceptions. This can facilitate future teacher interventions and the development of teaching and laboratory materials and methods. Understanding students' preconceptions is vital for effective teaching (Duit et al., 2012; Jáč et al., 2019) Clarifying disciplinary concepts involves simplifying and restructuring scientific content for school instruction (Duit et al., 2012; Jáč et al., 2019; Kattmann et al., 1997). Finally, creating a structured learning environment helps bridge the gap between students' and disciplinary conceptions, facilitating conceptual reconstruction (Duit et al., 2012; Jáč et al., 2019; Kattmann et al., 1997). A key feature of MER is its recursive nature, where the three domains reinforce one another (Duit et al., 2012).

Children's ideas develop from their experiences, reasoning, and abilities, shaped by a mix of preconceptions. As they learn, these preconceptions evolve as they encounter new information. However, it is important that students' preconceptions remain functional and applicable (Ozgur, 2013). With more advanced knowledge, students refine their initial ideas, helping bridge the gap between their prior knowledge and scientific concepts. This approach cultivates a generation of informed and innovative citizens in an ever-evolving field, like biotechnology is.

16.5 Conclusions

The analysis of the research results shows that students, regardless of their school level, face significant challenges in understanding key concepts related to biotechnology, genetic engineering, genetically modified organisms, and cloning. While the data does not reveal a significant difference in understanding between lower and upper secondary school students, misconceptions are prevalent across both groups. The research categorized students' responses into distinct areas, highlighting the difficulties they encounter.

In the field of biotechnology, both lower and upper secondary students struggled to define the term properly, often associating it with robotics or machinery instead of its actual link to biology and living organisms. This confusion suggests that students focus more on the "technology" aspect of the term rather than understanding its biological applications. These misconceptions highlight a critical gap in their foundational knowledge of biotechnology, which is essential for understanding its real-world applications.

Similarly, genetic engineering was another area where students showed limited understanding. Although many students recognized its connection to medicine and

DNA manipulation, they often confused it with genealogy or artificial insemination. This indicates that students lack a clear understanding of the scientific principles that underling genetic engineering. This confusion between related yet distinct concepts further complicates their grasp of biotechnology and genetic engineering as interconnected fields.

In the case of GMOs, students had a slightly better grasp of the concept, understanding that genetic modification involves altering the genes of an organism for specific benefits. However, many still harboured misconceptions, especially about how GMOs are created. For instance, some students believed GMOs result from sexual reproduction or crossbreeding, which reflects a lack of understanding of the scientific processes behind genetic modification.

Cloning, often perceived as a science fiction concept, was another area where students displayed significant confusion. Although most students understood the idea of cloning as producing two identical organisms, they struggled to explain how cloning works, often relying on ideas from popular media rather than scientific knowledge. This demonstrates that students' understanding of cloning is heavily influenced by fictional portrayals, leading to misconceptions about the process and its ethical implications.

The study underscores the need for targeted educational approaches that address these preconceptions and foster conceptual change. To achieve meaningful learning, teaching strategies must incorporate cognitive conflict. The Model of Educational Reconstruction is particularly effective in promoting conceptual change by using students' preconceptions in a way that bridges gaps between student and disciplinary knowledge. Thus, educators must not only provide accurate information but also create learning environments that encourage open dialogue and critical thinking. Ultimately, fostering a deeper understanding of biotechnology and related fields is essential for preparing students to become informed decision-makers in an increasingly complex scientific environment.

Acknowledgements The study was prepared under the institutional support of Faculty of Education, Palacký University in Olomouc.

References

Alanazi, F. H. (2021). Saudi students' and science teachers' knowledge of and attitudes towards biotechnology. *Journal of Biological Education, 57*(1), 196–213.

Bal, S., Samancı, N. K., & Bozkurt, O. (2007). University students' knowledge and attitude about genetic engineering. *Eurasia Journal of Mathematics, Science and Technology Education, 3*(2), 119–126.

Casanoves, M., González, Á., Salvadó, Z., Haro, J., & Novo, M. (2015). Knowledge and attitudes towards biotechnology of elementary education preservice teachers: The first Spanish experience. *International Journal of Science Education, 37*(17), 2923–2941.

Clark, D. P., & Pazdernik, N. J. (2012). *Biotechnology.* Elsevier.

Chi, M. T., Slotta, J. D., & De Leeuw, N. (1994). From things to processes: A theory of conceptual change for learning science concepts. *Learning and Instruction, 4*(1), 27–43.

Dawson, W., & Schibeci, R. (2003). Western Australian school students' understanding of biotechnology. *International Journal of Science Education, 25*(1), 57–69.

Dawson, V., & Soames, C. (2006). The effect of biotechnology education on Australian high school students' understandings and attitudes about biotechnology processes. *Research in Science & Technological Education, 24*(2), 183–198.

Drobník, J. (2008). *Biotechnologie a společnost* [Biotechnology and Society]. Karolinum.

Duit, R., & Treagust, D. F. (2003). Conceptual change: A powerful framework for improving science teaching and learning. *International Journal of Science Education, 25*(6), 671–688.

Duit, R., Gropengiesser, H., Kattmann, U., Komorek, M., & Parchmann, I. (2012). The model of educational reconstruction - A framework for improving teaching and learning science. In *Science education research and practice in Europe: Retrospective and prospective* (pp. 13–37). Sense Publisher.

Harms, U. (2002). Biotechnology education in schools. *Electronic Journal of Biotechnology, 5*, 5–6.

Horniaková, M. (2024). Moderné biotechnológie v školskom vzdelávaní z pohľadu zahraničných výskumných šetrení. [Modern biotechnology in school education from the perspective of foreign research] *Scientia in Educatione, 15*(2), 12–28.

Jáč, M., Kopecká, J., Morris, M., & Vránová, O. (2019). *Didaktické kazuistiky výuky přírodopisu a biologie* [Educational Case Studies of Biology Instruction.]. Univerzita Palackého v Olomouci.

Jeřábek, J., & Tupý, J. (Eds.) (2007). *The framework educational programme for basic education.* Ministry of Education Youth and Sport Czech Republic.

Jones, P., Clarke-Hill, P., Hillier, D., & Shears, P. (2000). Food retailers' response to GM controversy within the UK. *British Food Journal, 102*(5/6), 441–448.

Kattmann, U., Duit, R., Gropengießer, H., & Komorek, M. (1997). Das Modell der Didaktischen Rekonstruktion [The model of the didactic reconstruction]. *Zeitschrift Für Didaktik der Naturwissenschaften, 3*(3), 3–18.

Kooffreh, M. E., Ikpeme, E. V., & Mgbado, T. I. (2021). Knowledge, perception, and interest regarding biotechnology among secondary school students in Calabar, Cross River State, Nigeria. *Biochemistry and Molecular Biology Education, 49*(4), 664–668.

López, R. C., & Carrau, J. G. (2002, August 28–31). The GMO Regulation in the EU and the Commercial Conflict with the United States [Conference presentation]. International Congress, Zaragoza, Spain.

Mayring, P. (2014). *Qualitative content analysis: Theoretical foundation, basic procedures and software solution.* Springer.

Maes, J., Bourgonjon, J., Gheysen, G., & Valcke, M. (2018). Variables affecting secondary school students' willingness to eat genetically modified food crops. *Research in Science Education, 48*(3), 597–618.

OECD. (2013), Biotechnology. In *OECD Factbook 2013: Economic, environmental and social statistics.* OECD Publishing.

Ozgur, S. (2013). The persistence of misconceptions about the human blood circulatory system among students in different grade levels. *International Journal of Environmental and Science Education, 8*(2), 255–268.

Paš, M., Vogrinc, J., Raspor, P., Udovč Knežević, N., & Čehovin Zajc, J. (2019). Biotechnology learning in Slovenian upper-secondary education: Gaining knowledge and forming attitudes. *Research in Science & Technological Education, 37*(1), 110–125.

Posner, G. J., & Gertzog, W. A. (1982). The clinical interview and the measurement of conceptual change. *Science Education, 66*(2), 195–209.

Posner, G. J., Strike, K. A., Hewson, P. W., & Gertzog, W. A. (1982). Accommodation of a scientific conception: Toward a theory of conceptual change. *Science Education, 66*(2), 211–227.

Prokop, P., Lešková, A., Kubiatko, M., & Diran, C. (2007). Slovakian students' knowledge of and attitudes toward biotechnology. *International Journal of Science Education, 29*(7), 895–907.

Slavík, J., Janík, T., Najvar, P., & Knecht, P. (2017). *Transdisciplinární didaktika: o učitelském sdílení znalostí a zvyšování kvality výuky napříč obory*. [Transdisciplinary didactics: on teachers' knowledge sharing and improving teaching in different subjects]. Masarykova univerzita.
Strauss, A., & Corbin, L. (1990). Grounded theory research: Procedures, canons and evaluative criteria. *Zeitschrift Für Soziologie, 19*(6), 418–427.
Strike, K. A., & Posner, G. J. (1982). Conceptual change and science teaching. *European Journal of Science Education, 4*(3), 231–240.
Suryanti, D., Sinaga, P., & Surakusumah, W. (2018). Improvement of students' environmental literacy by using integrated science teaching materials. *IOP Conference Series: Materials Science and Engineering, 306*(1), 012031.
Šorgo, A., & Ambrožič-Dolinšek, J. (2010). Knowlege of, attitudes toward, and acceptance of genetically modified organisms among prospective teachers of biology, home economics, and grade school in Slovenia. *Biochemistry and Molecular Biology Education, 38*(3), 141–150.
Tekkaya, C. (2002). Misconceptions as barrier to understanding biology. *Hacettepe Üniversitesi Eğitim Fakültesi Dergisi, 23*(23), 259–266.
Toulmin, S. E. (1972). *Human understanding*. Cambridge University Press.
Veličković, V., Jović, M., Nalić, E., Višnjić, A., Radulović, O., Šagrić, Č., & Ćirić, M. (2015). Knowledge, attitudes toward, and acceptability of genetic modification among Western Balkan University students of life sciences (AGREE Study). *Journal of the American College of Nutrition, 35*(2), 150–162.
Welser, J. R. (1991). An industrial perspective on biotechnology issues. *Food Technology, 45*(4), 102–109.
Wisch, J. K., Farrell, E., Siegel, M., & Freyermuth, S. (2018). Misconceptions and persistence: Resources for targeting student alternative conceptions in biotechnology. *Biochemistry and Molecular Biology Education, 46*(6), 602–611.
Witzig, S. B., Freyermuth, S. K., Siegel, M. A., Izci, K., & Pires, J. C. (2013). Is DNA alive? A study of conceptual change through targeted instruction. *Research in Science Education, 43*(4), 1361–1375.

Chapter 17
Understanding the Pathway of Water Through the Body: Student Teachers' Knowledge and Essential Concepts for Future Pupils

Pernilla Granklint Enochson

17.1 Introduction

Knowledge about organs and organ systems is taught in schools all over the world. In Sweden, where the present study took place, this focus on the human body forms a large part of the curriculum and the content in textbooks at all stages of the mandatory schools (Skolverket, 2018), as well as in some of the upper secondary school courses (Skolverket, 2013). However, in most countries (if not all), much of the biology curricula focuses on students' understanding of the function of the human body.

By the time students reach secondary or upper secondary school, most have learned about the body on several occasions during their school years. Examples include the three curricula in the mandatory schools (grades 9–10): in the National Curriculum in England (2014, 2015), the curriculum for Secondary Level I—Biology in Baden-Württemberg, Germany (Bildungspläne Baden-Württemberg, 2016), and the National Agency for Education in Sweden (Skolverket, 2013, 2018). All three curricula aim to ensure that the students receive knowledge about the circulation system, the lungs and respiratory system, the excretory system, the function of the cells, and so on. However, in Sweden, biology can be studied as a standalone subject at upper secondary schools and as part of the more general subject of science. The general science subject includes biology, chemistry, physics, and geology, and the course is required by all upper secondary students in Sweden. The prospective teachers who aim to teach grades 1–6, need at least two courses in general science. The second science course and the biology course include a unit with a focus on the human body (Skolverket, 2013).

P. G. Enochson (✉)
Malmö University, Malmö, Sweden
e-mail: pernilla.granklint-enochson@mau.se

O. Morin et al. (eds.), *Challenges of a Changing World in Biology Education*, Contributions from Biology Education Research,
https://doi.org/10.1007/978-3-032-05346-6_17

In the science courses at the university, one aim is to gain knowledge of the organ systems and to set the subject-matter in an education setting so that the students gain the skills they need to teach the subject-matter to pupils. It is a challenge for adults to describe what happens to water from the time a person drinks it until they urinate. The problem is that the description of the process itself, even at a macro level, requires that the person understands the connections between several organ systems (for example, digestive, circulatory, and excretory systems) (Granklint Enochson, 2024).

17.2 Background

17.2.1 General Knowledge of Organs and Organ Systems

People cannot see what is happening inside the body, which makes it difficult to comprehend the processes that take place in it. One example is the connection between the input (via the mouth) and the output (urine) of the excretory system; what happens in between is not entirely clear. For instance, how sweating is linked to water excretion is seldom discussed in everyday conversations. Biology textbooks use the term "organ system", but what do we know about people's knowledge of organ systems?

Preschool children often draw individual organs without showing the connections between them (e.g. Cuthbert, 2000). The older the child becomes, the more he or she should understand how organs are comprised as systems that can be interpreted as organ systems (e.g., Andersson et al., 2020; Dempster & Stears, 2014; Garcia-Barrosa et al., 2011). The organ system that most people are familiar with is the digestive system, but what this knowledge means in this context requires explanation from a macro perspective.

The digestive system, with its input and output (that is, food enters through the mouth and comes out in the form of feces), is not particularly problematic to understand. The explanation that we ingest food through the mouth, then it goes to the throat, down to the stomach, and then further on to the intestinal system, is also relatively well-known from middle school (e.g. Carvalho et al., 2004; Garcia-Barrosa et al., 2011; Granklint Enochson & Redfors, 2012; Óskarsdóttir et al., 2011; Reiss et al., 2002; Tunnicliffe, 2004). Explaining the digestive system at this level includes being able to define the path of food at the macro level. However, it is much more difficult to connect cells or cell structures to reasoning about the digestive and circulatory systems (e.g., Granklint Enochson & Redfors, 2012).

17.2.2 The Body—Connections Between Organ Systems

In relation to explaining the "tube" in the digestive system, a more complex explanatory model is required to describe what happens in the body when a person drinks water (that is, from the moment of drinking a glass of water to when the water leaves the body as urine). To explain this process at the macro level, it is necessary to use an explanation that refers to at least three organ systems (the digestive system, the circulatory system, and the excretory system). Several studies have shown that it is difficult for students (and adults) to describe which systems are involved and how they are interconnected (Canlas & Molino-Magtolis, 2024; Carvalho et al., 2004; Clément, 2003; Cuthbert, 2000; Manokore & Reiss, 2003; Prokop et al., 2009).

One reason for this difficulty involves the student teachers' struggle to link the various organizational levels (Assaraf et al., 2013; Hmelo-Silver et al., 2008). Clément (2003) also pointed out some of the causes of the difficulties—one of which is ontological. In everyday speech, we rarely talk of intestinal walls as being permeable. Understanding the permeability in the connection between the different organ systems in the body is a prerequisite for explaining the processes of nutrients and water moving from one organ system to another (Clément, 2003). It is also significant that understanding the breaking down of nutrients is required to understand the transition between the gut and the circulatory system (Jahic Pettersson et al., 2021). A further problem with understanding the pathway of water through the body is that everyday perceptions about the process are based on explanatory models that are not scientifically based; for example, descriptions that contain explanatory models that describe tubes from the throat linking directly down to the kidneys (Clément, 2003; Granklint Enochson & Redfors, 2012) or that the water makes its way past the lungs before reaching the stomach (Granklint Enochson et al., 2015). It is relatively common for people to hold scientific misconceptions. Among ninth graders (Granklint Enochson & Redfors, 2012), a quarter of the participants exhibited various types of misconceptions, while in another study (Granklint Enochson, 2024) involving adults (pre-service teachers in preschool), nearly half held such scientifically unverified beliefs. Two dominant misconceptions emerged in these studies: one involves the idea of a direct tube from the mouth to the kidneys, as described above, while the other suggests that water is transported from the stomach directly into the body—sometimes with a connection to the circulatory system, but more often simply diffusing into the body. However, additional types of misconceptions exist. In a study conducted in South Africa, approximately half of the ninth-grade students assumed that water must pass through the lungs before reaching the stomach (Granklint Enochson et al., 2015). Notably, the students in this study were able to articulate their reasoning, with several stating that "water purifies the body" as an explanation for why they believed that water must pass through the lungs. Another finding in Granklint Enochson and Redfors (2012) was that the students with non-scientific descriptions also had difficulties explaining the function of the kidney.

Research has shown how students and adults use several differing non-scientific explanatory models to explain what happens to water in the body. Other studies show that the transitions between organ systems make understanding problematic, particularly because the explanation requires knowledge of the invisible (that is, entities and activities at the micro level). The present study focuses on the knowledge the pre-service student teachers bring with them into the science course at the university about the pathway of water through the body and what knowledge the students think they should teach to the pupils. In addition, this study examines what the students expect will be challenging to teach about the water pathway.

17.2.3 Learning About Organs and Organ Systems

Learning Science

According to Ausubel (1968), knowledge is built upon by adding further knowledge. In other words, knowledge that we have previously acquired is the basis for generating new knowledge. This can be linked to how pre-service student were taught about the human body in upper secondary school, specifically whether the instruction focused on individual organs or organ systems. Ultimately, this has implications for how universities address and build upon students' prior knowledge. However, understanding the fundamental functions of the body can be viewed as a characteristic of scientific literacy, particularly as it is stipulated in the school's curriculum.

Citizens need to be scientifically literate in order to be able to encounter, for example, the news. However, scientific literacy comes with understanding the epistemology of science (Glaze, 2018) (for example, how we know about the processes that happen in the body). Nevertheless, understanding about the body could also be explained by misconceptions, which may be related to how the science knowledge that the students are taught in school does not harmonize with ideas that are prevalent in their everyday culture (Aikenhead, 2005; Lemke, 2000; Phelan et al., 1991). It is also possible that the students have not been introduced to the topic in a way that allows them to develop a scientific approach to the content (Vygotsky, 1934/1986). Learning is also connected to interest in the subject. Potvin and Hasni (2014) showed that the problem of low student interest persists; interest declines the longer the students have been in primary school, with a large gap between what students want to know and what subject content they are taught.

Knowledge Required to be a Teacher

For effective teaching, three aspects are important: the teachers' own knowledge of the subject, the teachers' knowledge of the students, and the context in which the learning takes place (Shulman, 1990). How a teacher chooses to use their subject knowledge and the teacher's pedagogical awareness are also crucial for the students' knowledge development (e.g., Park & Oliver, 2008; Shulman, 1990). Pedagogical content knowledge (PCK) is a crucial aspect of effective teaching that encompasses a teacher's knowledge of instructional strategies, students' understanding, assessment, the curriculum, and teaching orientation (Tuithof, 2017; Williams & Lockle, 2012). It is a unique blend of content and pedagogical knowledge that develops over time and through experience (Van Driel & Berry, 2010). The alignment of a teacher's goals and beliefs with the curriculum goals is a key determinant of their PCK (Tuithof, 2017). PCK is seen as the integration of pedagogy and content and is essential for elevating teaching to a professional status (Shing et al., 2015).

17.3 Aim and Research Questions

The aim of this study is to investigate pre-service teachers' ideas about organs and organ systems, in particular the pathway of water through the body. The focus lies on pre-service teachers who will teach grades 1–3 and 4–6. Furthermore, the study examines what pre-service teachers perceive as challenges in teaching this subject-matter to their students in the future.

Research Questions (RQ):

What ideas do pre-service student teachers have about organs and organ systems, specifically regarding the pathway of water through the body?
What ideas do pre-service student teachers have about the function of water in the body?
What do pre-service student teachers perceive as challenges in future teaching situations?

17.4 Research Design and Method

Sample

The sample of this study consisted of 48 pre-service teachers. All participants were students enrolled at a Swedish university and in pursuit of a Master of Arts degree in primary education, either at the level of pre-school, school grades 0–3 (0 is pre-school class for children aged six years), or school grades 4–6. One-third of the participants were male (see Table 17.1). All the students were in the third year of a four-year course. The data were collected on their first day of the science course, with none of the students having previously taken university-level biology courses. What makes this particular teacher education unique compared to other Swedish universities is that, from the outset, all the students chose to enter a degree program with a focus on science education in one of the semesters. The students who would teach grades 0–3 had already had four weeks of teaching practice in schools, which was part of the course plan for the semester.

Table 17.1 The student teacher participants in the study

Pre-service teachers with the aim to teach in:	Number of participants	Female/Male
Grades 0–3	25	20 female, 5 male
Grades 4–6	23	13 female, 10 male

Data Collection

For this study, the data source was a questionnaire with two parts. In the first part, the students were asked to draw and write. For the drawing, an outline of a non-gender-specific human body figure was provided to them. The students were instructed to use the outline for drawing the pathway of water through the body. The figure used in this study had been used in several other studies before (e.g., Dempster & Stears, 2014; Granklint Enochson, 2024 used a similar figure) and was originally created by Pernilla Granklint Enochson in 2008 (Granklint Enochson, Helldén & Lindahl, 2008) (see Table 17.2). In the second part of the questionnaire, three open-ended questions were presented. These questions explored students' beliefs regarding the necessity of drinking water, as well as their expectations for future teaching in this subject.

- Why is it important to drink water?
- What do you expect your pupils to learn about water in the human body?
- What difficulties/limitations may arise when teaching about water in the human body, i.e., what problems might occur in the teaching situation?

And finally, a questionnaire with multiple-choice questions was used to examine whether or not there is a connection between the students' knowledge of the pathway of water through the body and the function of the kidneys. The following is an example of a multiple-choice question used:

What is the main function of the kidneys? (Put a cross)

- To produce antibodies against diseases
- To break down food
- To circulate blood
- To produce red blood cells
- To purify the blood of waste products.

Data Analysis

Teacher students' drawings and writings using the outline of the non-gender-specific human body were categorized according to coding instructions, as described by Granklint Enochson and Redfors (2012) and Granklint Enochson et al. (2015). The analysis is content-based and focuses on the level of complexity in the students' explanations, in terms of connecting different organ systems. The pre-service teachers' sketches and writings were analyzed according to the following categories:

A. No answer or answer not related to the question

B. Non-scientific descriptions based on alternative ideas of the organ system
C. Descriptions following a scientific explanatory model—important parts are missing
D. Descriptions following a scientific explanatory model—important parts included (at least three organ systems—digestive, circulation, excretion)

The analysis of the open-ended followed a more classic content analysis, which focused on how pre-service teachers expressed their knowledge in the text they wrote.

Zhang and Wildemuth (2017) outlined an eight-step process for conducting content analysis: preparing data, defining the focus of analysis, identifying categories, testing categories, coding the data, assessing the coding reliability, drawing conclusions, and reporting the findings. The category system in this study was developed following a similar process as described by Zhang and Wildemuth (2017). The data collection in this study utilized the same template and categorization system as those employed in previous research (e.g., Granklint Enochson & Redfors, 2012). Nevertheless, all templates were tested by university colleagues during a seminar to ensure the accuracy of the categorization of the figures. There was disagreement about only one template, which was ultimately placed in Category A instead of Category B.

17.5 Results

What ideas do pre-service student teachers have about organs and organ systems, specifically regarding the pathway of water through the body?

The results from the students' categorization of the drawings and writings on the template showed that two of the categories each accounted for one-third of the answers. These two categories were Category B (non-scientific descriptions based on alternative ideas of the organ system) and Category C (descriptions following a scientific explanatory model—important parts were missing) (see Table 17.2). A noteworthy aspect of Category B is that nearly all the student teachers had the understanding that water is distributed to other parts of the body through the stomach, thus lacking the understanding that water uptake takes place in the intestines. Also, according to Category C, either the circulation system or the kidneys were missing (see Table 17.2).

The open-ended question "Why is it important to drink water?"

This question has been examined from two perspectives: first, the pre-service teachers' understanding of the necessity of water consumption, and second, the extent to which their beliefs are associated with their knowledge of organ systems. The student teachers' responses to the open-ended question indicated that no direct link could be found between their knowledge of the water pathway and their understanding of the importance of drinking water. However, three different categories of answers to the question of why it is important to drink water were found. Eleven

Table 17.2 The categorization of the student teachers' knowledge

Pre-service teachers with the aim to teach in:	Category A	B	C	D	Sum
Examples of drawings			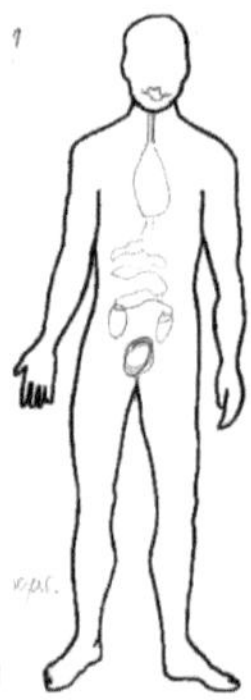	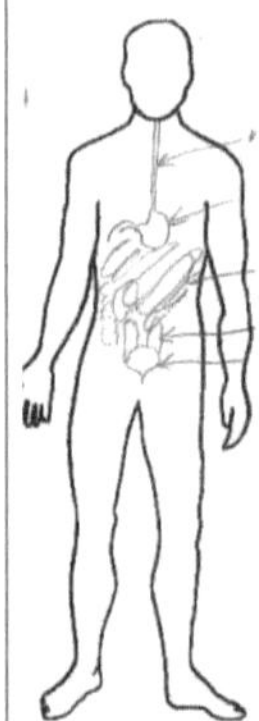	
Grades 0–3	4	7	7	7	25
Grade 4–6	5	7	9	2	23
In total	9 (19%)	14 (29%)	16 (33%)	9 (19%)	48 (100%)

pre-service student teachers gave answers that added nothing beyond the question (for example, it is important to drink) or the students wrote very generally about water (for example, water is good for the body). Twenty-one students answered with sentences like "filling up with water" (hydration). Twelve of the 48 students gave a more scientific explanation concerning water as carrier of nutrients (most of them) or as a regulator of temperature (Table 17.3). Three of the students had the misconception that water was part of the process of breaking down nutrients. One student did not give an answer.

Multiple-choice question "What is the main function of the kidneys?"

Regarding the multiple-choice question, "What is the main function of the kidneys?", 23 of the pre-service student teachers who intended to teach grade 4–6 and 19 pre-service student teachers out of 25 who intended to teach grade 0–3 selected the answer "to purify the blood of waste products". The percentage of students who gave such

Table 17.3 Some of the student teachers' knowledge about *Why is it important to drink water*

Water breaks down nutrients	The function of water in the body, such as transportation of nutrients (or regulation of temperature)
A: 0 B: 1 student C: 1 student D: 1 student In total, 3 students	A: 2 students B: 4 students C: 4 students D: 2 students In total, 12 students

an answer was equally distributed across categories A to D used to classify drawings and texts about the pathway of water through the body. This suggests that all students were aware that the kidneys filter waste products from the blood.

The open-ended question: What challenges may arise when teaching about the pathway of water through the human body, i.e., what problems might occur in the teaching situation?

When the pre-service student teachers elaborated on difficulties and challenges of teaching about the pathway of water through the body, there were three dominating themes. The first was their own level of knowledge. A common response—"Much more than I know now. Specifically, what processes occur in the body when the body takes in fluid"—came from a student who had a knowledge level of Category C (descriptions following a scientific explanatory model—important parts were missing). The students were aware of their lack of knowledge and hoped to gain more knowledge in the science course in order to teach the subject-matter to primary school students in the future. The second theme was related to the characteristics of water and the following statements: "How important water is for humans and that water cannot be compared to other liquids such as soft drinks, juices, coffee and tea" and "What water contains". This type of statement indicates that the pre-service students believed their own students would face challenges in understanding that the water in soft drinks is the same as tap water. The third theme was about "The benefit of water for hydration".

Furthermore, the students responded to the question about difficulties and challenges of teaching the subject-matter to primary school students in the future at a rather general level. More specifically, most pre-service student teachers indicated general pedagogic reasons such as pupils' behavior, time for preparation, and so on. Quite a few were worried about their ability to teach. However, only a few students (two students in 0–3 and four students in F-6) indicated disinterest in the subject as a challenge.

17.6 Discussion and Conclusion

A primary objective of this study was to examine the prior knowledge that pre-service student teachers, who will teach primary school students (grades 1–6), bring to university-level science courses. This cohort is particularly significant, as these individuals will be responsible for introducing pupils to formal science education. Understanding their conceptual approaches and scientific knowledge is therefore essential.

In comparison to previous research, the explanations provided by students in this study were less scientifically accurate than those of Grade 9 students in compulsory school (Granklint Enochson & Redfors, 2012) but more scientific than those observed in a study involving pre-service student teachers with an average age of 37, who were presumed to have prior experience teaching in preschool settings (Granklint Enochson, 2024). Notably, the students in the present study—having completed upper secondary school more recently than those in Granklint Enochson (2024)—did not exhibit the common misconception, documented in earlier research, of conceptualizing a direct tube from the mouth to the kidneys. However, a different type of misconception emerged in their drawings, which depicted water traveling from the stomach into the circulatory system rather than passing through the intestines. Despite these findings, no clear relationship was identified between students' understanding of the pathway of water through the body and their knowledge of kidney function or their awareness of the importance of water consumption for pupils. This contrasts with previous studies that have indicated such connections (e.g., Granklint Enochson & Redfors, 2012). It seems that students have not sufficiently problematized the water pathway or the diverse functions of water in the body in general, during their previous education. Some even mistakenly perceive water as a nutrient in itself. This gap in knowledge may stem from either inadequate exposure to the subject-matter (Vygotsky, 1934/1986) or a lack of interest in science during their schooling (Potvin & Hasni, 2014). Several studies have highlighted that students often show less interest in science compared to other school subjects (e.g., Potvin & Hasni, 2014) and that teachers play a crucial role in shaping students' engagement with scientific topics (e.g., Chen et al., 2024).

Given this, it is essential for university educators to assess and address the prior knowledge of pre-service teachers, ensuring they understand the connections between different organ systems and the interconnections linking them, as highlighted by Clément (2003). Additionally, university instructors must recognize and address common misconceptions among students, as these can significantly impact their future teaching practices. A well-structured teacher education program should not only provide a robust understanding of human biology at both the organ and system levels but also equip future educators with pedagogical strategies to address student misconceptions effectively (e.g., Park & Oliver, 2008; Shulman, 1990). Interestingly, the students in this study did not identify a lack of interest as a potential challenge in their future teaching. However, it remains crucial for educators to establish a baseline understanding of students' prior knowledge to facilitate the construction of new scientific concepts (Ausubel, 1968). Given that most participants began their academic studies within five years of completing upper secondary school, their secondary education experiences can be considered relatively recent. This study may therefore have a direct impact on how teacher education programs address the misconceptions and prior knowledge that students bring into their university studies.

The limitations of this study include the relatively small sample size and the fact that all participating students were admitted to the university in the same year, which may have influenced the results. The dropout rate among students was minimal and was limited to those who were absent for legitimate reasons. All students who were present at the university on the day of the study voluntarily chose to participate. Furthermore, the author of this article did not teach the participating group about human biology either before or after the data collection, ensuring that the responses were not influenced by a potential assessment situation.

References

Assaraf, O.B.-Z., Dodick, J., & Tripto, J. (2013). High school students' understanding of the human body systems. *Research in Science Education, 43*(1), 33–56. https://doi.org/10.1007/s11165-011-9245-2

Andersson, J., Löfgren, R., & Tibell, L. A. E. (2020). What's in the body? Children's annotated drawings. *Journal of Biological Education, 54*(2), 176–190. https://doi.org/10.1080/00219266.2019.1569082

Aikenhead, G. S. (2005). *Science education for everyday life: evidence-based practice.* Teachers College Press.

Ausubel, D. (1968). *Educational psychology – A cognitive view.* Holt, Reinhart & Winston.

Bildungspläne Baden-Württemberg. (2016). *Sekundarstufe I – Biologie.* Ministerium für Kultus, Jugend und Sport. Retrieved February 16, 2024, from https://www.bildungsplaene-bw.de/site/bildungsplan/get/documents/lsbw/export-pdf/depot-pdf/ALLG/BP2016BW_ALLG_SEK1_BIO.pdf

Canlas, I. P., & Molino-Magtolis, J. (2024). Reflecting on the relevance of drawing as a tool in eliciting pre-service teachers' preconceptions of human organs and organ systems. *Journal of Biological Education, 58*(4), 951–971. https://doi.org/10.1080/00219266.2022.2147207

Carvalho, G. S., Silva, R., Lima, N., Coquet, E., & Clément, P. (2004). Portuguese primary school children's conceptions about digestion: Identification of learning obstacles. *International Journal of Science Education, 26*(9), 1111–1130. https://doi.org/10.1080/0950069042000177235

Chen, C., Said, T., Sadler, P. M., Perry, A., & Sonnert, G. (2024). The impact of high school science pedagogies on students' STEM career interest and on their ratings of teacher quality. *Journal of Research in Science Teaching, 61*(10), 2535–2569. https://doi-org.proxy.mau.se/10.1002/tea.21948

Clément, P. (2003). Situated conceptions and obstacles: The example of digestion and excretion. In D. Psillos, P. Kariotoglou, V. Tselfes, E. Hatzikraniotis, G. Fasspooupoulos, G., & M. Kallery (Eds.), *Science education research in a knowledge-based society* (pp. 89–98). Kluwer Academic.

Cuthbert, A. J. (2000). Do children have a holistic view of their internal body maps? *School Science Review, 82*(299), 25–32.

Dempster, E., & Stears, M. (2014). An analysis of children's drawings of what they think is inside their bodies: A South African regional study. *Journal of Biological Education, 48*(2), 71–79. https://doi.org/10.1080/00219266.2013.837401

Garcia-Barrosa, S., Martínez-Losadaa, C., & Garridoa, M. (2011). What do children aged four to seven know about the digestive system and the respiratory system of the human being and of other animals? *International Journal of Science Education, 33*(15), 2095–2122. https://doi.org/10.1080/09500693.2010.541528

Glaze, A. (2018). Teaching and learning science in the 21st century: Challenging critical assumptions in post-secondary science. *Education Sciences, 8*(1), 12. https://doi.org/10.3390/educsci8010012

Granklint Enochson, P., Helldén G., & Lindahl, B. (2008). Student understanding about water transport in the human body and why water is healthy. In Allyson Macdonald (ed.), *Proceedings of the 9th Nordic Research Symposium on Science Education* (pp. 131–133), 11th–15th June 2008 University of Iceland.

Granklint Enochson P., & Redfors, A. (2012). Students' ideas about the human body and their ability to transfer knowledge between related scenarios. *European Journal of Health and Biology Education, 1*(1 & 2), 3–29. https://doi.org/10.20897/lectito.201202

Granklint Enochson, P., Redfors, A., Dempster, E. R., & Tibell, L. A. (2015). Ideas about the human body among secondary students in South Africa. *African Journal of Research in Mathematics, Science and Technology Education, 19*, 199–211. https://doi.org/10.1080/10288457.2015.1050804

Granklint Enochson, P. (2024). Pre-service teachers' ideas about the path of water through the body and their intentions about explaining it to preschool children. *Journal of Biological Education, 58*(4), 772–781. https://doi.org/10.1080/00219266.2022.2108106

Hmelo-Silver, C. E., Jordan, R., Liu, L., Gray, S., Demeter, M., Rugaber, S., Vattam, S., & Goel, A. (2008). Focusing on function: Thinking below the surface of complex natural systems. *Science Scope*. Summer, 27–35.

Jahic Pettersson, A., Tibell, L., & Löfgren, R. (2021). The brain needs nutrition': Pupils' connections between organizational levels. *Nordic Studies in Science Education, 17*(1), 49–63. https://doi.org/10.5617/nordina.7930

Lemke, J. L. (2000). Articulating communities: Sociocultural perspectives on science education. *Journal of Research in Science Teaching, 38*(3), 296–316. https://doi.org/10.1002/1098-2736(200103)38:3%3c296::AID-TEA1007%3e3.0.CO;2-R

Manokore, V., & Reiss, M. (2003). Pupils' drawings of what is inside themselves: A case study in Zimbabwe. *Zimbabwe Journal of Educational Research, 15*, 28–43.

Oskarsdottir, G., Stougaard, B., Fleischer, A., Jeronen, E., Lützen, F., & Kråkenes, R. (2011). Children's ideas about the human body – A Nordic case study. *Nordic Studies in Science Education, 7*(2), 179–189. https://doi.org/10.5617/nordina.240

Park, S., & Oliver, S. (2008). Revisiting the conceptualisation of pedagogical content knowledge (PCK): PCK as a conceptual tool to understand teachers as professionals. *Research in Science Education, 38*(3), 261–284. https://doi.org/10.1007/s11165-007-9049-6

Phelan, P., Davidson, A. L., & Cao, H. T. (1991). Students' multiple worlds: Negotiating the boundaries of family, peer, and school cultures. *Anthropology & Education Quarterly, 22*(3), 224–250. https://www.jstor.org/stable/3195764

Potvin, P., & Hasni, A. (2014). Interest, motivation and attitude towards science and technology at K-12 levels: A systematic review of 12 years of educational research. *Studies in Science Education, 50*(1), 85–129. https://doi.org/10.1080/03057267.2014.881626

Prokop, P., Fancovicová, J., & Tunnicliffe, D. (2009). The effect of type instruction on expression of children's knowledge: How do children see the endocrine and urinary systems? *International Journal of Environmental and Science Education, 4*(1), 75–93.

Reiss, M. J., Tunnicliffe, S. D., Möller Andersen, A. M., Bartoszeck, A., Carvalho, G. S., Chen, S.-Y., & Jarman, R. (2002). An international study of young people's drawings of what is inside themselves. *Journal of Biological Education, 36*(2), 58–64. https://doi.org/10.1080/00219266.2002.9655802

Shing, C. L., Saat, R. M., & Loke, S. H. (2015). The knowledge of teaching – Pedagogical content knowledge (PCK). *Malaysian Online Journal of Educational Sciences, 3*, 40–55.

Skolverket. (2013). *Curriculum for the upper secondary school*. Fritzes customer service Stockholm.

Skolverket. (2018). *Curriculum for the compulsory school, preschool class and school-age educare, revised 2018*. Norstedts Juridik kundservice Stockholm.

Shulman, L. S. (1990). Research in teaching and learning: A project of the American educational research association. In *Paradigms and Programs* (Vol. 1, pp. 3–36). Macmillan.

The national curriculum. (2014, 2015). Department of Education in the United Kingdom. https://www.gov.uk/national-curriculum

Tuithof, J. I. G. M. (2017). *The characteristics of Dutch experienced history teachers' PCK in the context of a curriculum innovation*. ISBN: 978-94-6103-061-0 Utrecht University Repository (Dissertation).

Tunnicliffe, S. D. (2004). Where does the drink go? *Primary Science Review, 85*, Nov/Dec, 8–10. ISSN-0269-2465

Van Driel, J. H., & Berry, A. (2010). Pedagogical content knowledge. In P. Penelope, E. Baker, & B. McGaw (Eds.), *International encyclopedia of education* (Third Edition) (pp. 656–666). Elsevier, ISBN 9780080448947.

Vygotsky, L. (1934/1986). *Thought and language* (A. Kozulin, Trans.). MIT Press.

Williams, J., & Lockle, J. (2012). Using CoRes to develop the pedagogical content knowledge (PCK) of early career science and technology teachers. *Journal of Technology Education, 24*(1). https://doi.org/10.21061/jte.v24i1.a.3

Zhang, Y., & Wildemuth, B. M. (2017). *Qualitative analysis of content. Applications of social methods to questions in information and library science* (pp. 318–329) (Barbara M. Wildemuth, red.). ISBN: 9781440839047 1440839042.

Chapter 18
Biology Teachers' Beliefs About the Social Certification and Dissemination of Biology Education Research

Alexander Georg Büssing and Bianca Reinisch

18.1 Introduction

In today's changing world, people need to be equipped with the required competencies, for example, to make informed decisions about new technologies such as vaccination using messenger ribonucleic acid (mRNA). To prepare citizens to act competently, there is a need for the public to be literate in basic biological concepts (Winterbottom, 2022). In this regard, school biology needs to play an important role for preparing future citizens for decision-making in challenging times, also considering the complex dynamics of living systems (Zeyer & Dillon, 2019).

However, transferring scientific knowledge from research studies into teaching practice represents a central problem of education (Kennedy, 1997). This also concerns biology education, because a gap between the results from biology education research and the implementation of these results in school teaching practice may hinder educational improvement and the preparedness of future citizens (Furman Shaharabani & Yarden, 2019). Therefore, there is a need for a better understanding of how the results from biology education research may enter school teaching practice and how this transfer can be fostered (Randi & Corno, 2007). Previous studies (e.g., Smith & Sela, 2005) have proposed different possibilities to narrow the gap between the results from biology education research and school teaching practice.

A. G. Büssing (✉)
Institute for Science Education, Technische Universität Braunschweig, Braunschweig, Germany
e-mail: alexander.buessing@tu-braunschweig.de

B. Reinisch
Universität Potsdam, Potsdam, Germany
e-mail: bianca.reinisch@uni-potsdam.de

O. Morin et al. (eds.), *Challenges of a Changing World in Biology Education*, Contributions from Biology Education Research,
https://doi.org/10.1007/978-3-032-05346-6_18

Some researchers proposed to provide teachers with research experiences to enhance their understanding of science education research, which could, in turn, improve their practice in lesson planning and teaching (Smith & Sela, 2005; Yosief et al., 2022). Other studies investigated the implementation of curricular innovations (Mpate, 2023) or the enactment of professional knowledge (Aumann & Weitzel, 2024). As some studies also integrated how teachers' practical knowledge and beliefs affect reforms (Van Driel et al., 2001), teachers' beliefs could be crucial in bridging the gap between research results and school teaching practice. Nonetheless, prior studies have not explored teachers' more general views on biology education research and their views on the transfer of research findings into school teaching practice.

The current paper addresses this gap by applying the Family Resemblance Approach (FRA), which represents an established approach to conceptualize the Nature of Science (NOS). We use this theory due to its analytical openness to explore biology teachers' beliefs of the social certification and dissemination of the results from biology education research as one selected category of the FRA. This dimension seemed to be most relevant for understanding the social-institutional dimension of transferring results from biology education research into school teaching practice. For this purpose, we present results from a qualitative interview study with eleven biology in-service teachers, which pre-service teachers conducted in the context of an innovative teacher education setting. Before further describing the methodological approach, we elaborate on the theoretical background including the transfer of results from biology education research, the adaptation of the FRA to biology education research, and teachers' beliefs.

18.2 Theoretical Background

18.2.1 Transferring Biology Education Research Results into Practice

For biology education research, conducting studies to improve school teaching represents a central aim of researchers, particularly in challenging times (Winterbottom, 2022). However, in contrast to other scientific fields such as law, architecture, or medicine, the implementation of research findings into school teaching practice seems to be more complex (Furman Shaharabani & Yarden, 2019). Possible reasons include the role of local traditions in schools and the missing understanding of how research results can be interpreted correctly (Cochran-Smith, 2005). In this regard, in-service teachers may experience a gap between their initial teacher education, which is often more oriented towards coherent theories in line with research results, than their later experiences from school practice (Korthagen & Kessels, 1999).

In prior studies, researchers described reasons such as the persuasiveness, relevance, and accessibility of research findings as possible reasons for teachers to experience a gap between research results and teaching practice (Kennedy, 1997). To improve the implementation of results from research into teaching practice, Van Driel et al. (2001) suggested learning in networks, peer coaching, collaborative action research, or the use of cases. Other researchers further underlined the social nature of implementing research into practice. For example, Allen (2009) used interviews with first-year graduate teachers about their experiences of teaching compared to their pre-service teacher education. The results showed that they valued practice-based experiences as more important than the theoretical subject-matter from their teacher education courses (Allen, 2009). Reasons for this may also include the different social contexts, as beginning teachers reported to struggle with taking up the social role as a teacher (Allen, 2009). Therefore, existing studies may have underestimated the role of social-institutional and subject-specific characteristics of schools, which could be a way to understand the transfer of results from biology education research into school teaching practice (Opfer & Pedder, 2011).

This is underlined by Larkin (2022), who argued for a stronger inclusion of social-institutional aspects by integrating the values of science education and the question of what constitutes "good" science education into reform efforts. More specifically, Larkin (2022) proposed that improving science education "requires thinking systemically about all of its aspects, including its goals, inputs, outcomes, measures, and reckoning with the social context in which such teaching and learning is situated" (p. 12). To meet this requirement, the FRA could be a valuable framework to investigate such subject-specific aspects.

18.2.2 The Family Resemblance Approach (FRA)

As a coherent framework to investigate the social-institutional dimensions of biology education research, the present study utilized the FRA. This framework represents one theory to understand NOS content and has been applied to different empirical investigations in science education (Cheung & Erduran, 2022; Reinisch & Fricke, 2022). The FRA defines specific categories, which can be used to compare scientific disciplines (Erduran & Dagher, 2014). In this way, the FRA allows researchers to investigate how "sciences are similar or dissimilar, thereby building up from scratch polythetic sets of characteristics for each individual science" (Irzik & Nola, 2011, p. 595).

A detailed overview of the categories of the FRA is provided in the result section of this contribution. The categories can be differentiated into a cognitive-epistemic and a social-institutional dimension (Cheung & Erduran, 2022). Given the focus on the gap between biology education research and school teaching practice in the present study, we specifically focussed on the FRA-category *social certification and dissemination* of the social-institutional system. This category describes how "scientists' work is reviewed and critically evaluated by their peers" (Erduran et al., 2019, p. 316). Furthermore, "this form of social quality control aids in the validation of new scientific knowledge by the broader scientific community" (Erduran et al., 2019, p. 316). This broader scientific community could also include teachers. Even if they do not perform formal evaluations such as peer-review, their inclusion of new results from biology education research into their teaching practice may represent an informal way of evaluating the research findings. Therefore, we investigated how teachers believe results from biology education research are evaluated and disseminated. This could lay the foundation for improving the communication of results from biology education research to teachers.

The present study applies the FRA, as this framework provides a robust theoretical foundation for a nuanced approach to the various aspects of the social and institutional dimensions of science, which may be of importance for understanding the practices of teachers and their beliefs (Larkin, 2022). Furthermore, it provides a heuristic for empirically investigating how teachers might perceive such dimensions in their discipline. Currently, other NOS-frameworks (e.g., the consensus view; Abd-El-Khalick & Lederman, 2000) seem not suitable for this, as they require a consensus about what aspects constitute the characteristics of biology education research in the first place. Therefore, the application of an analysis-based framework such as the FRA allows to lay the foundation for such subsequent studies.

18.2.3 Teachers' Beliefs

In prior research, teachers' views on science education have been examined from different theoretical perspectives, which often are interrelated with each other. Exemplary constructs include knowledge, attitudes, conceptions, mental representations, dispositions, or worldviews (Jones & Park, 2023). In the present study, we focus on teacher beliefs. Our understanding of beliefs is informed by their functions for teachers' personality structures. In this way, beliefs may act as filters, which shape experiences and content as relevant, function as frames that define problems or tasks, and later guide action (Fives & Buehl, 2012).

As the present study reports data from an interview study at only one timepoint, we conceptualize teachers' beliefs about the evaluation and dissemination of the results from biology education research as explicitly available and rather stable (Fives & Buehl, 2012). This is a specific view, that may be in contrast with other studies, which conceptualized beliefs within larger and dynamic systems (Jones & Park, 2023). Nonetheless, this view is required for achieving our research aims.

18.2.4 Aims of the Present Study

As described before, the FRA represents a framework with specific categories to analyse scientific disciplines. In the present chapter, we use this framework and apply it to biology education research to exploratively investigate in-service teachers' beliefs about the nature of biology education research (RQ_1) using semi-structured interviews. In a second step, we focused on the category *social certification and dissemination*, due to the connection of this category to knowledge transfer (RQ_2). As a final research question, we investigated if sections of the interviews included codings for the categories of *social certification and dissemination* and other categories of the FRA, so called code co-occurrences (RQ_3). This would give a deeper look into how specific aspects of the *social certification and dissemination* may be related to the other dimensions of the FRA on a qualitative basis:

Research question 1 (RQ_1): What are biology teachers' beliefs on the nature of biology education research?
Research question 2 (RQ_2): What are biology teachers' beliefs of the social certification and dissemination of results from biology education research?
Research question 3 (RQ_3): What code co-occurrences emerge between biology teachers' beliefs of the social certification and dissemination of results from biology education research and other categories of the FRA?

18.3 Methods

18.3.1 Research Design and Sample

The present paper presents results from a qualitative interview study in Germany. The German school system is structured into several school subjects (e.g., Biology, History). As a part of their studies at the university, teachers decide on one, two or even three different school subjects. In line with our research aim, we interviewed teachers, who had studied biology and currently taught biology classes at secondary schools. The interviews were done as part of a seminar with eleven pre-service teachers on the role of research and school teaching practice in biology education. Every pre-service teacher interviewed one in-service teacher about his or her views on biology education research. All teachers were individually interviewed using a semi-structured interview guide based on the aim of the investigation.

The pre-service teachers who conducted the interviews studied to become secondary school teachers ("Gymnasium" in the German school system) and were in their Master study. As a part of their teacher education, the seminar was intended to reflect on the dynamics between research and school teaching practice. The pre-service teachers were trained to conduct the interviews. The present paper is based on these interviews, but the authors of this paper analysed them independently of the pre-service teachers, who used the interviews for small poster presentations in the seminar.

Overall, eleven in-service biology teachers were interviewed ($M_{age} = 38.10$ years, $SD_{age} = 10.55$ years, $M_{experience} = 9.36$ years of experience). All teachers were informed about the purpose of the study beforehand using an introductory text in the interview guide and had the possibility of declining their participation at any time. Furthermore, all data were analysed anonymously, and the present paper uses pseudonyms for better comprehension. The interviews took between 35 and 60 min,

Table 18.1 Overview of the sample with pseudonyms, years of experience, and association with biology education

Pseudonym (age)	Experience	Ideas associated with "biology didactics" (German "Biologiedidaktik")
Adam (56)	28 years	"Which topics should be dealt with in a specific class, so students can understand them"
Bernd (48)	12 years	"To design classes, that biological research can be understood by students"
Clara (37)	7 years	"A basic framework for planning student-oriented teaching"
Daniela (33)	8 years	"I had no biology education research in my studies, but it is important for teachers"
Emily (49)	6 years	"Research results give insights and tools for designing their classes"
Frieda (43)	14 years	"Oh, that was a long time ago, about 12 to 14 years"
Gertrud (25)	1 year	"How biology is taught and its educational backgrounds for designing lessons in a specific way"
Hannah (25)	1 year	"Biology education is the science of learning and teaching of biology"
Ingrid (47)	20 years	"To allow students for understanding contents not too abstract"
Katja (30)	2 years	"Theory on good biology education"
Laura (26)	2 years	"How to prepare content to specific ages"

already showing that there also existed a large variation in length due to the different interviewers. Table 18.1 gives an overview of the sample and the teachers' first ideas associated with the term biology education as a view into their spontaneous beliefs.

18.3.2 Interview Design

The interviews were structured into several parts. First, the in-service teachers were asked about background characteristics, such as their age and teaching experience. To start their reflection process on the nature of biology education research, the interviewees were asked about the ideas they associated with the term biology didactics (German "Biologiedidaktik"). While it is difficult to standardize the teachers' responses, their associations provide insights into their global views of biology education research. Furthermore, the question about what teachers associated with biology didactics was a possibility to ensure that they talk specifically about biology education research (results are shown in Table 18.1).

After this rather broad introduction, the next parts focused on the social-institutional aspects of biology education. In particular, the interview guide was informed by the social-institutional category of *social certification and dissemination*. The interview questions from this section were: How do you think biology education research is evaluated? How do you think are biology education research results disseminated? What publications do you use to stay in touch with research results? Through which sources are the results from biology education research most often disseminated? Do you communicate with your colleagues about the results of biology education research? Do you think that school practice should be more often included in the dissemination of the results of biology education research?

In the final section, the interviewers asked the in-service teachers about their evaluation of their teaching and how the results of biology education research may improve their school students' learning. This part was not analysed for the present paper, even if connections of beliefs on biology education may be interesting for future studies.

18.3.3 Data Analysis

Before the analysis, the interview data was transcribed using the transcription software F4. The resulting text files were imported into MAXQDA24 software. A stepwise approach was used to analyse the data. In line with the structured qualitative content analysis following Mayring (2015), we first proposed deductive categories in line with the 11 FRA-categories (Table 18.2). After the first coding step of the material according to these categories to answer RQ_1, we inductively analysed the passages within the category *social certification and dissemination* to answer RQ_2. This particularly included the development of the superordinate categories of *social certification* and *dissemination.* Within these higher-order categories, we created new categories whenever a new subcategory could be differentiated. The results tables illustrate the nested nature of the categories by referring to a numbered system of codes. To answer RQ_3, we report exemplary categories, which were coded simultaneously with the *social certification and dissemination* to illustrate the co-occurrence of two categories in one section (called "code co-occurrences"). In this way, we were able to interpret more specific aspects of teachers' beliefs towards the *social certification and dissemination*, even if the dynamic relation between selected beliefs should be part of further research.

Several practices were followed to ensure the quality of data analysis (Göhner & Krell, 2020). First of all, the settings of the interviews as well as the analyses were standardised and documented. This concerned, for example, the training of the interviewers, documentation of interview questions in a semi-structured interview guide, and following established standards for the analysis. Furthermore, parts of the category system were discussed within the seminar and afterwards with another researcher, while the final analysis was done by a researcher trained in qualitative content analysis. When discussing the category system and codings, problematic points were addressed and settled.

Given the interpretative nature of the analysis and its focus on qualitative insights from individual cases, an interrater reliability calculation was not appropriate. As Schreier (2014) discusses, qualitative content analysis does not necessarily require

Table 18.2 Overview of FRA-categories with definitions from Erduran et al. (2019, p. 316), exemplary interview quotes, and description of the relevance for teachers beliefs

Category (definition)	Exemplary quote	Description
Cognitive-epistemic system dimensions		
1. Aims and values (Adherence to a set of values that guide scientific practices)	"(…) the validity and credibility [of results from biology education research]" (Hannah)	Validity and credibility represent important aims of biology education research
2. Scientific practices (Cognitive, epistemic, and discursive practices underpinning science)	"I could do surveys or tests at my school" (Katharina)	Surveys and tests are appropriate practices used by biology education researchers
3. Scientific knowledge (Scientific knowledge is holistic and relational, and TLM are conceptualized as a coherent network, not as discrete and disconnected fragments of knowledge)	"Of course it is important to learn new things (…), but I do not learn something new weekly or daily. (…) Because I learned it already sometime." (Daniela)	Results from biology education research may be perceived as static and teachers may be satisfied with basic knowledge
Social-institutional system dimensions		
4. Professional activities (Scientists engage in a number of professional activities to enable them to communicate their research)	"For example, concerning the digitalization, to design materials for making them usable." (Bernd)	Biology education researchers prepare learning materials for usage in classrooms
5. Scientific ethos (Scientists are expected to abide by a set of norms both within their own work, and during their interactions with colleagues and scientists from other institutions)	"(…) because it also has privacy problems [when doing research or learning with digital media]." (Hannah)	Privacy represents an important norm for biology education researchers
6. Social certification and dissemination (review and critical evaluatuation of scientists' work)	"Yes, the certification is in a classical way via articles in journals." (Bernd)	Reviewing and critical evaluation of work by others and the dissemination of research results
7. Social values of science (Social values within the scientific endeavour such as social utility, addressing human needs, and equality of intellectual authority)	"In my opinion, [results from biology education research] don't have such a high value (…) at least for teachers, so I would consider them rather unimportant." (Laura)	Results from biology education research may be of minor value for in-service teachers
8. Social organizations and interactions (The social organization and interactions within science)	"We visited the university [for a research project], that was really a highlight." (Clara)	The organizational differentiation of universities and schools play a major role

(continued)

Table 18.2 (continued)

Category (definition)	Exemplary quote	Description
9. Political power structures (The scientific enterprise operates within a political environment that influences the direction and use of science)	"What research from biology education said, that was of course implemented in class." (Daniela)	Research results include an authority, due to the political decisions to value them
10. Financial systems (Economic factors underlying science)	"It would be very fruitful (...) to have very good lessons (...), even if this would cost something, not everything can be for free." (Daniela)	Biology education research and its dissemination underlies economic constraints

statistical reliability measures but rather emphasizes intersubjective validation, transparency in category development, and systematic coding procedures to ensure rigor. Since our study was not designed to produce quantifiable generalizability, but rather to capture context-sensitive insights, quality assurance was ensured through systematic peer discussions, category validation, and methodical documentation (Schreier, 2014).

18.4 Results

18.4.1 *Teachers' Beliefs on the Nature of Biology Education Research*

Table 18.2 gives an overview of the categories resulting from the FRA and their respective definitions based on Erduran et al. (2019, p. 316), which were coded based on the first deductive coding step to answer the first research question (RQ_1). Except for *methods and methodological rules*, all categories could be coded in the transcripts.

Overall, the teachers often had a rather practical view on results from biology education research and researchers' activities. For example, the design of materials was mentioned as a central part of researchers' *professional activities (4.)*. The practical view was also visible in the category *social values* (7.), which included the utility of the results for effective learning in biology class, even when some teachers were rather critical about the overall usefulness of these results. This was also contradicted by other interviewees, which mentioned a political authority over affecting school practice implied by results from biology education research (9.). Other interesting insights were reported by the in-service teachers concerning the role of privacy as a part of the *scientific ethos* (5.), including the necessity to protect students' privacy when using data from them. Finally, also economic considerations were taken into account (10.). The further results will dive deeper into the *social certification and dissemination* (6.).

18.4.2 Beliefs About Social Certification and Dissemination

Table 18.3 shows the results for the second research question, including the inductive subcategories of (6.1) *social certification* and (6.2) *dissemination.* Concerning the *social certification* of results, some teachers reported sharing experiences about their teaching with different groups (6.1.1), even if other teachers mentioned not to discuss results with their peers (6.1.2). While this may reflect the social nature of their certification, many also mentioned a rather critical perspective on research results (6.1.3). Reasons include the abstract nature of the results, which are not specific for the classroom situations faced by the in-service teachers. Finally, this also can be derived as success criteria for a good certification, if teachers perceive the results from biology education research as recent (6.1.4) and situation-specific (6.1.5).

Within the category *dissemination*, teachers mentioned different *methods of dissemination* (6.2.1), which could be further inductively coded into more classical ways such as *public media* (6.2.1.1), *books* (6.2.1.2), and *(practitioner) journals* (6.2.1.3). Also, *colleagues* (6.2.1.4) and *digital opportunities* (6.2.1.5) are possibilities for disseminating results from biology education research. Finally, the teachers also mentioned the integration in *research projects* (6.2.1.6) or initial and ongoing *teacher education* (6.2.1.7). Regardless of this, teachers also reported *missing time* (6.2.2) and *teacher age* (6.2.3) as hindering aspects for the dissemination of results. Successful dissemination of results, using one of the mentioned methods, should therefore be as concise as possible, to make use of teachers' time, and keep teachers' personal characteristics in mind.

18.4.3 Code Co-occurrences with the Other Categories

Finally, we investigated which of the superordinate categories from the FRA were coded in the same sections as quotes from the category *social certification and dissemination.* These code co-occurrences may allow an insight into more specific characteristics of teachers' beliefs on the transfer of results from biology education research. As shown in Table 18.4, sentences coded for *social certification and dissemination* had code co-occurrences with the categories *aims and values, scientific knowledge, professional activities, scientific ethos, social values of science*, and *social organizations and interactions.*

As shown in the table, teachers reported that the application of research findings into school practice represent an important aim, which also could integrate the point of view of teachers as research subjects more strongly. Finally, the separation of biology education researchers and schools is experienced as a challenge for the

Table 18.3 Identified categories regarding *social certification and dissemination*

(Sub-)categories	Exemplary quote	Description
(6.1) Social certification	"Yes, it happens [that we talk about biology education research results]" (Daniela)	Teachers reported to discuss results from biology education research
(6.1.1) Sharing experiences	"You exchange experiences about what you tested with the others and what you learned from that." (Gertrud)	Experiences about pedagogy are exchanged
(6.1.2) No discussion of results	"Of course you talk automatically about different pedagogical perspectives, but to really look into recent results, it is not happening." (Katharina)	Disagreement with the view that discussions are happening
(6.1.3) Critical perspective	"[In research,] there are always perfect conditions. But we are in the third year of crisis and we are in crisis-mode between the need to teach specific things, because there are the diploma tests coming and the learner orientation. There is such a large gap, that I am unable to implement the research results." (Emily)	Critical perspectives on viability
(6.1.4) Topicality	"(…) but they also have to be recent (…)." (Emily)	Results have to be reçent
(6.1.5) Situation-specificity	"You need to be able to adapt the results to your situation." (Daniela)	Results have to be situation-specific
(6.2) Dissemination	"[If there is a new result], it has to be published somewhere." (Frieda)	Results from biology education research should be disseminated
(6.2.1) Dissemination methods	"Generally, it makes sense to publish [research findings] in journals or books." (Daniela)	Various methods for disseminating results from biology education research
(6.2.1.1) Public media	"If not in journals (…) or in the public media, no, [I have no contact to research results]." (Frieda)	Public media outlets (e.g., newspaper)
(6.2.1.2) Books	"I use books and sometimes also journals." (Laura)	Books on biology education
(6.2.1.3) (practitioner) journals	"(…) for example, *Unterricht Biologie*, this educational journal." (Clara)	Journals about research and practice
(6.2.1.4) Colleagues	"Also, the stuff, which you get from colleagues." (Gertrud)	Colleagues in schools distribute new results
(6.2.1.5) Digital opportunities	"(…) digitally in the internet, on the webpages of the researchers, if you have the possibility of downloading articles on their homepages." (Bernd)	Digital outlets such as homepages and newsletters

(continued)

Table 18.3 (continued)

(Sub-)categories	Exemplary quote	Description
(6.2.1.6) Research projects	"Sometimes, there is a proposal for implementing a survey for students for research. Then we are involved, otherwise not" (Frieda)	Participating in research projects
(6.2.1.7) Teacher training	"When researchers [from biology education] propose something in professional development settings." (Adam)	Teacher training for specific topics
(6.2.2) Missing time	"But again, time is the deciding resource." (Emily)	Not enough time for looking for new results
(6.2.3) Teacher age	"I believe, the results are being received more positively by younger teachers, because in former times, there was no such discipline (…)." (Daniela)	Dissemination may depend on age

usefulness of biology education research results. This may also be one reason for the teachers' experience of results from biology education research to be static. Both findings may be in contrast with researchers' experiences and demands, as will be discussed later.

18.5 Discussion

18.5.1 Exploration on Categories of a Nature of Biology Education Research

For the first research question (RQ_1), we were able to identify selected beliefs from the teachers for each of the dimensions of the FRA except for *methods and methodological rules*. The results illustrated some categories, which showed similarities of biology education research to other scientific disciplines. For example, the beliefs concerning the categories aims and values, political power structures, and financial systems are similar to the contents of these categories for biological science (Reinisch & Fricke, 2022). However, other categories included aspects specific for biology education research.

These more specific aspects include scientific practices such as taking surveys and tests in schools or the static nature of scientific knowledge from biology education research. While the practices may be in line with what researchers from other disciplines do, particularly the static nature of knowledge needs to be discussed critically. Here, future contrasting studies with biology education researchers would be interesting to investigate if they perceive biology education research results in the same way or as dynamic and in response to societal movements and grand

Table 18.4 Code co-occurrences of *social certification and dissemination* with other FRA-categories including an exemplary quote and the description of the relation between the categories

Co-occurred category	Exemplary quote	Description of relation with *social certification and dissemination*
Aims and values	"Many research approaches are not always entirely applicable or understandable for me. In general, I would prefer a closer connection to actual classroom teaching." (Laura)	Biology education research should be viable for school teaching practice to be certified by teachers
Scientific knowledge	"[No], because I learned it at some point before, and in the end, not much has really changed about it." (Daniela)	Teachers may view knowledge from biology education research as static
Professional activities	"But in general, it might be an option if the [researchers] offered an invitation -like a small seminar or a training session, yes, a training session specifically for this—to establish something like that and say, we are interested in exchanging ideas with schools, involving them in our research purposes. After all, they also want to ensure they have data in some way." (Adam)	Several activities of researchers such as offering teacher education or data collection are possibilities for social certification and dissemination
Scientific ethos	"So, yeah, since I didn't really get to learn much about the actual process of research itself, I didn't really feel affected or included, to be honest." (Hannah)	Teachers could be further included in the research process to adhere to norms such as openness, intellectual honesty, and respect for the research subject
Social values of science	"So, one should be able to receive these results in a way that is as stress-free as possible and suited to their situation." (Daniela)	Researchers should notice the necessities of practice and make their results easily accessible and specific
Social organizations and interactions	"When it comes to biology education, [researchers] should occasionally be involved in classroom activities, meaning actively participate—whether through observations or by teaching themselves" (Ingrid)	The institutional separation between research and teaching practice may be a hinderance for the transfer of biology education research results into school teaching practice

challenges (Winterbottom, 2022). Concerning the social-institutional dimensions, the teachers mentioned other specific aspects, illustrating how the application of the FRA-framework allows to differentiate biology education from other scientific disciplines. These first findings lay the foundation for a further investigation of characteristics from biology education research. Even when the focus of the present paper was on the social certification and dissemination, further studies could investigate these other dimensions more thoroughly to identify specific aspects for biology education research. This would add to other research, which was rather of explorative nature, for example by investigating pre-service teachers drawings of biology classrooms (Subramaniam, 2014).

18.5.2 Categories of Social Certification and Dissemination

Concerning our second research question (RQ_2), the teachers reported a broad range of different categories. Overall, the teachers interestingly reported to see a role for themselves in evaluating the results from biology education research. In line with this, they made references to the topicality and specific nature of the results, which may point to a dilemma for researchers. On the one hand, researchers strive for the best publication in a reputable reviewed journal. On the other hand, these journals may neither be accessible nor understandable for teachers (Eilks, 2018). For researchers, it could be important to consider the different roles of internal and external communication (Hanauska, 2019). That is, researchers communicate with each other via scientific journals (internal), but also with their practice field and the public (external).

Concerning ways for this external communication, the results for the methods of dissemination gave a good insight into several possibilities, including traditional, but also more innovative approaches. While the "classical" dissemination of results in books and journals for practitioners was mentioned regularly, particularly new media outlets and research projects emerged as interesting new avenues for disseminating results. These possibilities may also be more engaging, as research projects could be a possibility to better integrate teachers' views. In this way, projects not only be designed in a deficit-oriented way, but rather in a more dialogue-centred and participation-centred way (Lorke et al., 2024). As pointed out by Lorke et al. (2024), researchers should not only think about the missing knowledge of teachers and strive for the filling of these knowledge gaps (deficit-model), but rather see knowledge exchange as a constructive endeavour by both sides (dialog-model) or include teachers actively in the research process (participation-model).

18.5.3 Connections Between the Categories

The results for the third research question (RQ_3) showed that there were several code co-occurrences between the social certification and dissemination with the other categories, highlighting possible relations between the categories. Particularly, the transfer of research may be bound to the utility of the results, the understanding of what researchers do, and the way organisations such as schools and universities function and interact with each other (Hinton & Fischer, 2008). For example, the insights of researchers into teaching practices at school were mentioned as important for the viability of the research. This may be explainable by the different values adhered to research results (i.e., accuracy and objectivity) in contrast to school practice (i.e., viability and utility), which would also be in line with other studies. For example, a prior study found how the switch of pre-service teachers into early school practice may change the perceived relevance of research results and educational theories (Allen, 2009). Nonetheless, there also may be a need to further reflect the understanding of scientific theories by teachers (Reinisch & Krüger, 2018), because this understanding may affect teachers' expectations (Furman Shaharabani & Yarden, 2019).

For the future of transfer, researchers should keep in mind teachers' perspectives on results from biology education research and how strongly the reception of these results is bound to the utility of the results. This extends prior studies by more strongly taking the social-institutional aspects of biology education research into account (Larkin, 2022). However, these results can only be the foundation, as they only represent qualitative insights into selected aspects. In how far these could lay the ground for a further investigation of a general *nature of biology education research* may be up to further studies.

18.5.4 Limitations

Within our explorative study, we were confronted with some content-related and methodological limitations. First of all, the qualitative nature of the data cannot be generalized to all teachers, but rather represents beliefs of selected teachers in specific interview situations. As discussed, we believe that the results allow interesting insights into the transfer of research results, but need to be replicated and should also be quantified to get a more coherent picture. Such quantitative studies could also include control variables such as the role of teaching experience or personal identity, which were mentioned in other studies (Akerson et al., 2014). Furthermore, the role of the different interviewers may have had an effect on the results, as the interviews likely included more variance, than if the interviews were performed by one interviewer only. Concerning the content-related limitations, there sometimes occurred problematic differentiations between the categories. One example refers to the categories *scientific practices* and *methods and methodological rules*, as it may be unclear, which of these categories included data gathering in the field using surveys or tests. This is also in line with other studies applying the FRA (Reinisch & Fricke, 2022). Nonetheless, also understanding the category of *social certification and dissemination* may be subject to discussion. For example, the FRA includes certification as well as dissemination in one category. Based on our results, it may be possible to further differentiate this into two categories. Furthermore, teachers are currently not included in traditional ways of evaluating research such as peer-review, which is why this may be in contrast to the initial definition of this category. Particularly this limitation leads to the need for a better understanding of how an application of the FRA may be helpful for understanding the characteristics of biology education research.

18.6 Conclusion and Outlook

The present explorative study was able to give insights into in-service teachers' beliefs about the social certification and dissemination of results from biology education research. Further including these views of teachers may be the foundation for a more bidirectional way of communicating research results. Nonetheless, there is only a limited understanding of what the application of the FRA-categories to biology education research really includes and what constitutes the content within the categories, which is why further work in this regard is needed.

A better understanding of the social and institutional context of biology education may help teacher educators to steer teacher education and improve the connection between research and teaching practice (Korthagen, 2016). Finally, it may be fruitful to further theoretically elaborate on the conceptualization of beliefs and their possible differentiation from related constructs. For example, there may be a cognitive knowledge base ("How is biology education research done"), but also a more affective and normative dimension ("What should biology education research do"). Both may be relevant for teacher education (Büssing, 2024). Nonetheless, it also needs to be discussed if there can be a consensus about cognitive-epistemic and social-institutional aspects of biology education research, or if it should be viewed in the heterogeneity it comprises (van Dijk, 2011). A further investigation into a possible consensus may be a requirement for testing the explorative assumptions from this study more quantitatively, as this would make the different aspects more feasible and therefore measurable. This new work may spread further light on bringing results from biology education research into school practice and therefore also prepare future citizens for a changing world.

References

Abd-El-Khalick, F., & Lederman, N. G. (2000). Improving science teachers' conceptions of nature of science: A critical review of the literature. *International Journal of Science Education, 22*(7), 665–701. https://doi.org/10.1080/095006900500044044

Akerson, V. L., Pongsanon, K., Weiland, I. S., & Nargund-Joshi, V. (2014). Developing a professional identity as an elementary teacher of nature of science: A self-study of becoming an elementary teacher. *International Journal of Science Education, 36*(12), 2055–2082. https://doi.org/10.1080/09500693.2014.890763

Allen, J. M. (2009). Valuing practice over theory: How beginning teachers re-orient their practice in the transition from the university to the workplace. *Teaching and Teacher Education, 25*(5), 647–654. https://doi.org/10.1016/j.tate.2008.11.011

Aumann, A., & Weitzel, H. (2024). Exploring a theory-practice gap: An investigation of pre-service biology teachers' enacted TPACK. In K. Korfiatis, M. Grace, & M. Hammann (Eds.), *Shaping the future of biological education research* (pp. 311–323). Springer International Publishing. https://doi.org/10.1007/978-3-031-44792-1_22

Büssing, A. G. (2024). Die Natur der Naturwissenschaftsdidaktiken als Reflexionsmöglichkeit für die Lehrkräftebildung: Ein Vorschlag zur Anwendung des Family Resemblance Approach. In B. Reinisch, D. Krüger, & D. Mahler (Eds.), *Biologiedidaktische nature of science-forschung: zukunftsweisende praxis* (pp. 45–58). Springer Berlin Heidelberg. https://doi.org/10.1007/978-3-662-68409-2_5

Cheung, K. K. C., & Erduran, S. (2022). A systematic review of research on family resemblance approach to nature of science in science education. *Science & Education, Online First Article.* https://doi.org/10.1007/s11191-022-00379-3

Cochran-Smith, M. (2005). Studying Teacher education: What we know and need to know. *Journal of Teacher Education, 56*(4), 301–306. https://doi.org/10.1177/0022487105280116

Eilks, I. (2018). On the role of publications in science education and the question of their impact and evaluation. *Action Research and Innovation in Science Education, 1*(2), 19–22. https://doi.org/10.51724/arise.11

Erduran, S., & Dagher, Z. R. (2014). *Reconceptualizing the nature of science for science education* (Vol. 43). Springer Netherlands. https://doi.org/10.1007/978-94-017-9057-4

Erduran, S., Dagher, Z. R., & McDonald, C. V. (2019). Contributions of the family resemblance approach to nature of science in science education. *Science & Education, 28*(3–5), 311–328. https://doi.org/10.1007/s11191-019-00052-2

Fives, H., & Buehl, M. M. (2012). Spring cleaning for the "messy" construct of teachers' beliefs: What are they? Which have been examined? What can they tell us. In *APA educational psychology handbook* (Vol. 2, pp. 471–499).

Furman Shaharabani, Y., & Yarden, A. (2019). Toward narrowing the theory–practice gap: Characterizing evidence from in-service biology teachers' questions asked during an academic course. *International Journal of STEM Education, 6*(1), 21. https://doi.org/10.1186/s40594-019-0174-3

Göhner, M., & Krell, M. (2020). Qualitative Inhaltsanalyse in naturwissenschaftsdidaktischer Forschung unter Berücksichtigung von Gütekriterien: Ein Review. *Zeitschrift Für Didaktik der Naturwissenschaften.* https://doi.org/10.1007/s40573-020-00111-0

Hanauska, M. (2019). 27. Historical aspects of external science communication. In A. Leßmöllmann, M. Dascal, & T. Gloning (Eds.), *Science communication* (pp. 585–600). De Gruyter. https://doi.org/10.1515/9783110255522-027

Hinton, C., & Fischer, K. W. (2008). Research schools: Grounding research in educational practice. *Mind, Brain, and Education, 2*(4), 157–160. https://doi.org/10.1111/j.1751-228X.2008.00048.x

Irzik, G., & Nola, R. (2011). A family resemblance approach to the nature of science for science education. *Science & Education, 20*(7–8), 591–607. https://doi.org/10.1007/s11191-010-9293-4

Jones, M. G., & Park, S. (2023). Science teacher attitudes and beliefs: Reforming practice. In *Handbook of research on science education* (pp. 1101–1122). Routledge.

Kennedy, M. M. (1997). The connection between research and practice. *Educational Researcher, 26*(7), 4–12. https://doi.org/10.3102/0013189X026007004

Korthagen, F. (2016). Inconvenient truths about teacher learning: Towards professional development 3.0. *Teachers and Teaching, 23*(4), 1–19. https://doi.org/10.1080/13540602.2016.1211523

Korthagen, F., & Kessels, J. P. A. M. (1999). Linking theory and practice: Changing the pedagogy of teacher education. *Educational Researcher, 28*(4), 4–17. https://doi.org/10.3102/0013189X028004004

Larkin, D. B. (2022). Getting to a good place with science instruction: Rethinking an appropriate conception of teaching science. *Science Education, 106*(5), 1054–1070. https://doi.org/10.1002/sce.21742

Lorke, J., Ballard, H. L., & Robinson, L. D. (2024). More complex than expected—Mapping activities and youths' experiences at BioBlitz events to the rosette model of science communication. *Frontiers in Environmental Science, 11*, 1270579. https://doi.org/10.3389/fenvs.2023.1270579

Mayring, P. (2015). *Qualitative Inhaltsanalyse: Grundlagen und Techniken.* Beltz.

Mpate, H. (2023). Biology teachers' implementation of the competence based curriculum in Tanzania: Challenges and opportunities. *Journal of Biological Education*, 1–21. https://doi.org/10.1080/00219266.2023.2282428

Opfer, V. D., & Pedder, D. (2011). Conceptualizing teacher professional learning. *Review of Educational Research, 81*(3), 376–407. https://doi.org/10.3102/0034654311413609

Randi, J., & Corno, L. (2007). Theory into practice: A matter of transfer. *Theory into Practice, 46*(4), 334–342.

Reinisch, B., & Fricke, K. (2022). Broadening a nature of science conceptualization: Using school biology textbooks to differentiate the family resemblance approach. *Science Education*, 1–33. https://doi.org/10.1002/sce.21729

Reinisch, B., & Krüger, D. (2018). Preservice biology teachers' conceptions about the tentative nature of theories and models in biology. *Research in Science Education, 48*(1), 71–103. https://doi.org/10.1007/s11165-016-9559-1

Schreier, M. (2014). Varianten qualitativer Inhaltsanalyse: Ein Wegweiser im Dickicht der Begrifflichkeiten. *Forum: Qualitative Social Research, 15*(1). https://doi.org/10.17169/FQS-15.1.2043

Smith, K., & Sela, O. (2005). Action research as a bridge between pre-service teacher education and in-service professional development for students and teacher educators. *European Journal of Teacher Education, 28*(3), 293–310. https://doi.org/10.1080/02619760500269418

Subramaniam, K. (2014). Student teachers' conceptions of teaching biology. *Journal of Biological Education, 48*(2), 91–97. https://doi.org/10.1080/00219266.2013.837405

van Dijk, E. M. (2011). Portraying real science in science communication. *Science Education, 95*(6), 1086–1100. https://doi.org/10.1002/sce.20458

Van Driel, J. H., Beijaard, D., & Verloop, N. (2001). Professional development and reform in science education: The role of teachers' practical knowledge. *Journal of Research in Science Teaching, 38*(2), 137–158. https://doi.org/10.1002/1098-2736(200102)38:2%3c137::AID-TEA1001%3e3.0.CO;2-U

Winterbottom, M. (2022). Biological education in the age of Covid: Taking a global perspective. *Journal of Biological Education, 56*(2), 121–121.

Yosief, A., Sulieman, M. S., & Biede, T. (2022). Improving the practices of teacher educators through collaborative action research: Challenges and hopes. *Educational Action Research, 32*(2), 1–18. https://doi.org/10.1080/09650792.2022.2066147

Zeyer, A., & Dillon, J. (2019). Science|Environment|Health – The emergence of a new pedagogy of complex living systems. *Disciplinary and Interdisciplinary Science Education Research, 1*(1), 1–10. https://doi.org/10.1186/s43031-019-0014-9

Part IV
Teaching Strategies and Learning Environments

Chapter 19
Is Learning Always Linear, Uniform, and Parallel to Teaching? Rethinking Common Assumptions About Learning

François Lombard, Marie Merminod, and Laura Weiss

19.1 Introduction

Common educational practice often rests on assumptions about how learning occurs and how teaching drives learning. These tacit assumptions have been challenged by a substantial body of research including Astolfi (2008), Good (2008), Hattie and Yates (2013), Millar (2013), and Taber (2019). Indeed "the taken-for-granted assumptions of both a disciplinary tradition and a teaching tradition can make solutions look deceptively self-evident" (Goodyear, 2015, p. 31).

Building on these concerns, this article aims to uncover and discuss three implicit assumptions, drawing on data that reflects students' learning progression—during learning.

As part of a long-term Design-Based Research (DBR) project, we developed methods to assess students' learning progressions at a fine-grained time scale, *during* the learning process. The analysis of our data, combined with insights from other research, in particular Zabel and Gropengiesser's (2011) study 'Learning Progress in Evolution Theory: Climbing a Ladder or Roaming a Landscape?', led us to identify and uncover these assumptions. Goodyear (2015), among numerous others, suggests what the student does is actually more important in determining what is learned, than what the teacher does. Such evidence during progression is required, but students' understanding cannot be directly assessed, and is based on students' productions

F. Lombard (✉)
Université de Genève, TECFA, Geneva, Switzerland
e-mail: Francois.lombard@unige.ch

M. Merminod
Geneva, Switzerland

L. Weiss
Université de Genève, Geneva, Switzerland
e-mail: laura.weiss@unige.ch

O. Morin et al. (eds.), *Challenges of a Changing World in Biology Education*,
Contributions from Biology Education Research,
https://doi.org/10.1007/978-3-032-05346-6_19

(Bennett, 2011), either from instruments specifically produced such as questionnaires, or from informal data (such as traces of productions while learners perform learning tasks, that interfere less with learning as used here) Mottier Lopez (2015). She also reminds us that the validity of any form of assessment must be considered in relation to the decision it helps to take. The aim of these methods was not to compare interventions, as in the classical experimental research paradigm, but rather to produce evidence that could assist teachers in guiding learning while it happens and to improve teaching designs. These methods were applied to data gathered from multiple small cohorts of higher secondary school students (2006–2020) and we argue they provide unique evidence to discuss these assumptions.

We will first present three selected assumptions; illustrate how they manifest in education and refer to a brief selection of research. Then these assumptions will be discussed with data from a sample design, put into perspective with other studies and finally educational implications will be considered.

19.1.1 Assumption A1: "Learning Paths Are Linear"

Considerable time spent by many educators organizing slides or sequencing activities strongly suggest they act on this assumption and that learning could efficiently be considered as linear. Organizing the sequence of lectures, learning activities, experiments and social interactions seems a widely followed approach to lesson planning and to prescribe teaching (Altet, 1993). Curricula were often structured as a program—a sequenced list of knowledge, concepts, and chapters e.g., Plan d'études Collège de Genève (2016)—suggesting they were elaborated under this assumption. However, major initiatives proposed assessment of curricula based on other assumptions. For example, NGSS (2013) is structured around core ideas, scientific practices, and crosscutting concepts.

Also, research on learning progressions (LP) presents hypotheses about the pathways students are likely to follow while learning concepts and practices. The knowledge students are expected to start with is called the lower anchor, and the upper anchor represents what students are expected to understand (Furtak & Tayne, 2019). Because learning is often observed before and after a sequence of activities, many educators assume from LP research that learning happens in a similarly linear fashion. However, research summarized by Taber (2019) suggests students' learning occurs in small steps and is iterative rather than linear.

19.1.2 Assumption A2: "There Is One Learning Path That All or Most Learners Follow"

A2 supposes all learners will learn in the same sequence, overlooking possible heterogeneity. Teachers' strong reliance on lesson planning, sequence organization, and time allocation (Taber, 2019) also implicitly relies on this assumption.

Although LP research may appear to support A2, Duncan and Gotwals (2015, p. 411) caution that LPs, which focus on broad chapter-level progressions, measure performances of students that attained large steps in progressions and average out great numbers, tend to overlook the complexity of what happens in what they name "messy middles". While discussing the strength constraints on the learning process, they suggest it might be relatively loose, and that many different learning paths could be possible leading to the same outcome. Also, these paths may be difficult to define because of apparent inconsistencies in students' reasoning. In these messy middles, students may have some, but not all, of the necessary pieces of knowledge and would thus be able to respond to some assessment tasks but not to others. The authors caution that these patterns of responses would differ across students, creating multiple messy middles. In such cases, defining a path, or paths, between the lower and upper anchors is tricky. The extent to which use of a learning progression can support teachers' ability to interpret and respond to student ideas, is an issue raised by Furtak and Tayne (2019). Indeed, most of the research on student LPs lacks the finer granularity needed to fully understand what happens during learning (Furtak & Tayne, 2019), and checking this assumption is often overlooked. Actually, there is little use by teachers of assessment tools that capture students' current understanding during the learning process (Furtak & Tayne, 2019).

Furthermore, this assumption and large-scale studies that average out individual differences (Taber, 2019, p 130) hinder addressing the needs of diverse learners, cultural or gender differences, a long-overlooked equity issue.

19.1.3 Assumption A3: Teacher-Organized Activity Is a Good Predictor of What Will Be Learnt and Learning Paths

This assumption has led to frustration among teachers when students fail to learn as expected: "But I told them!" Research has long shown that teaching sequences do not necessarily predict students' learning progression. Bachelard (1934) discussed how teachers don't understand why learners don't understand. Schoenfeld (1988) highlighted this in "The Disasters of Well-Taught Mathematics Courses". While teachers often assume the way they organize lessons predicts student learning sequence, teaching is not equivalent to understanding. Teaching can set in place epistemic, material and social structures that guide, but does not determine what students do (cognitively) and learn (Goodyear, 2015). This A3 (and A1) could result from a long-lasting confusion between teacher's necessary stimulation of learning and development of learning, a process shown to happen inside each learner (Piaget, 1937; Vygotsky, 1962).

Assessment of learning progressions must refer to clear learning goals (Hattie & Clarke, 2019; Mottier Lopez, 2015; Taber, 2019). In the currently taught molecular paradigm of biology (Entress, 2022), historian of biology Morange (2021) defines accepted questions as molecular causal explanations of underlying mechanisms. Understanding how a phenomenon was or is causally produced is referred to as causal mechanical explanations by Ariely et al. (2024). However, assessment is often reduced to descriptions. Factors include tradition, avoiding problems with parents or authorities, social pressure for selecting elites (Hadji et al., 2015), time pressure, assessment ease (checking simple terms, structures, categories against textbook answers is easier and cause little dispute). After Zhang et al., (2007, p. 24) we refer to students' efforts to produce not only descriptions of the material world but also theoretical explanations and articulation of hidden mechanisms central to the nature of science as epistemic complexity.

Scientific explaining is a process by which more mechanistic details hidden in simple explanations are revealed, and is referred to as unpacking "black boxes" (Haskel-Ittah, 2023). Explanations in science education will inevitably contain some detailed mechanisms and some black boxes where deeper understanding is not yet required. Haskel-Ittah (2023) distinguishes hidden black boxes (where some steps are not mentioned), and implicit black boxes (used as filler terms—hidden by a magic word), that can be fruitful because they stand as placeholders for gaps in explanations. These gaps can be masked by terms, such as activate, cause, encode, inhibit, produce, process, and represent (often used to indicate a kind of activity in a mechanism without providing any detail about how that activity is carried out) (Craver & Darden, 2013). For them, these placeholders can stand for hypothetical phenomena needing to be investigated, or where the gain in detail makes no difference in understanding the phenomenon for the person receiving the explanation.

While it is impossible to open all these black boxes (since every description of a mechanism bottoms out at some point), leaving them all closed to avoid problems and rushing toward the "correct solution" prevents in-depth understanding (Orange & Orange Ravachol, 2017). However, deciding which questions need to be developed (i.e., which black boxes need to be unpacked) is not easy for learners—according to what each individual already understands. Unpacking is required for explanations whose absence will "change the predicted behaviour of the mechanism in a way that matters for one's purpose" (Craver & Darden, 2013, p. 35).

While to the expert many naïve explanations might seem incoherent, they could well reflect steps in the progression (Astolfi, 2008). These naïve explanations might seem sufficient to students, but they can prevent seeking deeper understanding, as students are often unaware of gaps in their knowledge. This "illusion of explanatory depth" becomes evident only when students are asked to provide a detailed explanation (Haskel-Ittah, 2023). Furthermore, they need to understand what qualifies as a good answer in the specific curricular context.

Hattie and Yates (2013) show that knowledge gaps motivate learning but knowledge chasms demotivate, so black boxes can be placeholders until learners are ready to deepen their understanding (Craver & Darden, 2013). When students will feel the need to unpack explanations depends on their growing understanding, which might vary considerably from student to student and along time. Thus, it is an important responsibility of teachers to help each learner realize when and which explanations need to be unpacked to gain the deeper understanding required in their curriculum. To this effect teachers require fine grained evidence at the time of learning. Here, we will use the lens of explanatory black boxes for analysing students' explanations and learning progressions, and explore the possible messy middles that Duncan and Gotwals (2015) suggest.

To sum up, our RQs are:

RQ1 Does our data support that learning paths are linear?
RQ2 Does our data support that there is one learning path that all or most learners follow?
RQ3 Does our data support that presented material and teacher-organized activity is a good predictor of learning paths?

19.2 Research Design and Method

19.2.1 Data Source: Sample Design on Which Assumptions Were Tested

Data was collected after Lombard et al. (2018) from multiple small cohorts (N = 12–16) of higher secondary school students (2008–2019) during their yearlong learning in a computer-supported collaborative learning (CSCL) environment. Groups of 3–4 students collaborated to produce together one wiki document which was used to collectively prepare for exams (see Fig. 19.1). Since this text focuses on three specific assumptions, a brief overview of the design in which they were tested is provided. In each inquiry cycle, the chapter under study (e.g., immunology) was divided into four subchapters, with each assigned to a different group.

A reference map (RM) was built by the teacher to visualize the knowledge structure of the learning goals in the curriculum. See Fig. 19.2 and 19.4 for examples (Topic: Immunology, mechanisms of allergy, humoral immunity). As suggested by Taber (2019), concept mapping is an effective tool for clarifying the conceptual structure of a subject. The concept maps used in this study were constructed according to Novak and Cañas (2008), where each item is a concept or a causal link between concepts. While concept maps are often used as conceptual tools to support learning (Bereiter, 2002), in this study they were not produced by students, nor provided to students. They were used as a reference for researchers, designers, and teachers to track student progress.

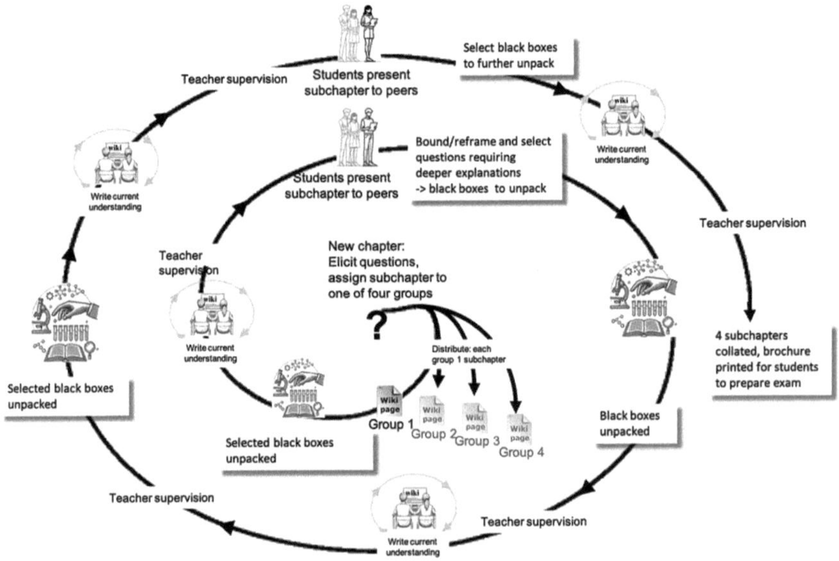

Fig. 19.1 Outline of the design illustrating the iterative activities: students inquire along questions they are responsible for, write in a wiki page their understanding as it progresses, present to peers, refocus on questions that now emerge as important, deepen their inquiry, unboxing black boxes and improving explanations. The teacher supervises and insures they can find answers in experiments or resources. The four wiki pages are collated into a brochure that is critical for all students to prepare the exam.

Throughout their inquiry, students wrote their evolving understanding on a dedicated wiki page (see Fig. 19.1). Twice during the process, each group presented their subchapter explanations to peers. With questions from peers and teacher supervision, questions still needing deeper explanations were identified. Groups then deepened their inquiry and refining their explanations (unboxing black boxes). Teacher supported this process by guiding students to relevant experiments or resources. Ultimately, the four wiki pages were compiled into a comprehensive brochure, typically spanning 65 printed pages, which served as a critical study aid for all students in preparing for exams. The data comprises time-stamped student productions recorded in these wiki pages, consisting of multiple text versions over time. All versions of the documents were automatically recorded—typically 110 versions over a three-week period, encompassing six 90-min lessons; all cohorts amounted to ~3 million words.

19.2.2 Assessment Methods

Process variables were developed to capture the progress of students' explanations all along the learning process and inform teacher guidance. These sequential

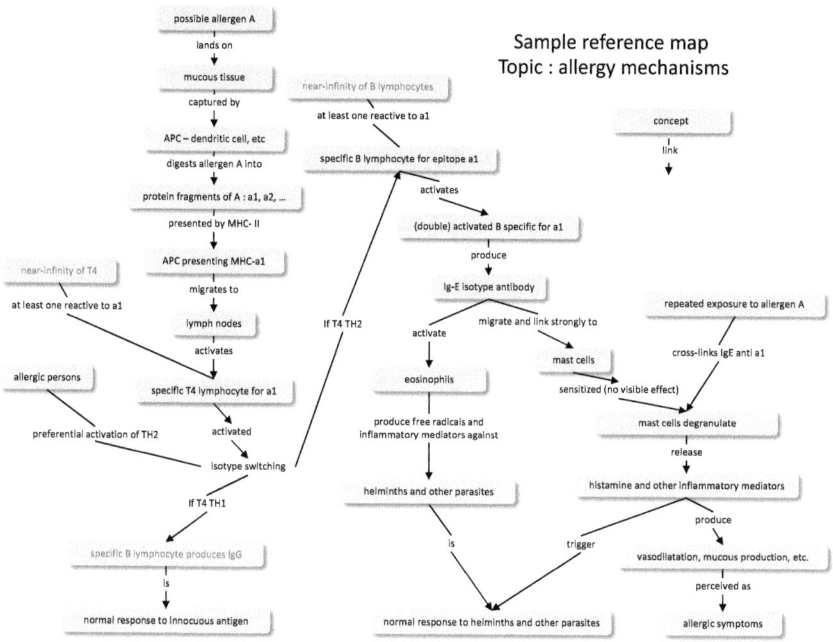

Fig. 19.2 Sample reference map (RM), structuring the explanations as concepts and causal links. Successive versions of student explanations along learning were mapped onto these items to visualize (RMCC) progressions and further analysis (cPI) Light grey boxes are concepts treated in other wiki pages

versions offer unusually fine insight about the students' evolving understanding, during learning progression.

To analyse individual learning paths, we employed the Reference Map Change Coding (RMCC) method (Lombard et al., 2018). This method maps students' successive text productions against the reference map (RM) for each subchapter, to collect evidence about the progression of understanding. See Fig. 19.2, 19.3 and 19.5 for examples.

Each version was examined and, in the significant versions (average: 12) where items of the RM changed (not only language corrections, etc.) each semantic unit was analysed for the presence of concepts or causal links and mapped onto the RM. The comparison of these coded RMs—during each inquiry cycle and across cohorts—offered fine-grained evidence of students' progress (see Fig. 19.4 and 19.5 for examples). Successive, progressively more complete maps reflected students' understanding. See Fig. 19.3 for a visualisation of selected stages in progression. Comparing these maps across cohorts allowed analysis of variability in progression. See Fig. 19.4 for a comparison of different cohorts at equivalent stage in progression. All coding was first done by two researchers not involved in the teaching until >80% intercoder agreement.

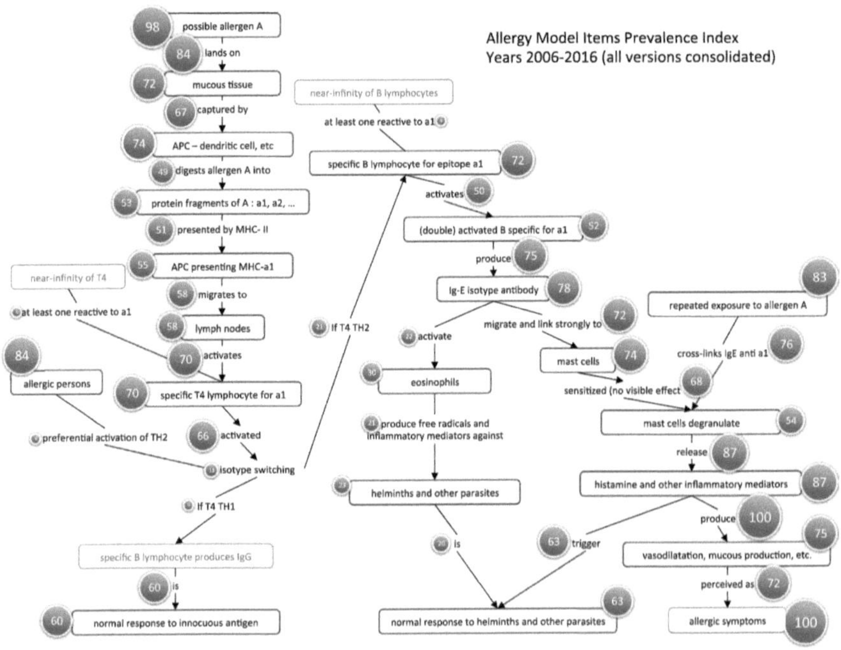

Fig. 19.3 Sample cPI results: The mean occurrence of each item in student productions across all version and all cohorts is expressed as percentage and size of the badge next to each item. Small values are slow spots

To identify patterns and trends in students' attainment of learning goals across cohorts—and to explore learning heterogeneity—we used an extension of RMCC, the consolidated Prevalence Index (cPI) (Lombard et al., 2018). The Prevalence Index (PI) represents the mean occurrence of specific concepts or items across significant versions of the students' productions. For instance, if the item "B lymphocyte" appears in six out of ten versions, its PI is 60%. The cPI calculates the average PI across cohorts. Items that appear consistently early across cohorts have a high cPI, while those appearing later—or not at all—have a low cPI, (termed here "slow spots") (see Fig. 19.3 for the allergy topic example).

The RMCC method was used to test assumptions A1 and A2. To test assumption A3, we used RMCC and cPI.

To further examine if progression of students' productions from descriptions to explanations paralleled the learning objectives, we assessed their Epistemic Complexity (EC). Students' productions were rated on a four-point scale adapted from Zhang et al. (2007): unelaborated facts (UF), elaborated facts (EF), unelaborated explanations (UE), and elaborated explanations (EE) as described in Lombard and Schneider (2013).

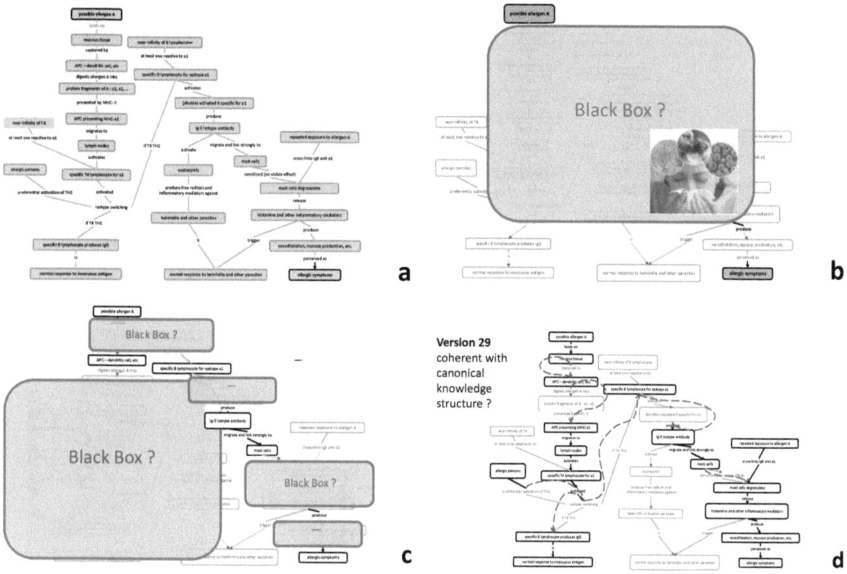

Fig. 19.4 Sample student productions during learning mapped onto the reference map of learning goals; **a** and **b** version 2, **c** version 4, **d** version 29. Time sequence suggests a mosaic-like rather than linear progression. Intermediate understanding often appears incoherent in relation to canonical knowledge structure and is insufficient for fully explaining natural phenomena required. **d** near final version appears closer to canonical knowledge structure

19.3 Results

19.3.1 Does Our Data Support or Refute Assumption A1: "Learning Paths Are Linear" (RQ1)?

Analysis by RMCC method revealed students' understanding that did not follow a linear, canonical knowledge structure, as typically found in textbooks or teaching. Instead, learning progressed in a mosaic pattern (A brief example in Fig. 19.4). During progression, students displayed understanding of some items of the learning goals but without all the expected logical connections (Fig. 19.4c). Over time, explanations deepened, and connections were progressively made.

To sum up, our results did not show a linear progression aligned with the planned sequence of canonical knowledge in textbooks. Nevertheless, many students, despite following different paths, eventually achieved the learning goals ("full mosaic"). Patterns in progressions were revealed that could inform guidance and design improvement. Further examination of these results revealed where in the RM and when students reached the deep explanation expected.

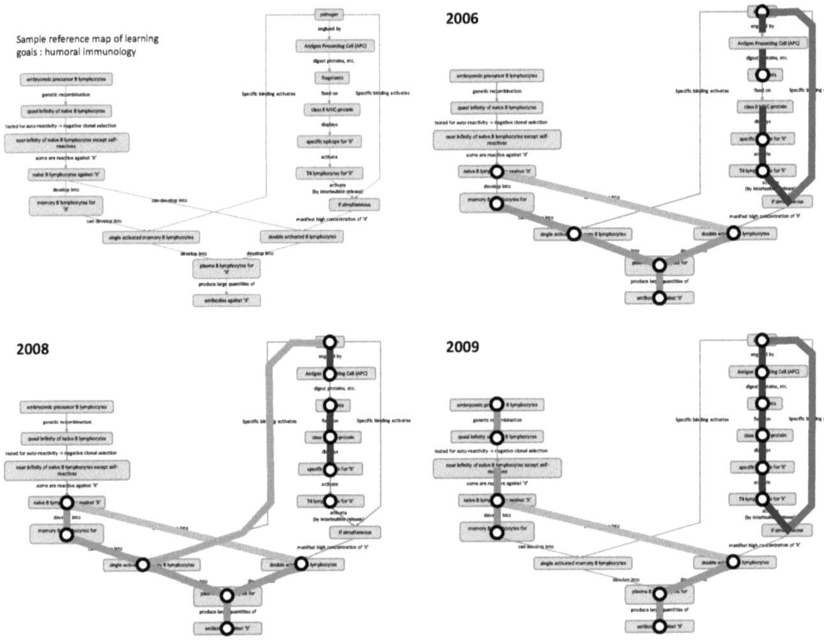

Fig. 19.5 Reference map of learning goals for humoral immunology (top left). Selected maps representing student understanding from three representative cohorts (2006, 2008, 2009) at equivalent stages show quite different incomplete understanding. Arbitrary colors were chosen to help visualize equivalent sections of explanations

19.3.2 Does Our Data Support or Refute Assumption A2: "There Is One Learning Path That All or Most Learners Follow" (RQ2)?

Comparing learning progress across different cohorts using the RMCC method showed considerable variation in the sequence of attainment of learning goals (Fig. 19.5).

Closer examination showed some items of the learning goals (mainly in the centre of the learning goals as illustrated in Fig. 19.5) were attained in quite variable sequences, suggesting multiple messy middles. Despite this variability, most students finally reached the learning goals. In this respect our findings challenge assumption A2.

However further analysis with cPI revealed that across cohorts some learning goals were reached on average earlier (high cPI)—mainly on both ends of the learning goals as illustrated in Fig. 19.4c, while others, slow spots, were reached later or inconsistently.

Together these findings suggest that assuming a single learning path might overlook important influences (heterogeneity of learners, internal and external constraints) other than teacher organization, and miss crucial guidance opportunities.

19.3.3 Does Our Data Support or Refute Assumption A3: "Teacher-Organized Activity Is a Good Predictor of What Will Be Learnt and Learning Paths" (RQ3)?

The RMCC method produced evidence of students' progressive understanding that often did not align with teacher-organized activities and canonical structure (Fig. 19.4). Interestingly, apparent inconsistencies between students' reasoning and expert reasoning persisted throughout most of the learning process, until students reached the learning goals ("full mosaic"), at which point they were able to solve complex problems in exams.

Further analysis using cPI (Fig. 19.3) revealed that some of the most central explanations (e.g., "How do T4 lymphocytes stimulate antigen-reactive B lymphocytes to produce antigens") often had low cPI (i.e., they appeared late on average across cohorts), indicating that students needed more time to master these learning goals. See for example in Fig. 19.3, items "isotype switching" (cPi 11) and "if T4 TH2" (cPi 21).

Analysis for Epistemic Complexity (EC) showed growth mostly at the end of the learning process; students were able to express the most complex explanations later than simpler explanations and descriptions (Fig. 19.6).

Taken together, these RMCC, cPI, and EC results suggest that understanding did not clearly follow the textbook structure or the sequence of teacher-led activities, but also that it seemed under other influences that will be discussed in Sect. 19.4.

19.4 Discussion

19.4.1 Discussing Assumption A1: "Learning Paths Are Linear" (RQ1)

When analysed using the RMCC method, our data revealed that learning followed an irregular, mosaic-like pattern, challenging assumption A1. Instead of a linear progression, students' progress appeared as "variable messy middles" as Duncan and Gotwals (2015) suggested.

During most of the progression, they could answer some assessments but not others that were hidden in large, vague explanations. These intermediate explanations can be seen as black boxes: temporary understandings that need to be unpacked to form accurate predictions and explanations of natural phenomena (Haskel-Ittah,

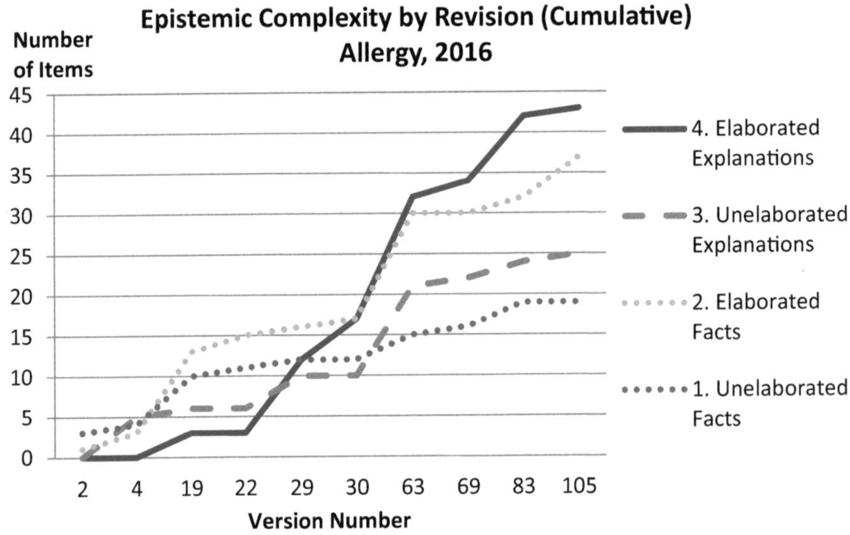

Fig. 19.6 Epistemic complexity (EC) vs. inquiry progression (version number of student productions) for year 2016. Count of semantic units coded in each category. Elaborated explanations in student texts increases mostly toward the end of inquiry cycle as can be seen in level 4 (blue solid line). Grey dotted or dashed lines visualize levels 1–3

2023). Figure 19.4 illustrates some of this process: as students progress, a naïve explanation (Fig. 19.4b) gets unpacked, revealing more black boxes (Fig. 19.4c). Smaller black boxes indicate minor gaps in understanding, while a larger black box (bottom left) conceals a critical mechanism required by the curriculum (clonal selection). Students need some prodding to realize there is a knowledge gap. For example, asking how the explanation "allergens cause allergic reactions" accounts for the fact pollen triggers sneezing in some individuals but not in others can lead students to further unpack their understanding, revealing more hidden black boxes that will have to be unpacked (Fig. 19.4c). Teachers require evidence, such as that provided by RMCC, to identify when students need to be stimulated to question their explanations, and helped to unpack these black boxes.

The increase in epistemic complexity (EC) towards the end (Fig. 19.6) also suggests that deeper black boxes (possibly the most difficult explanations) cannot be unpacked until others (easier explanations and descriptions) have been unpacked. This suggests that in these areas developmental constraints and affordances shaped learning, allowing for multiple paths to reach the same outcome, as Duncan and Gotwals (2015) proposed. Although the learning paths did not align with canonical sequences, many students still reached the learning goals through divergent pathways. Some learning goals appear loosely constrained at this fine-grained level, explaining their observed variability, while others could be more constrained, accounting for the less variable items (high and low cPI).

Overall, our results do not support a simple linear progression. Instead, they suggest a combination of messy middles and developmental constraints, leading to a mosaic progression: early for some items and delayed unpacking of others.

19.4.2 *Discussing Assumption A2: "There Is One Learning Path That All or Most Learners Follow." (RQ2)*

Learning paths varied considerably across cohorts, with no single path evident, as shown in Fig. 19.5 challenging assumption A2. Similar to assumption A1, our findings align with Duncan and Gotwals' (2015) multiple messy middles. This also suggests that student heterogeneity was widespread rather than exceptions in homogeneous classes. The cPI analysis also revealed areas with high variability but some "slow spots", indicating that learning paths for some goals are influenced not only by teacher interventions but also by internal or external constraints (see Sect. 19.5).

While our data involved groups of 3–4 students rather than individuals, the variability would likely increase with individual-level data, further questioning this assumption. Differences in teacher interventions across cohorts could also contribute to this variability, highlighting a common yet overlooked issue when comparing educational interventions. Taken together, our results suggest that assuming a single learning path may overlook the diversity of learners, and is in line with Taber's (2019, p. 130) concern that variation within each condition is often averaged out many classical large randomized studies (2019, p. 130). This raises concerns about equity, since considering students as normally homogeneous in their progression could lead to inequities in the support they receive in their diversity (Hadji et al., 2015). Fine-grained evidence, like that provided by RMCC, can inform feedback to ensure all students receive appropriate guidance.

19.4.3 *Discussing Assumption 3: Teacher-Organized Activity Is a Good Predictor of What Will Be Learnt and Learning Paths (RQ3)*

RMCC analysis showed that students' learning paths did not consistently align with teacher-led sequences or textbook knowledge structure. While some patterns, such as consistent slow spots, did emerge, the heterogeneity in learning progression could not easily be attributed solely to teacher activity. This may also reflect constraints within the knowledge structure such as conceptual dependencies (Sweller, 2010), suggesting that complex explanations are delayed until students develop partial understandings, allowing them to unpack more difficult black boxes. In line with Duncan and Gotwals (2015), our data suggests that for some learning goals, it may be that the constraints are strong and there may be only a few paths, or perhaps just one for learning. The slow spots identified in our data likely reflect areas of learning that are constrained by such internal or external influences.

On the other hand, the heterogeneity we found in other areas could therefore be explained by some areas of the RM that are loosely constrained and mostly influenced by a mix of internal (e.g., cognitive limitations) and external factors (e.g., curriculum design, social and cultural influences).

To sum up, our results imply teaching and learning did not follow a co-linear path, and relying on A3 is problematic. Teachers need real-time evidence of students' progress to offer effective feedback (Brookhart, 2014)—for example in the form of raising awareness of gaps in understanding, to guide unpacking (Haskel-Ittah, 2023).

19.5 Conclusion

While the methods used in this study produced results that mostly do not support assumptions A1, A2, and A3 within this specific learning design, they lack the statistical power to definitively disprove these assumptions. It is important to note that this was not the intended purpose of the study. As Mottier Lopez (2015) pointed out, the relevance of assessment methods must be considered in relation to the decisions they inform. The methods applied here were designed to guide teachers and improve learning designs, and they have proven valuable in that context. Using these methods to provide evidence for decisions related to experimental research would first require further validation.

Answering RQ1, RQ2, RQ3 our findings align with Taber (2019), describing student learning as incremental, interpretive, and iterative. Furthermore, these methods confirmed assessing during learning can reveal important evidence about learning processes (including identifying slow spots where learning requires special attention) and issues (the influence of messy middles and constraints on explanation refinement), to uncover the real heterogeneity and guide students.

Our results contribute to the body of research emphasizing the importance of evaluating the impact of learning including Brookhart (2014), Hattie and Clarke (2019), Mottier Lopez (2015), Ruiz-Primo and Furtak (2007), Taber (2013). Such methods could also help discuss the extent to which use of a learning progression can support teachers' ability to interpret and respond to student ideas, an issue raised by Furtak and Tayne (2019). Furthermore, they specifically highlight the role and means of assessment in science education, as underscored by Millar (2013), and in particular to the need for fine-grained assessments of explanations during the learning process. Our results also support Altet's (1993) argument that collecting evidence of student understanding during learning, coupled with adaptive teacher guidance, is more effective than focusing mainly on the timing and sequencing of activities. This challenges frequent traditional practices in vocational training that focus on rigidly sequencing activities and lectures. In a wide spectrum of pedagogies, evidence about effectiveness of learning during progression helps identify what aspects of the learning environment are less effective. This could lead to reconsider the way teaching and learning environments are prepared, independently of pedagogical strategies.

Our results also align with Hattie and Clarke's (2019) emphasis on the value of feedback during learning. They also show that feedback given at the same time as grades—although it is required to justify those grades—does not improve learning. Instead, formative feedback is most effective when provided during learning (Black & Wiliam, 2009; Brookhart, 2014; Furtak & Tayne, 2019; Ruiz-Primo & Furtak, 2007; Taber, 2019). Interestingly Hattie and Clarke (2019) show that effective teacher feedback involves both the giving and receiving (by teachers and/or by students) of relevant insights into student understanding, misconceptions, and errors during learning.

Fine-grained evidence is essential for teachers to raise awareness of knowledge gaps (Hattie & Clarke, 2019) and in order to focus learning effort on black boxes when students are ready to unpack them (Haskel-Ittah, 2023). Even when considering a wider spectrum of pedagogies, she argues for the importance of raising awareness by students (and possibly by educators) of black boxes where some steps are not mentioned (hidden) or black boxes used as filler terms—hidden by a magic word (implicit). When lectures are the chosen method of teaching, using black boxes in explanations is also inescapable, however explicitly mentioning those that will not be unpacked (Haskel-Ittah, 2023) can help students later deepen their explanations, whereas implicit black boxes could hinder further understanding.

We suggest that the evidence discussed here could be used to analyse students' explanations, models, and arguments, as well as to track learning progressions in most educational designs, not only the ones illustrated in our example. While some teachers might argue that they lack the time to analyse student productions in such detail, simplified versions of criteria such as EC could be used to produce feedback guiding towards explanations rather than descriptions only. Establishing a reference map could help build consensus among teachers on the learning objectives. Ensuring that learners understand what constitutes a good answer throughout the learning process is crucial for effective learning (Hattie & Clarke, 2019). They also show that simply stating learning objectives at the outset may not be sufficient. Indeed, activities that enhance students' awareness of what the learning goals really are can yield strong learning gains (Kapur & Bielaczyc, 2012). The criteria in these methods offer a theoretical foundation (Lombard et al., 2018) to justify specific biology learning goals.

While immediate feedback during learning can efficiently convey what a good explanation should include, this requires evidence of students' current understanding and prior knowledge, which can both be derived from traces of students' productions. The increasing use of computer-supported learning platforms offers new opportunities to collect real-time data on learning, a development that was rare at the start of this project.

Fine-grained assessment of learning—revealing slow spots, variable areas in our case—could be used to shed light on the issue raised by Duncan and Gotwals (2015) regarding the developmental constraints and affordances that could shape learning paths. Such methods could provide evidence to discuss the possible causes of the slow spots they revealed (i) external: knowledge field constraints such as conceptual dependencies, element interactivity (Sweller, 2010), or shortcomings in the teaching design; (ii) internal: constraints such as cognitive construals (Betz et al., 2019). Investigating cognitive load theory (Sweller, 2010) and the influence of cognitive construals—such as teleological, essentialist, and anthropic thinking (Betz et al., 2019) on persistent misconceptions may provide deeper insights into the role of constraints and affordances.

Further research analysing the data from this project could help uncover these potential influences. More research is also needed to justify what qualifies as a good answer in biology.

Acknowlegments The authors thank Marie Sudries for stimulating comments and language improvement. The data processing was done by Marie Merminod then a researcher at Université de Genève.

References

Altet, M. (1993). Préparation et planification. In J. Houssaye (Ed.), *La Pédagogie, une encyclopédie pour aujourd'hui* (pp. 78–88). ESF.

Ariely, M., Nazaretsky, T., & Alexandron, G. (2024). Causal-mechanical explanations in biology: Applying automated assessment for personalized learning in the science classroom. *Journal of Research in Science Teaching, n/a*(n/a). https://doi.org/10.1002/tea.21929

Astolfi, J.-P. (2008). *La saveur des savoirs. Disciplines et plaisir d'apprendre.* ESF.

Bachelard, G. (1934). *La formation de l'esprit scientifique.* Vrin.

Bennett, R. E. (2011). Formative assessment: A critical review. *Assessment in Education: Principles, Policy & Practice, 18*(1), 5–25. https://doi.org/10.1080/0969594X.2010.513678

Bereiter, C. (2002). *Education and mind in the knowledge age* (Second). Lawrence Erlbaum Associates.

Betz, N., Leffers, J. S., Thor, E. E. D., Fux, M., de Nesnera, K., Tanner, K. D., & Coley, J. D. (2019). Cognitive construal-consistent instructor language in the undergraduate biology classroom. *CBE—Life Sciences Education, 18*(4), ar63. https://doi.org/10.1187/cbe.19-04-0076

Black, P., & Wiliam, D. (2009). Developing the theory of formative assessment. *Educational Assessment, Evaluation and Accountability (Formerly: Journal of Personnel Evaluation in Education), 21*(1), 5–31.

Brookhart, S. (2014). *The essence of formative assessment* [Presentation].

Craver, C. F., & Darden, L. (2013). *In search of mechanisms: Discoveries across the life sciences.* University of Chicago Press.

Duncan, R. G., & Gotwals, A. W. (2015). A tale of two progressions: On the benefits of careful comparisons. *Science Education, 99*(3), 410–416. https://doi.org/10.1002/sce.21167

Entress, C. (2022). The disappearance of natural history, fieldwork, and live organism study from American biology teacher education. *Science & Education.* https://doi.org/10.1007/s11191-022-00351-1

Furtak, E. M., & Tayne, K. (2019). *Affordances and constraints of learning progression designs in supporting formative assessment* (pp. 241–256). https://doi.org/10.1007/978-3-030-17219-0_15

Good, T. L. (2008, April). *Forty years of research on teaching 1968–2008: What do we know that we didn't know then?*

Goodyear, P. (2015). Teaching as design. *HERDSA Review of Higher Education, 2*, 27–50.

Hadji, C., Bentolila, A., Meirieu, P., & Raulin, D. (2015). *L'évaluation à l'école: Pour la réussite de tous les élèves*. Nathan.

Haskel-Ittah, M. (2023). Explanatory black boxes and mechanistic reasoning. *Journal of Research in Science Teaching, 60*(4), 915–933. https://doi.org/10.1002/tea.21817

Hattie, J., & Clarke, S. (2019). *Visible learning: Feedback*. Routledge.

Hattie, J., & Yates, G. (2013). *Visible learning and the science of how we learn* (1. publ.). Routledge.

Kapur, M., & Bielaczyc, K. (2012). Designing for productive failure. *Journal of the Learning Sciences, 21*(1), 45–83. https://doi.org/10.1080/105084406.2011.59

Lombard, F., Merminod, M., Widmer, V., & Schneider, D. K. (2018). A method to reveal fine-grained and diverse conceptual progressions during learning. *Journal of Biological Education, 52*(1), 101–112. https://doi.org/10.1080/00219266.2017.1405534

Lombard, F., & Schneider, D. K. (2013). Good student questions in inquiry learning. *Journal of Biological Education, 47*(3), 166–174. https://doi.org/10.1080/00219266.2013.821749

Millar, R. (2013). Improving science education: Why assessment matters. In D. Corrigan, R. Gunstone, & A. Jones (Eds.), *Valuing assessment in science education: Pedagogy, curriculum, policy* (pp. 55–68). Springer.

Morange, M. (2021). *A history of biology* (T. L. Fagan & J. Muise, Trans.). Princeton University Press.

Mottier Lopez, L. (2015). *Evaluations formative et certificative des apprentissages: Enjeux pour l'enseignement*. De Boeck.

Next Generation Science Standards. (2013). *Next generation science standards: For states, by states* (p. 18290). National Academies Press. https://doi.org/10.17226/18290

Novak, J. D., & Cañas, A. J. (2008). The theory underlying concept maps and how to construct and use them. *Florida Institute for Human and Machine Cognition.*

Orange, C., & Orange Ravachol, D. (2017). Problématisations scientifiques fonctionnalistes et historiques en éducation relative à l'environnement et au développement durable: Le cas de l'évolution climatique. In P. Roy, A. Pache, & B. Gremaud (Eds.), *Formation et pratiques d'enseignement en questions* (pp. 21–37).

Piaget, J. (1937). *La naissance de l'intelligence chez l'enfant*. Delachaux & Niestlé.

Plan d'études Collège de Genève, 59 (2016).

Ruiz-Primo, M. A., & Furtak, E. M. (2007). Exploring teachers' informal formative assessment practices and students' understanding in the context of scientific inquiry. *Journal of Research in Science Teaching, 44*(1), 57–84. https://doi.org/10.1002/tea.20163

Schoenfeld, A. H. (1988). When good teaching leads to bad results: The disasters of 'well-taught' mathematics courses. *Educational Psychologist, 23*(2), 145–166. https://doi.org/10.1207/s15326985ep2302_5

Sweller, J. (2010). Element interactivity and intrinsic, extraneous, and germane cognitive load. *Educational Psychology Review, 22*(2), 123–138. https://doi.org/10.1007/s10648-010-9128-5

Taber, K. (2013). *Classroom-based research and evidence-based practice: An introduction* (2nd ed.). SAGE.

Taber, K. (2019). *MasterClass in science education: Transforming teaching and learning.*

Vygotsky, L. (1962). *Thought and language* (E. Hanfmann & G. Vaka, Trans.). MIT Press.

Zabel, J., & Gropengiesser, H. (2011). Learning progress in evolution theory: Climbing a ladder or roaming a landscape? *Journal of Biological Education, 45*(3), 143–149.

Zhang, J., Scardamalia, M., Lamon, M., Messina, R., & Reeve, R. (2007). Socio-cognitive dynamics of knowledge building in the work of 9-and 10-year-olds. *Educational Technology Research and Development, 55*(2), 117–145.

Chapter 20
Learning About Nutrition Through Play

Martha Georgiou

20.1 Introduction

It is well known that a healthy diet is extremely important for human health and development. It is also well known that the nutritional needs of children and adolescents are specific (Brown, 2016), and in addition to their healthy development, nutrition also affects students' school performance (Worobey & Worobey, 1999). However, according to World Obesity Federation data in 2022, over 390 million children and adolescents aged 5–19 years were overweight, including 37 million children under the age of 5, and 160 million who were living with obesity (Okunogbe et al., 2022). If we consider that young people of these ages often do not eat healthily, we can obviously understand how these findings can be derived. According to Pirouznia (2001), adolescents in particular make food choices that are strongly influenced by the habits of their peers, and choices may sometimes be conscious while other times are not. The World Health Organization (WHO, 2018) states that the choice of junk food, which usually has high caloric value and low nutritional value, is among the causes of eating disorders, but such food is highly popular among adolescents during their outings. Furthermore, eating disorders are also caused by the attempts of adolescent girls, in particular, to achieve a certain silhouette through exhaustive diets, due to similar social norms. In any case, wherever these food choices come from, they can lead to health-threatening conditions and diseases. It is therefore obvious that supporting adolescents' behavior around healthy eating is extremely important for their present and future health.

Nutrition knowledge plays a crucial role in shaping adolescents' attitudes toward healthy eating. However, research in many different countries, for example, the US, Poland, Turkey, and Iran, have revealed deficiencies in nutritional knowledge among

M. Georgiou (✉)
Department of Biology, National and Kapodistrian University of Athens, Athens, Greece
e-mail: martgeor@biol.uoa.gr

O. Morin et al. (eds.), *Challenges of a Changing World in Biology Education*,
Contributions from Biology Education Research,
https://doi.org/10.1007/978-3-032-05346-6_20

adolescents (Barzegari et al., 2011; Kostanjevec et al., 2012; Mizia et al., 2021; Pirouznia, 2001; Saribay & Kirbas, 2019). Similarly, a previous study (Georgiou & Moshogianni, 2024) revealed that Greek adolescent students' knowledge about nutrition is limited and fragmented concerning all nutrition groups: Students mainly seek information on the internet and from their parents, and much less often from school, where appropriate educational programs are only rarely offered, and even then, do not provide solid knowledge since the curricula include few relevant concepts. Furthermore, Greek students prefer unhealthy foods to fruit and vegetables, with the result that Greece has very high rates of childhood obesity (Bebetsos et al., 2015). At the same time, the Greek curricula are poor in nutritional concepts, where their teaching time is limited to only four hours in compulsory education, and those hours are part of the Home Economics course rather than Science courses.

For all the above reasons, we decided to design educational material that can easily be used in the classroom. We created an educational game, as the integration of games in the classroom environment helps to enhance learning motivation and active participation by students, while at the same time enhancing students' social and problem-solving skills (Misra et al., 2022). In addition, game-based learning has been found to be effective in teaching nutrition knowledge (Viggiano et al., 2015; Waters et al., 2011) and to influence adolescents' eating habits and behaviors (Baranowski et al., 2003). We focused on card games because they are entertaining, maintain direct interpersonal interaction both between teachers-learners and among learners (Liu & Chen, 2013), and are fairly easy and inexpensive to create. Thus, the research question was: "Is a card game effective in promoting adolescents' knowledge about nutrition and, by extension, their attitudes towards it?" The design of the educational game on nutrition and the results of its pilot implementation are described below.

20.2 Theoretical Background

Several researchers hold the view that nutritional knowledge is positively correlated with adolescents' eating attitudes (Barzegari et al., 2011; Mizia et al., 2021; Skinner & Woodburn, 1983), although others believe that this correlation is not clear enough (Contento et al., 2002). However, in their recent research, Mizia et al. (2021) argue that while the correlation of knowledge of nutrition concepts with healthy eating habits is positive, the correlation of knowledge with unhealthy habits is negative. In other words, if adolescents had sufficient knowledge about nutrition, they could probably make better choices about the foods they consume, without this implying that knowledge can always prevent unhealthy choices: Obviously, the phenomenon is multifactorial.

Game-based learning (GBL) has been widely recognized as an effective pedagogical approach for enhancing engagement, motivation, and knowledge retention in students (Misra et al., 2022). Among the various types of educational games, card games are particularly valuable due to their simplicity, accessibility, and ability to foster both individual and collaborative learning experiences. Research has shown that card games encourage active participation, peer interaction, and cognitive processing, making them an ideal tool for reinforcing conceptual understanding in educational settings (Liu & Chen, 2013).

Additionally, card games align well with the principles of social learning theory (Bandura, 1977), which emphasizes learning through observation, imitation, and interaction with others. When students play educational card games, they engage in discussions, negotiate meanings, and refine their knowledge through structured gameplay, which supports deeper learning and critical thinking (Gee, 2003). Furthermore, the interactive nature of card games allows for immediate feedback and iterative learning, helping students to reinforce correct information and adjust misconceptions in real time.

Another advantage of card-based educational games is their ability to integrate knowledge with real-life decision-making scenarios in a socio-scientific framework (Georgiou & Mavrikaki, 2013; Georgiou et al., 2020; Sá-Pinto et al., 2022). Research indicates that learning is more effective when students can apply new information in meaningful contexts (Georgiou, 2024; Mayer, 2005). Card games provide structured yet flexible opportunities for students to experiment with different strategies, make informed choices, and develop problem-solving skills in a risk-free environment. Moreover, studies have found that educational card games can positively influence students' attitudes and behaviors, particularly in subjects related to health education, such as nutrition and well-being (Amaro et al., 2006; Blakely et al., 2009).

The Theory of Planned Behavior (Ajzen, 1991) provides a useful framework for understanding how the game may influence students' dietary decisions. This theory posits that behavior change is driven by attitudes, subjective norms, and perceived behavioral control. By involving students in activities that require them to categorize foods and make healthier choices, the game can positively influence their attitudes towards nutrition and increase their awareness of the impact of their dietary habits.

Adolescents who are aware of the health consequences of poor nutrition are more likely to develop intrinsic motivation to adopt balanced eating habits. More precisely, research indicates that structured assessments focusing on meal frequency, water consumption, macronutrient diversity, and portion sizes provide valuable insights into students' dietary awareness and their ability to make balanced choices. These categories align with widely accepted frameworks in nutrition education, which emphasize the importance of practical knowledge application for long-term behavior change (Contento, 2011; Worsley, 2002). Assessments based on these criteria allow educators to identify specific gaps in students' knowledge and tailor interventions accordingly. However, knowledge alone is not always sufficient to change behavior; attitudes toward food are also influenced by social, cultural, and environmental factors. Educational interventions that integrate both knowledge dissemination and attitude formation, such as game-based learning, can be particularly effective in fostering long-term healthy eating behaviors (Hwang et al., 2012).

By leveraging these pedagogical benefits, card games serve as an engaging and practical tool for facilitating learning across diverse educational contexts. Their adaptability and ease of implementation make them an attractive option for educators seeking innovative methods to enhance student participation and comprehension.

20.3 Methodology

20.3.1 Constructing the Educational Game

In order to develop the educational game "Food for Good" (FG^2), we relied on previous research we had conducted to identify the strengths and weaknesses of students' knowledge about nutrition. Specifically, our study (Georgiou & Moshogianni, 2024) aimed to survey Greek adolescents' understanding of nutrition concepts, and the results revealed significant gaps in their knowledge. These findings directly informed our game design, ensuring that key misconceptions were addressed, and that all essential nutrition groups were included in the gameplay.

The survey of 249 Greek adolescent students was conducted using the GeSNK questionnaire (Calella et al., 2017). The results revealed that there are weaknesses in students' knowledge about all nutrition groups, without exception. For example, our research found that only 57.3% of students answered nutrition-related questions correctly, highlighting deficiencies in their understanding of key dietary components such as carbohydrates, proteins, and fats. Additionally, misconceptions about the health impact of certain foods were prevalent. A more specific example from the findings was the ignorance of 52.7% of students about omega-3 and omega-6. In fact, they did not know at all that these were fatty acids, despite the extensive marketing of these ingredients in foods. The survey also showed that there are misconceptions even about the connection between fitness/body function and nutrition. This important finding indicated that in the construction of the game, we should introduce data and information relating to these topics. Thus, we decided that it was important not to exclude any of the macronutrient groups from the educational game we were planning. On the contrary, we aimed to ensure that students would encounter all nutritional groups and their nutritional value. Furthermore, our research also showed that students often received nutrition information from unreliable sources such as the internet and peers, rather than from structured educational programs. This further reinforced our decision to create an engaging, scientifically accurate tool to support formal education. Given these goals, we designed FG^2 to reinforce accurate nutritional knowledge and encourage healthier decision making through interactive learning.

In planning the game, we considered it important that this game should be easy to play in the classroom so that it could function as a supplementary educational material, since the school curriculum is very poor in nutritional concepts. Finally, we decided to give a character of familiarity and everyday life to the scientific content by choosing to include foods that belong to the usual daily diet of Greek adolescents in order to attract their interest not only through the alternative educational approach offered by a game, but also through its content.

Thus, 40 cards were designed. The front side of the card has information about the qualitative composition of a food and the correlation between this food and human health, as well as similar suggestions. On the reverse side there are four pictures, one

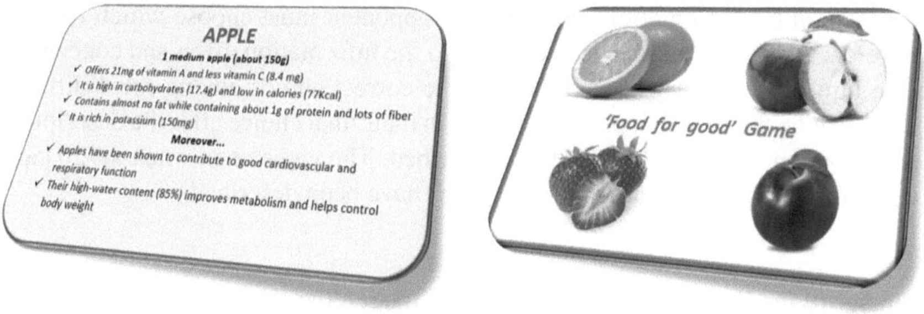

Fig. 20.1 "Food for Good" (FG^2) educational nutrition game card

of which matches the description on the front side. Thus, the cards per quartet had the same back side and different front sides (Fig. 20.1).

The rationale for designing the quads was to associate "similar" foods in the minds of students. This resulted in the following groups: fruits, dairy, protein, "drinkables", snacks, vegetables, and processed foods. We felt it was very important to include processed foods in the teaching content: If students are only exposed to categories of unprocessed foods, while they are likely to gain relevant information and knowledge about those, we may not ensure they will be able to make the necessary combinations in the case of processed foods, which is quite essential. Moreover, they often come into contact with processed foods in their daily life, so including them in the game could, on the one hand, increase the students' interest and, on the other hand, could provide useful information and knowledge about real-life cases of food combining.

20.3.2 *Rules of the Game and Play Procedure*

To ensure maximum participation and engagement of students, specific rules were established (Fig. 20.1). In addition, the rules were designed to help students/players engage with the content of the game and gain as much nutritional information as possible.

General instructions

This game can be played either with two players, or with two teams of two players each. One extra player also needs to take the role of referee. All players will need to go through the referee position in turn, so the game should be played for as many rounds as the number of the total players. The game has two parts, which are described below:

Part 1

Players have ten random cards in their hands out of a total of forty in the game. After rolling a die to determine who will play first, they begin to take turns, and their goal is to take their opponents' card. To do this, one player describes the food on the card and based on the information provided, the opponent must choose which food the card refers to. They need to listen carefully to the information given and concentrate on the information that will lead them to the correct choice among the 4 different ones shown on the card. If a player is wrong in their final choice after the description, then the opponent keeps the card they described. The game ends when all 20 cards (i.e., 10 from each player or from each team) have been described.

Part 2

An image of the Food Pyramid is provided by the referee. The winning student/team is asked to select and place on it the foods they consider provide a complete and healthy daily diet. The players have the right to take two cards from their opponent, if they deem it necessary to successfully complete the pyramid. On completion of this phase, the referee checks that the choices are correct with the help of a pyramid at their exclusive disposal which includes the necessary clues. If the referee judges that the choices are in accordance with the rules of healthy and balanced diet, then the winner is the player/team who made the correct placement. If not, the other player/team is given the right to make some changes and if they manage to correct the mistakes, then they win.

What we wanted to achieve from the design of the game was to help the student/players learn about the nutritional value of different foods (i.e., macronutrients, vitamins, minerals, etc.). In addition, we were interested in their being able to use this information in a specific context, i.e., in planning a healthy daily diet. This of course means taking into account all the main meals of a day, mentioning the number and size of portions for each food category, and including an appropriate amount of water. Therefore, it was important to allow the students to make their own personal choices, based on all the information on the cards and the pyramid, while adhering to the guidelines of a healthy and balanced diet.

20.3.3 Implementation of the FG^2

At this pilot stage, we were interested in getting a first insight into the educational value of FG^2 and its effectiveness in terms of managing the adolescents' daily food choices in combination with nutritional knowledge, in order to take the necessary corrective actions required for an improved version of the game. Hence, this approach allowed us to explore both students' knowledge and tendencies on a topic that involves health implications.

Participants

In this pilot phase, 50 K-10 students constituted the convenient sample of participants. They came from a suburban school in Athens with a moderate socio-economic status, and according to their teachers, they were performing moderately well. They had not previously been taught nutrition issues beyond the minimum teaching hours (4) required by K7, nor had they participated in extra-curricular activities/programs related to nutrition issues. The ratio of girls to boys was approximately 45–55% respectively (i.e., 22 girls and 27 boys). In this pilot stage of the game construction, the aim was to capture some first impressions about its effectiveness.

Procedure and data collection

Before the game was implemented, students were given a variety of pictures of different foods and asked to individually draw a one-day diet plan poster by choosing from among them. Items from all food groups were represented, as well as foods from students' daily lives. These preliminary responses would later be compared to post-game responses on the same task to determine whether their choices approached or deviated from a healthy-balanced diet after the game. This approach was chosen as it is known that drawing allows the expression of students' ideas without restriction (White & Gunstone, 1992) and since their knowledge would be assessed through the questionnaire, we wanted to examine their tendencies regarding food choice in a more open way.

Students were also given the GeSNK questionnaire (Calella et al., 2017) to complete before the game was implemented, and their answers would later be compared to their answers on the same questionnaire after the game had been completed.

The students played for a total of three teaching hours, i.e., for three consecutive days of one teaching hour per day. To conduct the game, five groups were formed, i.e., 2 opposing teams of 2 players and a referee.

Two weeks after the completion of the game, the students were again asked to draw a diet plan poster as they had done before the game, but the food items were not necessarily the same as those encountered in the FG^2. As mentioned above, the aim was to determine whether the game had any influence on their dietary choices.

At the same time, Students were also asked to complete the GeSNK questionnaire (Calella et al., 2017).

20.3.4 Data Analysis

In order to proceed with the analysis of the posters designed by the student/players, we needed to define the categories that we would evaluate in the content of the designs. Thus, the following categories were defined: (a) number of and reference (i.e., name or time) to daily meals, (b) inclusion of water and reference to the quantity, (c) inclusion of all necessary categories of food macronutrients, and (d) reference to recommended portion sizes. Points were awarded for each daily meal depicted on the poster, as shown in Table 20.1.

As previously mentioned, these criteria provide valuable information about students' nutritional awareness and their ability to make balanced choices (Contento, 2011; Worsley, 2002). The categories were not only used for evaluating the final student outputs, but also played a crucial role in the game design: Each category corresponded to specific game mechanics intended to reinforce these elements. For

Table 20.1 Data analysis scheme

Category		Points	Total possible points/ category
(a)	Each meal included	1	10
	Reference to name of meal/time consumed	1	
(b)	Inclusion of water	1	2
	Reference to quantity	1	
(c)	Inclusion of food categories/ macronutrients: fruits, vegetables, dairy, source of fat, source of carbohydrate, source of protein (according to Mediterranean food pyramid)	1 point for each group	6
(d)	Inclusion of correct recommended portion sizes for each meal	5	5

example, the inclusion of meal frequency was emphasized through the sequencing of game parts.

By aligning the assessment categories with the structure of FG^2, we ensured that students were consistently exposed to essential nutrition principles throughout the course of the game. This approach allowed for a more cohesive evaluation, linking the knowledge acquired during the game with tangible improvements in students' understanding of nutrition, as reflected in their post-game assessments.

Both the data obtained from the poster design task and from the questionnaire responses were evaluated and analyzed using the SPSS v.29 statistical package.

20.4 Findings

The results of the poster designs are shown through the illustrative Figs. 20.2 and 20.3, which indicate an improvement in the students' visualization of a daily diet. However, statistical analysis of the results presents a different conclusion: Paired sample t-tests showed that there is a statistically significant difference in the total score between the initial and final poster, i.e., before and after the game ($p < 0.001$). The same tests were performed separately for each of the categories, and the findings showed that statistically significant differences were found in the second and fourth categories (b) and (d), respectively, while in the other two categories (a) and (c), there was a predominance in the score after the game application without this being reflected statistically significantly. In other words, the number of daily meals and their time of consumption during the day did not appear to differ in relation to the content knowledge presented to them through the game. Furthermore, although students included more food categories and better-quality combinations in the posters they designed after playing the game, this was only reflected in terms of descriptive statistics. On the other hand, students' tendency to report and visualize their water intake and to depict in their drawings the portion sizes of each food appeared to differ significantly before and after playing the game.

The processing of the questionnaire responses showed a statistically significant improvement in the students' knowledge before and after playing FG^2. As already mentioned, the GeSNK questionnaire includes general nutrition questions. More information about the part of the questionnaire used is given in a previous article by the author (Georgiou & Moshogianni, 2024).

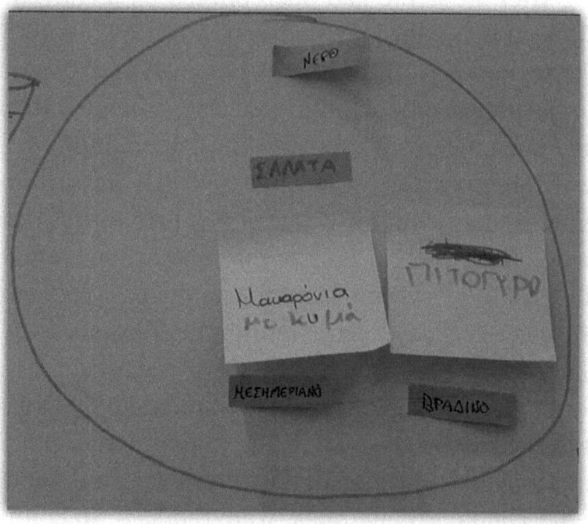

Fig. 20.2 Daily diet (before playing the game)

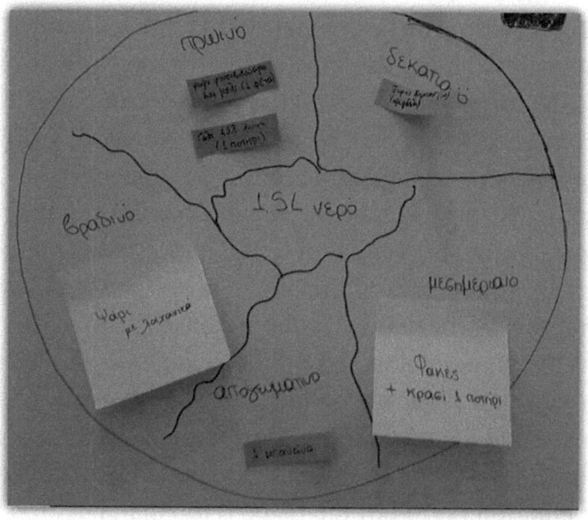

Fig. 20.3 Daily diet (after playing the game)

20.5 Discussion

This article describes the construction of FG^2, an educational game designed to inform students about nutrition in a playful way, and ultimately aiming to reinforce their sound and healthy food choices. It is a card game, easy to create and easy to use in different settings, since it can be played inside the classroom as well as outdoors.

Before playing the game, students were asked to complete a GeSNK questionnaire and to draw a poster which illustrated a healthy diet; they were also asked to complete the same two tasks after having played the game for purposes of evaluating how successful or not the game had been.

After the game had been piloted on a convenient sample of students, the results showed that their knowledge of nutritional concepts did, in fact, improve. Although the second part of the results obtained after evaluating the students' posters revealed some improvements, they did not indicate universal superiority compared to what had been the case before the implementation of FG^2.

More specifically, it appeared that the game succeeded in helping students to realize that drinking water is not just an optional procedure, but a concrete, daily need. Both the cards that referred, where possible, to the water content of food, and the food pyramid that the student/players were asked to complete at the end of the game, succeeded in highlighting the necessity of consuming sufficient water on a daily basis. The same was true with regard to the recommended quantities of different foods. The food pyramid, which refers to the portion sizes per food, certainly played a major role in this. Thus, the initial picture presented by the pre-game posters was significantly and statistically improved. Junk food options, dominated by meat (twice a day) and carbohydrates, were reduced. In contrast, inclusion of other food groups such as fruit appeared. Of course, the positive aspect was that food portion sizes per meal were reported to a much greater extent after the game, but nevertheless 'food qualities' (i.e., category (c)) showed improvement only in terms of descriptive statistics. Thus, fiber (fruit and vegetables) was often omitted, but less so compared to the original posters, and in some posters, junk food options combined with sweets continued to appear. What perhaps seemed familiar to the students was the number of meals per day and the timing of consumption. Although there was an increase in students' performance here with correct references to breakfast, lunch, etc., it was not significant compared to the impressions on the posters before playing FG^2.

Of course, all of the above raise questions and clearly lead to thoughts around improving the game. It is important to keep in mind that this project was a pilot study of the game conducted on a convenient sample; thus, its usefulness is in capturing trends and recording evidence rather than allowing generalizations to be made. We also need to consider the short implementation period of the game, as attitude modification takes more time than knowledge acquisition, and therefore a longer implementation period of the game (Holzmann et al., 2019). In addition, previous studies on game-based learning interventions in nutrition education have shown that long-term engagement and reinforcement activities are crucial for sustained behavior change (Viggiano et al., 2015). Future research should explore extended implementation periods and assess whether observed knowledge improvements translate into sustained dietary changes. However, in this pilot stage, our goal was only to determine whether the game 'works' and at what level, or if a global change in the design of the game is needed, without expecting to obtain definitive results. It is important to emphasize that we do not believe that capturing a daily meal on posters can equate to a change of attitude on the part of students, nor does it ensure that this will happen in real life when they are asked to choose what to eat. However, it is an initial indication of an understanding of healthy and balanced choices and, more importantly, a way for them to realize that they can choose between many different foods they enjoy, which at the same time meet criteria for good body function. Knowledge about foods containing proteins, carbohydrates, fibers, fats, etc. can help them to make the necessary choices and combinations in their daily life, while at the same time taking care of their health. It is, therefore, logical that in this short period of application of the game, a sufficient degree of behavioral change would be difficult to achieve to an absolute degree; even if it were possible, a way would have to be found to check whether any such change correlates with actual eating habits. This is also one of our next objectives.

Our results have twofold significance: they demonstrate the feasibility of developing students' interest in and knowledge of nutrition through implementation of our FG^2 game, and are in line with other research showing that game-based teaching interventions focused on nutritional topics were significantly effective in enhancing students' learning and attitudes (e.g., Amaro et al., 2006; Viggiano et al., 2015; Yien et al., 2011). Contento (2011) had similar outcomes studying applied Theory of Planned Behavior (Ajzen, 1991) to dietary interventions, reinforcing the idea that knowledge alone is not sufficient to effect a change in behavior, but structured experiential learning can facilitate it. Moreover, games are considered to be beneficial in terms of long-term knowledge retention due to their experiential and entertaining nature (Blakely et al., 2009). Thus, FG^2 could not only serve as an engaging learning tool, but also as a means to support long-term behavioral changes in eating habits, addressing both the cognitive and motivational aspects of nutrition education. However, to reach the final version of FG^2, in terms of achieving its educational objectives, some modifications and improvements are needed, such as enrichment with additional cards, allowing more time for its implementation, and perhaps redefinition of the rules at some points.

In future, it would be valuable to compare FG^2's effectiveness with other interactive nutrition education tools, such as digital serious games or mobile applications, which have been shown to enhance dietary knowledge and behavioral intentions in adolescents (Waters et al., 2011). Identifying the strengths and weaknesses of different methodologies could provide insights into optimizing game-based learning approaches for nutrition education.

20.6 Conclusion

FG^2 is an educational game designed to help students acquire knowledge about nutrition concepts with the aim of improving their eating habits and ultimately their overall health. It is a card game that can function as a supplementary educational material and is easy to use. By evaluating its first version through a pilot study, it appeared that it could achieve some of its objectives, such as enriching students' nutritional knowledge. However, there are areas where the game made no difference, such as students' correct choice of foods for a healthy daily diet. For this reason, improvements will be made both in the construction of the game and in the method and scope of its application in order to draw more solid conclusions. It is encouraging, of course, that positive aspects of this alternative and fun approach to learning have already been identified, and we hope that FG^2 can become an ally in nutrition education both at home in Greece and abroad.

References

Amaro, S., Viggiano, A., Di Costanzo, A., Madeo, I., Viggiano, A., Baccari, M. E., Marchitelli, E., Raia, M., Viggiano, E., Deepak, S., Monda, M., & De Luca, B. (2006). Kalèdo, a new educational board-game, gives nutritional rudiments and encourages healthy eating in children: a pilot cluster randomized trial. *European Journal of Pediatrics, 165*, 630–635.

Ajzen, I. (1991). The theory of planned behavior. *Organizational Behavior and Human Decision Processes, 50*(2), 179–211. https://doi.org/10.1016/0749-5978(91)90020-T

Bandura, A. (1977). *Social learning theory.* Prentice-Hall.

Barzegari, A., Ebrahimi, M., Azizi, M., & Ranjbar, K. (2011). A study of nutrition knowledge, attitudes and food habits of college students. *World Applied Sciences Journal, 15*(7), 1012–1017.

Baranowski, T., Baranowski, J., Cullen, K. W., Marsh, T., Islam, N., Zakeri, I., Honess-Morreale, L., & Demoor, C. (2003). Squire's Quest!: Dietary outcome evaluation of a multimedia game. *American Journal of Preventive Medicine, 24*(1), 52–61.

Bebetsos, E., Zorzou, A., Bebetsos, G., Kosta, G., & Karamousalidis, G. (2015). Children's self-efficacy and attitudes towards healthy eating. An application of the theory of planned behavior. *International Journal of Sports and Physical Education, 1*, 1–8.

Blakely, G., Skirton, H., Cooper, S., Allum, P., & Nelmes, P. (2009). Educational gaming in the health sciences: Systematic review. *Journal of Advanced Nursing, 65*(2), 259–269.

Brown, J. E. (2016). *Nutrition through the life cycle.* Cengage Learning.

Calella, P., Iacullo, V. M., & Valerio, G. (2017). Validation of a general and sport nutrition knowledge questionnaire in adolescents and young adults: GeSNK. *Nutrients, 9*(5), 439.

Contento, I. R., Randell, J. S., & Basch, C. E. (2002). Review and analysis of evaluation measures used in nutrition education intervention research. *Journal of Nutrition Education and Behavior, 34*(1), 2–25.

Contento, I. R. (2011). *Nutrition education: Linking research, theory, and practice.* Jones & Bartlett Publishers.

Gee, J. P. (2003). What video games have to teach us about learning and literacy. *Computers in Entertainment, 1*(1), 20.

Georgiou, M. (2024). Widening students' world views via the implementation of socioscientific issues in educational practice. In *A moral inquiry into epistemic insights in science education: Personal and global perspectives of socioscientific issues* (pp. 151–179). Springer Nature.

Georgiou, M., & Mavrikaki, E. (2013). Greek students' ability in argumentation and informal reasoning about socio-scientific issues related to biotechnology. In C. P. Constantinou, N. Papadouris, & A. Hadjigeorgiou (Eds.), *Proceedings of the 10th conference of the European Science Education Research Association* (pp. 1158–1166).

Georgiou, M., Mavrikaki, E., & Constantinou, C. P. (2020) Is teaching biology through socio-scientific issues enough for the development of argumentation skills? In B. Puig, P. B. Anaya, M. J. G. Quilez, & M. Grace (Eds.), *Biology education research. Contemporary topics and directions* (pp. 177–186). Servicio de Publicaciones [Publications Service], Universidad de Zaragoza. https://doi.org/10.26754/uz.978-84-16723-97-3

Georgiou, M., & Moshogianni, M. (2024). Mapping Adolescents' Nutritional Knowledge. In K. Korfiatis, M. Grace, & M. Hammann (Eds,), *Shaping the future of biological education research. Contributions from biology education research.* Springer. https://doi.org/10.1007/978-3-031-44792-1_14

Holzmann, S. L., Schäfer, H., Groh, G., Plecher, D. A., Klinker, G., Schauberger, G., Hauner, H., & Holzapfel, C. (2019). Short-term effects of the serious game "fit, food, fun" on nutritional knowledge: A pilot study among children and adolescents. *Nutrients, 11*(9), 2031.

Hwang, G.-J., Wu, P.-H., & Chen, C.-C. (2012). A game-based learning approach to improving students' learning achievements in a nutrition course. *The Turkish Online Journal of Educational Technology, 11*(2), 10–19.

Kostanjevec, S., Jerman, J., & Koch, V. (2012). The influence of nutrition education on the food consumption and nutrition attitude of schoolchildren in Slovenia. *Online Submission.*

Liu, E. Z. F., & Chen, P. K. (2013). The effect of game-based learning on students' learning performance in science learning–A case of "conveyance go." *Procedia-Social and Behavioral Sciences, 103*, 1044–1051.

Mayer, R. E. (2005). *The Cambridge handbook of multimedia learning.* Cambridge University Press.

Misra, R., Eyombo, L., & Phillips, F. T. (2022). Benefits and challenges of using educational games. In *Research anthology on developments in gamification and game-based learning* (pp. 1560–1570). IGI Global.

Mizia, S., Felińczak, A., Włodarek, D., & Syrkiewicz-Świtała, M. (2021). Evaluation of eating habits and their impact on health among adolescents and young adults: A cross-sectional study. *International Journal of Environmental Research and Public Health, 18*(8), 3996.

Okunogbe, A., Nugent, R., Spencer, G., Powis, J., Ralston, J., & Wilding, J. (2022). *Economic impacts of overweight and obesity.* 2nd Edition with Estimates for 161 Countries. World Obesity Federation.

Pirouznia, M. (2001). The association between nutrition knowledge and eating behavior in male and female adolescents in the US. *International Journal of Food Sciences and Nutrition, 52*(2), 127–132.

Sá-Pinto, X., Bennierman, A., Børsen, T. H., Georgiou, M., Jeffries, A., Pessoa, P., Soussa, B., & Zeidler, D. (2022) *Learning evolution through socio scientific issues.* UA Editora. https://doi.org/10.48528/4sjc-kj23

Saribay, A. K., & Kirbas, S. (2019). Determination of nutrition knowledge of adolescents engaged in sports. *Universal journal of educational research, 7*(1), 40–47.

Skinner, J. D., & Woodburn, M. J. (1983). Nutrition-related characteristics of high school teachers and student performance. *Journal of Nutrition Education, 15*(3), 99–104.

Viggiano, A., Viggiano, E., Di Costanzo, A., Viggiano, A., Andreozzi, E., Romano, V., Rianna, I., Vicidomini, C., Gargano, G., Incarnato, L., Fevola, C, Volta, P., Tolomeo, C., Scianni, G., Santangelo, C., Battista, R., Monda, M., Viggiano, A., De Luca, B., & Amaro, S. (2015). Kaledo, a board game for nutrition education of children and adolescents at school: Cluster randomized controlled trial of healthy lifestyle promotion. *European Journal of Pediatrics, 174*, 217–228.

Yien, J. M., Hung, C. M., Hwang, G. J., & Lin, Y. C. (2011). A game-based learning approach to improving students' learning achievements in a nutrition course. *The Turkish Online Journal of Educational Technology, 10*(2), 1–10.

Waters, E., de Silva-Sanigorski, A., Burford, B. J., Brown, T., Campbell, K. J., Gao, Y., Armstrong, R., Prosser, L., & Summerbell, C. D. (2011). Interventions for preventing obesity in children. *Cochrane Database of Systematic Reviews* (12).

White, R., & Gunstone, R. (1992). *Probing understanding*. The Falmer Press.

World Health Organization. (2018). Spotlight on adolescent health and well-being. Findings from the 2017/2018 Health Behaviour in School-aged Children (HBSC) survey in Europe and Canada. *International report. Volume 2. Key data.* Geneva; 2020. https://apps.who.int/iris/handle/10665/332104

Worobey, J., & Worobey, H. S. (1999). The impact of a two-year school breakfast program for preschool-aged children on their nutrient intake and pre-academic performance. *Child Study Journal, 29*(2), 113–113.

Worsley, A. (2002). Nutrition knowledge and food consumption: Can nutrition knowledge change food behaviour? *Asia Pacific Journal of Clinical Nutrition, 11*(Suppl 3), S579–S585.

Chapter 21
Student's Explanations of Evolutionary Processes Within the Narrative Framework of a Board Game About Chance in Evolution

Magali Coupaud, François Dessart, Fabienne Paulin, and Alice Delserieys

21.1 Introduction

This research studies the narratives of middle-school students after playing a board-game about evolution integrated into educational resources for secondary school. The board-game 'Darwinium' was designed by a team of researchers and teachers and places middle school students in the fictional situation of researchers observing the evolution of fictional animal populations placed in an experimental dome. The game mechanics and the story it tells are considered a "narrative framework" (Dessart, 2021). In other words, it sets the initial situation, a set of theoretical and methodological standards on which the players/students can base the reconstruction of past events. During a game play, players have to report on the evolution of their fictional animal population in verbal narrative form and graphic form. This project is based on previous work that has highlighted the difficulties in apprehending the Theory of Evolution (ToE) and the ideas of chance that are linked to it (Coupaud et al., 2023; Fiedler et al., 2019).

M. Coupaud (✉) · A. Delserieys
Aix Marseille Université, Marseille Cedex, France
e-mail: magali.coupaud@univ-amu.fr

A. Delserieys
e-mail: alice.delserieys@univ-amu.fr

F. Dessart · F. Paulin
Université Claude Bernard Lyon, Lyon, France
e-mail: francois.dessart@univ-lyon1.fr

F. Paulin
e-mail: fabienne.paulin@univ-lyon1.fr

O. Morin et al. (eds.), *Challenges of a Changing World in Biology Education*, Contributions from Biology Education Research,
https://doi.org/10.1007/978-3-032-05346-6_21

21.2 Theoretical Framework

21.2.1 Evolution and Chance

In this work, we refer to an updated ToE. The Darwinian theory has evolved and it has been transformed significantly since its inception. The development of the Synthetic Theory of Evolution (STE) in the 1930s was a key moment, providing genetic explanations for the fundamental principle of Darwinism: "descent with modification" (Dobzhansky, 1937; Huxley, 1942; Mayr, 1942; Simpson, 1944). Since then, starting in the 1950s, the development of new knowledge (biological, geological, astronomical, etc.) has led to questioning certain aspects of the STE, which proposed that "Evolutionary transformation is adaptive and occurs slowly and gradually, cumulatively, by adding minute changes under the influence of natural selection" (de Ricqlès & Padian, 2009, p. 349). The hegemony of natural selection, which minimizes the role of genetic drift, was reconsidered by evolutionists from the 1960s onwards. We owe Sewall Wright his theorization of genetic drift, in particular its importance for small populations. However, it was with Motoo Kimura (1968) and the discovery of neutral mutations that genetic drift became a central element of ToE. Gould and Eldredge challenged gradualism and natural selection with punctuated equilibria and the concept of exaptation (Eldredge & Gould, 1972). Similarly, the rise of cladistics made the processes and patterns of evolution more autonomous (Hennig, 1966).

These conceptual changes have led some authors to question the need to produce a new synthesis of evolutionary theory (de Ricqlès & Padian, 2009; Laland et al., 2014; Pigliucci, 2013).

Anouk Barberousse proposes a formulation of the updated ToE that we adopt for this work: "In a schematic way, we can say that the ToE takes as its starting point the fact that organisms can reproduce and that the offspring resulting from these reproductive events may differ from their parents. Furthermore, each organism has a limited existence in space and time. The ToE tells us that in this context, the structure of the genealogical network depends on random sorting processes (drift) and selective processes (natural selection) that act on organisms in each generation. The action of these two processes is determined both by the intrinsic characteristics of the organisms and by the spatio-temporal context in which they find themselves" (Barberousse, 2011, p. 369).

This recent proposition gives to the concept of chance a more important role in evolutionary explanations, which is supported by the following statement from Gayon (2009, p. 339): "Evolutionists have become more pragmatic, more pluralistic, and also more contingentist, meaning more attentive to the changing complexity of the circumstances that have shaped the history of life".

Several authors have explored the links between the theory of evolution and the concept of chance (Beatty, 1995; Desjardins, 2016; Gayon, 2005; Merlin, 2013; Powell, 2012). Gayon (2005) identifies "three possible meanings of the word 'chance': luck, probability, and contingency with respect to a given theoretical system" (p. 387). In an evolutionary process, chance in the sense of "luck" is associated with mutations leading to genetic innovations. The mutation of a gene can be considered fortuitous at the individual level, in the sense that it is not associated with any specific purpose. A priori, for individuals, mutations are neither an advantage nor a disadvantage but rather depend on environmental conditions.

The second sub-concept of chance that operates in the ToE, randomness, refers to a statistical approach and is introduced with random probabilities, notably associated with the mechanism of genetic drift at the population level or the mechanism of reproduction (Gayon, 2005; Merlin, 2013). Indeed, if we consider a generation of parents with known genetic characteristics, the allele frequency of the next generation is not strictly determined but depends on the laws of probability based on the size of the initial population. Taking randomness into account in the ToE allows us to move beyond a deterministic view.

The third sub-concept of chance, contingency, represents a mechanism that combines fortuitous events, contingent within a historical perspective. Indeed, the natural selection of a variation at the population level results from the interaction of independent causal series: the presence of external factors (such as changes in the environment) and the internal variation within the organism, variation due to mutations (Gayon, 2005; Merlin, 2013). This type of chance plays a major role in evolutionary processes by eliminating any idea of intentionality. This is an important aspect for overcoming the epistemological obstacle of finalism (Coupaud et al., 2023).

21.2.2 Understanding of Evolution

Many authors have highlighted the difficulty pupils and students have in grasping the concept of chance and genetic drift. There are many misconceptions about the process of genetic drift, and these do not improve much as a result of learning about the ToE (Andrews et al., 2012; Garvin-Doxas & Klymkowsky, 2008; Mead & Scott, 2010).

A study conducted with middle school students highlights correlations between their knowledge of evolution and their perception of chance (Coupaud, 2018). They associated chance with fate, luck or misfortune, inducing a strong finalistic idea of evolution. A majority of the surveyed students expressed strong agreement with a definition of chance associated with fate, luck, or misfortune. According to these authors, such a conception is linked to a low acceptance of evolution, resulting in a strong teleological view of evolution. This study highlights a connection between students' agreement with probabilistic thinking and their agreement with evolutionary thinking (Coupaud, 2018). However, there is a noted lack of awareness among students regarding probabilistic thinking in everyday situations. Furthermore, didactic research indicates that reasoning at the individual level may obstruct understanding evolution, which requires probabilistic population thinking among middle school students (Gobert, 2014). Another study reveals that while the concept of chance is not part of the core representations of evolution among students, it is recognized by teachers and researchers (Paulin, 2024). Students often use the term "aleatory" in relation to mutations and genetic drift, suggesting they perceive chance more on a molecular level than a macroscopic one. We believe that understanding the contingency of evolutionary events is essential for shifting students' finalistic views.

Considering these obstacles, we propose a historical approach to teaching the ToE through narrative forms. A functional approach without narrative, may not enable students to overcome these obstacles. Encouraging students to narrate the evolution of a fictional population based on a game could be a valuable strategy. Indeed, recent studies in the didactics of life sciences have demonstrated the potential of narrative tools for learning about evolution (Zabel et al., 2021). Chance is intrinsically linked to the game due to the player's uncertainty regarding the outcome (Solinski, 2015). A significant part of the pleasure of playing lies in this uncertainty. A game creates a context that encourages players to fully engage in the proposed activity, referencing the concept of flow in psychology (Alvarez et al., 2016; Zabel & Gropengießer, 2015), allowing for creativity to flourish.

21.2.3 Narrative Discourses

An interesting feature of the game is that it implements a narrative framework for the player. This process might be more or less conscious. A story of what happened is expressed to narrate each game. But also, "in the background", each game holds what did not happen and what could have happened. From this point of view, the story accounts for the contingency of the events of the game. It is this handling of contingency by the story that interests us here.

By considering the story depicted in each game as a narrative process, we analyzed students' texts likely to include a narrative framework. Following the proposition of Barthes (1966), we consider that "everything is a story" (Barthes, 1966) and that scientific texts often take a narrative form. A story is an integral part of human experience, and at the same time, has the power to give to the world (Bruner, 2002). As a result, students make frequent use of narrative processes. Previous research in science education has shown the role played by narrative frameworks in the reconstruction of past events (Dessart, 2019, 2021; Orange-Ravachol, 2012): it allows the construction of a framework of intelligibility within which events occur and, above all, make sense. But this narrative framework also participates in the construction of other stories that are inserted into it ("framed" stories) by structuring the narrative process: beginning/middle/end of the story (Ricœur, 1984), focus on the action/on the characters in particular (Brémond, 1973).

As part of the playful activity around the Darwinium game studied here, students (aged 13–14 years old) are asked to produce a text and a diagram of what happened regarding the evolution of a population in the game. These students' productions reflect the way they think about evolution and associated chance, and the way they report on it.

The narrative framework of the game does not impose a particular type of discourse. As a result, the students' writings take a variety of forms and can be part of different types of discourse: descriptive, narrative and argumentative. Descriptive discourse does not provide an explanation. Narrative discourse organizes characters involved in an action and a succession of events leading to the end of the story (Ricoeur, 1984). It can lead students to focus on one or more characters in the story, without allowing them to consider the whole picture. Argumentative discourse produces justified statements. Arguments can focus on the origin of facts, the reason for their existence. Arguments can also target a purpose, an intention (Figon & Dessart, 2022).

21.3 Research Question

We seek to answer the following question: What kinds of explanations and what kinds of discourses on biological evolution and associated chance do students produce after engaging with the narrative framework of the Darwinium game?

Our first hypothesis is that the explanations provided by students from the game's narrative framework vary from evolutionary thinking, in agreement with ToE, to fixist thinking, in disagreement with ToE including different forms of chance.

Our second hypothesis is that these different explanations may be related to the type of discourse employed:

- evolutionary processes and chance should not be expressed in descriptive discourse, but rather in narrative or argumentative discourse,
- the beginnings of population thinking could be expressed in narrative discourse, and perhaps more explicitly in argumentative discourse,
- finalist ideas could appear in narrative discourse.

21.4 The Board-Game Darwinium

The board game Darwinium, designed prior to this study by a multi-disciplinary research team, incorporates two concepts of chance: randomness and contingency. This board game is considered to be an artefact that may contain implicit knowledge related to evolutionary chance. It involves players in a fictitious simulation. The players study the changes occurring in a fictional experimental population in response to conditions imposed by the player, but not totally controlled. The game is based on a chaotic gameplay mechanic with random events.

21.4.1 What is the Story?

The story of the game is one of experimental evolution. Players take on the role of researchers documenting the evolution of a population of a fictional species placed in an experimental dome. Each team investigates the population's changes occurring on the game board in response to partially controlled conditions in the dome. This experimental dome immerses players in a scenario of experimental evolution.

21.4.2 Game Mechanics

The game features three observable characteristics (beak shape, skin color, feet type), each with five phenotypic traits. These phenotypic traits are represented on cards with sketches, descriptions, and symbols, along with corresponding trait-tokens (Fig. 21.1).

The players locate their populations on a central board representing the experimental dome. Each team has a player's board to place the trait tokens and track their population characteristics. This player's board illustrates the possible traits according to the trait cards drawn and indicates an affinity between a trait and an environment

Fig. 21.1 Example of cards representing a trait (here the webbed-feet variation of the feet type characteristic) with corresponding trait-tokens (C1 for the webbed-feet)

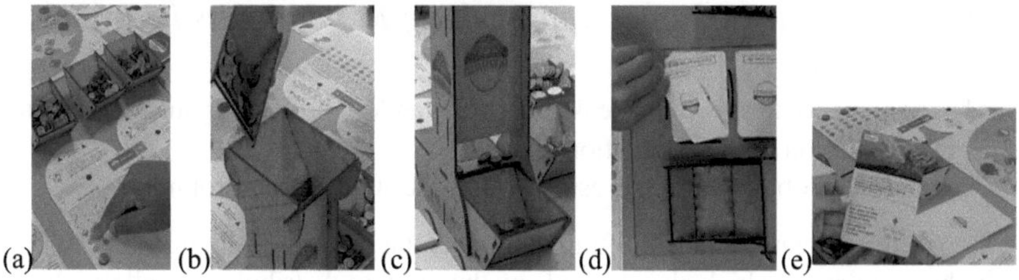

Fig. 21.2 Players add trait tokens according to the "affinity with the environment" (**a**), throw all trait tokens in the dice tower (**b**), retrieve a random portion of trait tokens (**c**), and draw a new trait card (**d**), "Event" card drawn (**e**)

(Fig. 21.2a). They also have a research notebook to report the changes of their population. Players modify the environmental conditions in the dome. These changes are triggered by an action of the player or undergone by an event card drawn.

The game mechanics allow for random mixing of traits, influenced by changing environmental conditions.

21.4.3 The Steps of the Game in Each Round

Darwinium is played in rounds consisting of seven steps in each round:

1. Moving territories: Players can move their population into a different environment.
2. Action in the dome: Players can initiate actions that alter the dome's physical conditions. The consequence of these changes on the populations themselves is known only once the action is played, and it can affect all players with positive or negative impact on their proportion of trait tokens.
3. Environmental Affinity: Each player successively throws all the trait tokens into a specially designed dice tower (Fig. 21.2b). The tower allows a random sorting of the pieces. Some are kept by the player, some are discarded, some are trapped in the tower.
4. Reproductive Shuffling: Players sort trait tokens in a dice tower for random distribution, determining which traits are retained, discarded or trapped in the tower (Fig. 21.2c).
5. Emergence of New Traits: New traits can emerge through random mutations when trait cards are drawn (Fig. 21.2d). Players adjust their player board to add the required tokens corresponding to the new traits drawn.
6. Event: An event card is drawn inducing changes in the dome's environment (Fig. 21.2e). All players apply the changes that might impact their population (gain or loss of tokens that change the proportion of traits in their population).
7. News conference: Players fill an observation notebook to report on what happened to their population during the round.

21.4.4 Darwinium Game: Approaching Randomness in ToE

This board game focuses on two ideas of chance: random probability and contingency to explain evolution at the population level.

The random probability is expressed with three steps in each round:

- Step 4, the "Reproductive shuffling", models the frequency changes of the traits in the population which also depends on the random distribution of alleles during reproduction.
- Step 5, the "Appearance of new traits", models the appearance of new variants in the population, coming from random mutations or from the random expression of recessive alleles when "Traits" cards are drawn.
- Step 6, the "Event", models potential environmental changes affecting the population.

Contingency is seen in the preparation of tokens (that are then, thrown in the dice tower) (step 3), where the number is influenced by environmental type and resources, allowing advantageous traits to increase in frequency. This step models the process of natural selection.

21.4.5 Darwinium Game as a Narrative Framework

The Darwinium game provides a narrative framework because:

- The rules of the game impose constraints (types of environments, natural disasters, etc.) on the progress of the game in a defined order from the beginning to the end of the game. The sequence of events affects the frequency of the traits in the animal populations. But, while constraining the story, they leave room for evolutionary possibilities: not everything is possible, and what is possible is contingent (it might not happen).
- The mechanics of the game leads the player to carry out a series of actions (taking tokens, putting them in the tower, drawing cards from the environment, etc.). Each action produces a game fact, each of which represents an evolutionary event. For example, the "take tokens" action corresponds to a game event (what happens in the game), but it also represents a random evolutionary event (reproductive chance). With each action, the player moves from the possible (what could happen) to a factual reality (what really happens).

In this way, the Darwinium game models evolutionary processes according to a scenario that acts as a metaphor, analogically substituting an evolutionary reality. According to Richard in Taylor and Dewsbury (2018), "a metaphor is a comparison between two seemingly dissimilar concepts that involves the "carrying over of a word from its normal use to a new use" (p. 1). This is the metaphorical dimension of the game, which each student is led to experience. The difficulty for the student is not to stick to the event of the game (taking tokens for example), but rather to see it as an evolutionary event (reproductive chance).

In line with the Psychologist Bruner, we consider that narrative frameworks such as stories are a way of thinking (Bruner, 1996) and organizing knowledge (Echterhoff & Straub, 2003). In our own works, we underline that narratives can allow the learners to engage in a work of scientific problematization and allow the expression of creativity and openness to new possibilities in the search for explanations.

Consequently, with this narrative framework, we think that students can structure their text in order to rebuild the future evolutive events and process.

The "narrative framework" of the Darwinium game is built on:

- populational thinking: the narrative of the game and texts found in the elements of the game never mention individuals but rather use groups of animals,
- generational change: a timeline is not expressed as such, but the game introduces a temporality through the succession of generations in the population of animals,
- material artefacts manipulation that introduces random processes: (1) Tokens, representing various traits of the populations of the game, thrown in a dice tower (random distribution of alleles during sexual reproduction) (Fig. 21.2b). (2) Preparation of the number of tokens to throw in the dice tower (contingency of natural selection). (3) "Traits" cards drawn depending on the sorting result of the dice tower (random mutations) (Fig. 21.2c). (4) "event" cards randomly drawn (see Fig. 21.2e),
- modification of the environment: players can alter environments and respond to event card impacts.

21.5 Research Design and Method

21.5.1 Data Collection

The data are collected during a debriefing session, after the game session of Darwinium game, conducted in a biology class. This phase is crucial for transferring the knowledge gained from the play activity (Sanchez, 2019). The teacher goes through the stages of the game with the pupils, going back and forth between reality (the mechanisms explaining evolution) and the model (the game elements). During this debriefing session, 57 students (13–14 years old) produced narrative writings. They were asked to write the story of their population, to express their understanding of what happened during the game and how it related to evolution. Students were encouraged to write a text using the notes they took during the game in a specifically designed booklet. This study analyses a selection of narrative texts (N = 30) that feature either ideas of chance in evolutionary processes or obstacles related to understanding chance in evolutionary processes (finalism), or both. Texts that did not feature these ideas are not part of this analyses. The aim is to highlight the ways

1 : 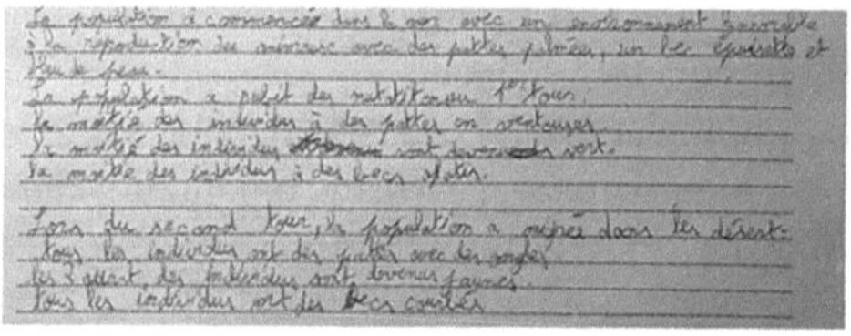

1: The population began in the sea, with an environment favourable to the reproduction of animals with webbed feet, a landing net beak and blue skin. The population underwent mutations in the first round: half of the individuals have legs with suction cups, half of the individuals have turned green and half of the individuals have flattened beaks.
In the second round, the population migrated to the desert: all the individuals have feet with nails; three quarters of the individuals have turned yellow; all the individuals have curved beaks.

2 : 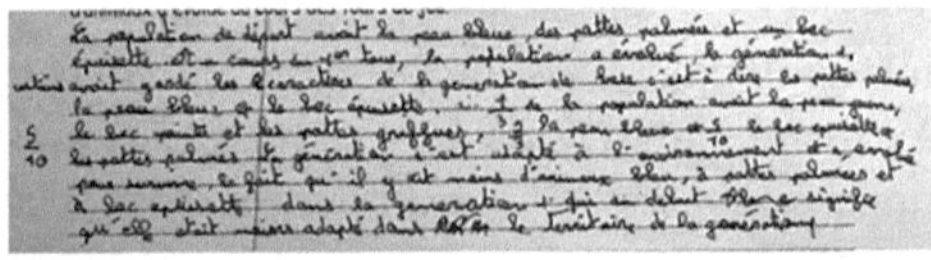

2: The starting population had blue skin, webbed feet and a landing net beak. During the first round, the population evolved; in generation 1, some had kept the characters of the base generation, i.e. webbed feet, blue skin and a landing net beak. One third of the population had yellow skin, a pointed beak and clawed feet, two thirds blue skin and 5/10 the landing net beak and webbed feet. The generation adapted to the environment and evolved to survive, so the fact that there are fewer blue animals with webbed feet and landing net beaks in generation 1 than there were at the beginning means that they were less adapted in generation 1's territory.

3 : 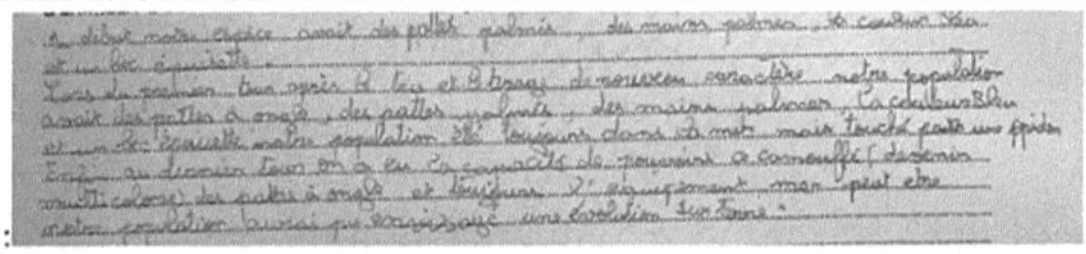

3: At the beginning our species had webbed feet, webbed hands, blue colour and a landing net beak. In the first round, after sorting and drawing the new characters, our population had nail feet, webbed feet, webbed hands, blue colour and a landing net beak. Our population was still in the sea but affected by an epidemic. Finally, in the last round, we had the ability to camouflage ourselves (become multicoloured), nail legs and still the sea equipment; perhaps our population could have considered evolving on land.

Fig. 21.3 Example of student's narrative writing with ideas of chance in evolutionary processes (2.1), finalism (2.2), both (2.3). Translation by authors

in which these selected texts fit into the narrative framework set up by the game (see Fig. 21.3).

The texts were analysed based on the framework described below with a qualitative approach, and, the resulting frequency of the categories of analysis was analysed using statistical tests.

21.5.2 Analytical Framework

The students' texts were analysed using a dual framework. The first framework was epistemological, focusing on evolutionary and chance concepts (see Table 21.1), while the second focused on discourse types (see Table 21.2). This analysis identified categories and subcategories in a deductive approach. We counted the number of texts demonstrating references to each categories and subcategories (see Tables 21.1 and 21.2).

Concerning evolution in the first analytical framework, students' texts were classified on their attitudes toward evolution as follows:

- evolutionist when they express the fact that populations change from generation to generation (considered as accepting evolution but with no certainty of understanding it) (for example "Our population has evolved as a function of chance and environment."),
- finalist, when evolutionary changes, are explained by an intention to adapt (for example "Our population has been forced to adapt."),
- fixist, when texts express the fact that populations don't change (no example in the data).

Furthermore, we analysed a related category: population thinking. This corresponds to students' awareness of the differences between individuals in the same

Table 21.1 Absolute frequencies of students' texts attributed to each category and subcateogy of analysis of the students' evolutionary or chance perception framework

Categories of analysis	Subcategories of analysis	Absolute frequency (N = 56 texts)
Evolutionist		54
Finalism		22
Fixist		0
Population thinking	Level 1	47
	Level 2	18
Chance	Random probabilities	23
	Contingency	6
	Incomplete knowledge	0
	Destiny	0

Table 21.2 Absolute frequency of students' texts attributed to each category of analysis relative to the nature of discourses framework

Category of analysis	Subcategories of analysis	Absolute frequency (N = 56)
A single type of discourse (11/56)	Descriptive	6
	Narrative	4
	Argumentative	1
Combination of discourses (45/56)	Descriptive + narrative	18
	Descriptive + argumentative	9
	Narrative + argumentative	9
	Descriptive + narrative + argumentative	9

population. It is opposed to typological or essentialist thinking, which prevents students from understanding that the future of a population is made up of each individual's particular and random history. We have identified two levels of understanding to characterize this population thinking:

- level 1 population thinking: texts refer to a population without discriminating between individuals (for example: "Our population has evolved, acquiring characteristics that we didn't have at the outset"),
- level 2 population thinking: texts refer to differences between individuals of the same population and generation (for example: "My population is now diverse: one third of it is multicolored, one fifth has suction-cup feet and three fifths have a pointed beak.").

Concerning student's explanations about chance, the students' texts are considered as:

- using random probabilities when a random event obeys a law of probability (for example: "[The change in animal characteristics over the course of générations], I know that this is due to genetic mixing, and therefore random."),
- using contingency when a contingent event occurs in a theoretical field of possibilities, following the encounter of two random and independent series of events (for example: "traits have changed thanks to genetic mixing and environmental change."),
- using incomplete knowledge when the texts express the idea that chance exists only because we don't know enough about deterministic processes (no example in the data),
- using destiny when the texts express the idea that explanation is found in a divine intervention and that randomness does not exist (no example in the data).

The subcategories of random probabilities and contingency are consistent with ToE and the categories of incomplete knowledge and destiny are in conflict with it.

The second analytical framework characterizes discourse types in the students' texts using specific subcategories:

- descriptive discourse is focused on the characteristics of each population and their distribution across generations, without explanations,
- narrative discourse is organized with a structure of story (beginning-middle-end), a personification and chronology,
- argumentative discourse is based on reasons for the changes in populations from one generation to the next.

21.5.3 Statistical Analysis

The data resulting from the texts' qualitative analysis were analyzed using the JASP 0.19.1 software. Chi-square tests of independence were used to assess the relationships between the nature of discourses (descriptive, narrative and argumentative) and evolutionary or chance perception used by students to explain the evolution of their fictional population. We checked that all the conditions of the chi-square test were met, in particular that the expected frequencies of the contingency tables were greater than 5.

21.6 Results

21.6.1 Quantitative Results

The quantitative results provide an initial descriptive analysis of the following aspects:

- Students' acceptance and understanding of evolution: the majority of the texts express an evolutionist idea, and no text denies evolution. The majority of the texts produced by the students after using the game express an evolutionist idea and no text denies evolution. The pupils talk in terms of population and not species. This would encourage population thinking. For example, they use percentages to describe the evolution of their population. However, some texts contain terms of individuals, differing from one another, or from one generation to the next. The results also highlight that from all of the texts, less than half used finalist formulations (Table 21.1).
- Students' chance perception: around one third of the texts put chance forward in the explanations given, in the sense of random probabilities. But some texts invoked the idea of contingency to explain the evolution of their population (Table 21.1).
- The nature of discourses used by students to write their text: most texts combine different types of discourses (Table 21.2): in particular, narrative and argumentative discourses in the same text or in most cases, descriptive and narrative discourses in the same text. For example, we can see a combination of descriptive discourse (Fig. 21.4a, in green) and narrative discourse (Fig. 21.4a, in blue).

These results partly confirm the first hypothesis: students expressed a variety of explanations for ToE using different discourses. Nevertheless, no fixist explanations for ToE or incomplete knowledge and destiny explanations for chance were produced by the students (Table 21.1).

Chi-square tests of independence were used to assess the relationships between the nature of discourses (descriptive, narrative and argumentative) and explanations of ToE and chance used by students to explain the evolution of their fictional population.

The results of the chi-square tests of independence partially confirm the 2nd hypothesis: There is a relationship between certain explanations of ToE and chance and certain types of discourse. The results showed that there was no significant relationship between the type of descriptive discourse and the set of categories accounting for the explanations of ToE and chance stated by the students ($p > 0.05$). On the other

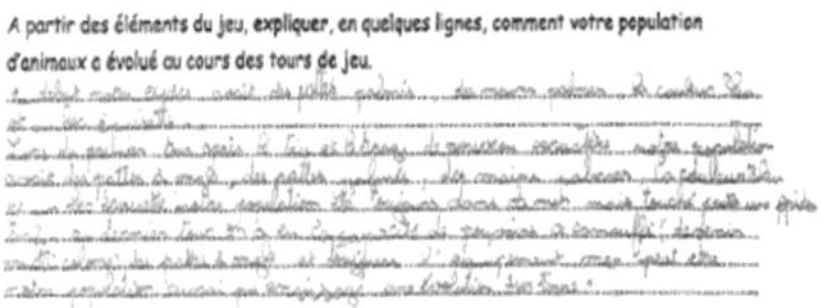
A partir des éléments du jeu, expliquer, en quelques lignes, comment votre population d'animaux a évolué au cours des tours de jeu.

(a)

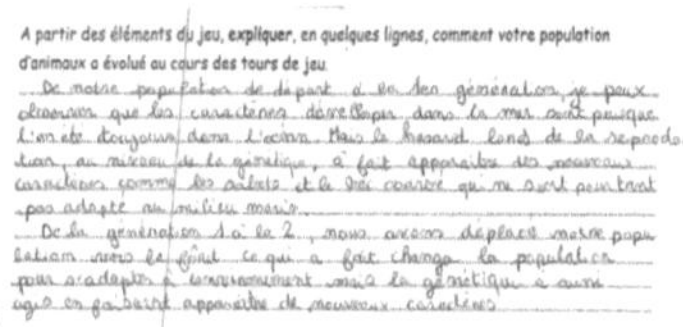
A partir des éléments du jeu, expliquer, en quelques lignes, comment votre population d'animaux a évolué au cours des tours de jeu

(b)

(a) "*At the beginning, our species had webbed feet, webbed hands, blue colour and a landing net beak. In the first round, after sorting and drawing new characters, our population had legs with toes with nails, webbed feet, webbed hands, blue colour and a landing net beak. Our population was still in the sea, but affected by an epidemic. Then, in the last round, we had the ability to camouflage ourselves (become multicoloured), legs with toes with nails and still the sea equipment [webbed feet, blue colour and a landing net beak], perhaps our population could have considered evolving on land.*"

b) *From our starting population to the 1st generation, I can see that the characteristics developed in the sea are still there, as it has always been in the ocean. But chance during reproduction, at the genetic level, has led to the appearance of new characteristics such as hooves or curved beaks, which are not adapted to the marine environment. From generation 1 to 2, we have moved our population to the forest, which has caused the population to change to adapt to the environment, but genetics has also acted to bring out new traits.*

Fig. 21.4 **a** Example of student's narrative writing with descriptive (in green) and narrative discourses (in blue), **b** example of student's narrative writing with use population thinking inconsistently (in red) and narrative discourse (in green). Translations by authors

hand, there is a relationship between narrative discourse and the students' population thinking level 1 and 2 (respectively $\chi^2(1, N = 56) = 4.67$, $p = 0.03$, $\chi^2(1, N = 56) = 4.81$, $p = 0.03$). It can be said that when students use population thinking (level 1) in their text, they tend to use a narrative discourse in which they personify certain elements of the game. For example, we can see that, in the text presented in Fig. 21.4, the student talks about the population as a central character (Fig. 21.4b, in red) and he uses a narrative discourse (Fig. 21.4b, in green). But, the chisquare test shows a negative relationship between narrative discourse and level 2 explanations. On the other hand, argumentative discourse is related to fixist and random-probability explanations (respectively, $\chi^2(1,N = 56) = 14.67$, $p < 0.001$; $\chi2(1,N = 56) = 8.93$, $p = 0.003$). Students who propose fixist explanations, in disagreement with the ToE, or explanations linked to random probabilities, in agreement with the ToE, preferentially use an argumentation discourse.

21.6.2 Qualitative Results

A more in-depth qualitative analysis of our corpus allows us to refine our conclusions and gain a clearer understanding of how students use narratives to describe the evolution of their population. The data reveal that students often struggle to move beyond simple chronological storytelling. For them, events appear to follow one another in sequence. For example, the student in Fig. 21.4a describes three main points in time corresponding to three rounds of the game ("At the beginning", "in the first round", "then in the last round"). For each point in time, there is a description of the population but there are no clear links between them. This observation aligns with the relatively small number of argumentative texts we observed in the corpus.

Moreover, when students reference specific moments in the game, such as drawing tokens or moving populations, they tend to focus solely on these game actions. There is little or no connection to the evolutionary events these actions are designed to represent. This suggests that students may have difficulty internalizing the objectives of the activity. They stay at the level of the model that the game represents but have difficulty to transfer to the actual concepts of evolution. This is a main challenge in the use of the game Darwinium in class, and the link between the Darwinium game and the broader concept of evolution needs to be better supported throughout the learning experience.

Finally, we noticed that students who provide explanations based on populations often describe these populations in a "personified" way, as if the population itself were the main character in their narrative. This tendency is especially pronounced in more narrative-driven texts, as quantitative results show. At this stage, we still need to explore this line of thinking further to determine to what extent students are engaging with the idea of populations in an evolutionary sense or whether their descriptions mimic this concept without fully grasping it.

21.7 Discussion

Most texts produced by students after playing Darwinium (54/56) contain elements of discourse that are in line with evolutionary thinking: they describe changes in populations over the course of generations. We sometimes found these relevant elements alongside other irrelevant elements such as finalist ideas. Some students thus produce mixed and contradictory explanations that reveal a partial understanding of evolution. In this study, however, finalist or teleological answers were in the minority (22/56), which is unusual (see, for example, Kampourakis & Zogza, 2008). There are two possible explanations for this favorable result. First, different scenarios were created during the game using identical populations. These scenarios were then presented to all students during the debrief. We can assume that the presentation of this diversity partly reduced the students' tendency to produce teleological explanations. On the other hand, certain stages of the game forced students to consider chance in the future of their population, which in turn may have thwarted their spontaneous finalist vision.

We expected to find a relationship between narrative discourse and finalism. However, this relationship only exists with argumentative discourse. This may be explained by the fact that, in the narrative texts, the pupils who experienced chance were less finalistic than expected.

The fact that there is a relationship between finalism and argumentative discourse suggests that argumentation is not a guarantee of progression towards evolutionary thinking in students. Indeed, arguments can be as much about the origin of facts as they are about finality (Figon & Dessart, 2022).

Furthermore, the students' use of narrative in the Darwinium game is reduced to the juxtaposition of events (game events or evolutionary events represented by the game). As a result, they give little emphasis to the role of contingency in evolutionary processes. But this work on chance, and especially contingency, which is fundamental to evolutionary sciences (Gould, 1990), relies on narrative, not just chronological reconstruction. Events cannot be thought of as mere occurrences: they take on meaning above all through the relationships they establish with one another (Dessart & Triquet, 2019). Working out these relationships between events is, in our view, a matter of inter-object thinking (Piaget, 1977; Resta-Schweitzer & Weil-Barais, 2009), which pupils have difficulty implementing here.

Finally, the difficulties encountered by the pupils could also stem from a lack of mastery of metaphorical thinking: the Darwinium game is a metaphorical representation of evolutionary processes and patterns. However, the students remain stuck on the game's events: they are unable to shift the metaphor from the game to evolutionary events. Debreafing could be a key moment for pupils to move from game metaphor to evolutionary process modeling. This underlines the work that could be done in science classes linking metaphorical thinking with scientific explanations (Ricœur, 1975; Taylor & Dewsbury, 2018).

21.8 Conclusion

The Darwinium game takes students on a temporal adventure in which they experience a diversity of population futures, over several generations, starting from the same initial situation. They also experience the role of chance in this field of possibilities.

This exploratory study shows that the narrative structure of the Darwinium game effectively engages students in evolutionary thinking. However, the concept of contingency where events are dependent on chance or external factors remains a challenging idea for pupils to grasp. While the game encourages students to move towards a more scientific understanding of evolution, their interpretations often blend different concepts, resulting in what could be described as a form of naïve syncretism, where elements of scientific and intuitive thinking coexist in a somewhat simplified or incomplete manner (Aptyka et al., 2022; Parraguez et al., 2023). One significant challenge for many pupils lies in the game's mimetic aspect its attempt to simulate evolutionary processes in a way that mirrors, or "mimes," real-world biology, akin to Aristotle's idea of mimesis. Many pupils seem to have difficulty fully internalizing this analogy, meaning they may not easily transfer what they learn within the game to broader evolutionary principles.

Despite these challenges, the imaginative and narrative elements of the Darwinium game stand out as a major strength. These aspects of the game offer a rich and engaging context that we believe can facilitate deeper learning, but they need to be explored and refined. The narrative, by allowing pupils to immerse themselves in the evolutionary story, seems to create a sense of involvement that could potentially lead to better retention and understanding of chance. However, to exploit the full educational potential of this imaginary framework, further research is needed to further analyze the impact of the narrative framework on explanations of ToE and chance in student writing. And by developing more educational strategies to help students connect the game's narrative to scientific concepts of evolution, educators may be able to foster a more nuanced and scientifically accurate understanding of evolution and chance. We believe the Darwinium game to be a relevant tool for bridging the gap between the naivety of intuitive thinking and the complexity of scientific thought.

References

Alvarez, J., Djaouti, D., & Rampnoux, O. (2016). *Apprendre avec les serious games ?* Réseau Canopé.

Andrews, T. M., Price, R. M., Mead, L. S., McElhinny, T. L., Thanukos, A., Perez, K. E., Herreid, C. F., Terry, D. R., & Lemons, P. P. (2012). Biology undergraduates' misconceptions about genetic drift. *CBE-Life Sciences Education, 11*, 248–259.

Aptyka, H., Fiedler, D., & Großschedl, J. (2022). Effects of situated learning and clarification of misconceptions on contextual reasoning about natural selection. *Evolution: Education and Outreach, 15*(1). https://doi.org/10.1186/s12052-022-00163-5

Barberousse, A. (2011). Pourquoi et comment formaliser la théorie de l'évolution? In T. Heams, P. Huneman, G. Lecointre, & M. Silberstein (dirs), *les mondes darwiniens - l'évolution de l'évolution* (pp. 417–439). Editions Matériologiques.

Barthes, R. (1966). Introduction à l'analyse structurale des récits. *Communications, 8*(1), 1–27. https://doi.org/10.3406/comm.1966.1113

Beatty, J. (1995). The evolutionary contingency thesis. In G. Wolters & J. Lennox (Eds.), *Theories and rationality in the biological sciences* (pp. 45–81). The University Pittsburg Press.

Brémond, C. (1973). *Logique du récit*. Seuil.

Bruner, J. S. (1996). *The culture of education.* Havard University Press.

Bruner, J. S. (2002). *Pourquoi nous racontons-nous des histoires ?* Retz.

Coupaud, M. (2018). *Évolution du vivant et hasard: étude quantitative des conceptions d'élèves de collège dans le contexte français* (Phd thesis). Université Aix-Marseille.

Coupaud, M., Gobert, J., Castéra, J., & Delserieys, A. (2023). Relations entre les idées de hasard chez des collégiens de 11 à 16 ans en France et leur degré de compréhension de l'évolution du vivant. In C. Fortin, & J. Gobert (Eds), *Recherches en didactique de l'évolution: Enseignement, apprentissage et formation.* International Science and Technology Editions.

Desjardins, E. (2016). Contingent evolution: Not by chance alone. In G. Ramsey & C. Pence (Eds.), *Chance in evolution* (pp. 223–243). University of Chicago Press.

Dessart, F. (2019). *Récit et médiation scientifique « sur le terrain » en géologie. Une approche épistémologique et didactique de la mise en récit dans les situations de médiation de la géologie in situ* (PhD thesis). Université Claude Bernard Lyon 1.

Dessart, F. (2021). Parcours narratif et parcours géologique: La mise en récit de l'histoire géologique d'un site naturel. *Recherches En Didactiques, 32*(1), 65–86. https://doi.org/10.3917/rdid1.032.0065

Dessart, F., & Triquet, E. (2019). Les analepses dans les récits de médiation géologique « sur le terrain »: Un outil pour mobiliser l'actualisme et la démarche historique ? *RDST, 19*, 117–140.
Dobzhansky, T. (1937). *Genetics and the origin of species*. Columbia University Press.
Echterhoff, G., & Straub, J. (2003). Narrative Psychologie: Facetten eines Forschungsprogramms. *Erster Teil: Handlung Kultur Interpretation, 12*(2), 317–342.
Eldredge, N., & Gould, S. J. (1972). Punctuated equilibria: An alternative to phyletic gradualism. *Models in Paleobiology, 82*, 115.
Fiedler, D., Sbeglia, G., Nehm, R., & Harms, U. (2019). How strongly does statistical reasoning influence knowledge and acceptance of evolution? *Journal of Research in Science Teaching, 56*(9), 1183–1206.
Figon, F., et Dessart, F. (2022). Prise en compte d'un obstacle didactique en formation initiale d'enseignants de SVT. Vers une typologie du finalisme dans les écrits d'étudiants en master MEEF. In *XIIe rencontres scientifiques de l'association pour la recherche en didactique des sciences et des technologies* (pp. 173–178).
Garvin-Doxas, K., & Klymkowsky, M. (2008). Understanding randomness and its impact on student learning: lessons learned from building the biology concept inventory (BCI). *CBE-Life Sciences Education, 7*(2), 227–233.
Gayon, J. (2005). Chance, explanation and causation in evolutionary theory. *History and Philosophy of the Life Sciences, 27*, 387–397.
Gayon, J. (2009). Mort ou persistance du darwinisme? regard d'un épistémologue. *Comptes Rendus Palevol, 8*(2–3), 321–340.
Gobert, J. (2014). *Apprentissage de raisonnements néodarwiniens en classe de sciences de la Vie et de la Terre* (Phd thesis). Université de Caen Basse-Normandie.
Gould, S. J. (1990). *Wonderful life: The burgess shale and the nature of history*. W. W. Norton & Company.
Hennig, W. (1966). *Phylogenetics systematics*. University of Illinois Press.
Huxley, J. (1942). *Evolution: the modern synthesis*. George Allen et Unwin Ed.
Kampourakis, K., & Zogza, V. (2008). Students' intuitive explanations of the causes of homologies and adaptations. *Sciences & Education, 17*, 27–47.
Kimura, M. (1968). Evolutionary rate at the molecular level. *Nature, 217*, 624–626.
Laland, K., Uller, T., Feldman, M., Sterelny, K., Muller, G. B., Moczek, A., Jablonka, E., Odling-Smee, J., Wray, J. A., Hoekstra, H. E., Futuyamad, J., Lensky, R. E., Mackay, T. F. C., Schluter, D., & Strassmann, J. E. (2014). Does evolutionnary theory need a rethink? *Nature, 514*, 161–164.
Mayr, E. (1942). *Systematics and the origin of species from the viewpoint of a zoologist*. Columbia University Press.
Mead, L. S., & Scott, E. (2010). Problem concepts in evolution Part II: Cause and chance. *Evo Edu Outreach, 3*, 261–264.
Merlin, F. (2013). *Mutations et aléas: le hasard dans la théorie de l'évolution*. Hermann.
Orange-Ravachol, D. (2012). *Didactique des sciences de la Vie et de la Terre*. Presses universitaires de Rennes.
Parraguez, C., Núñeza, P., Krüger, D., & Cofré, H. (2023). Describing changes in student thinking about evolution in response to instruction: The case of a group of Chilean ninth-grade students. *Journal of Biological Education, 57*(5), 1022–1038.
Paulin, F. (2024). The "conceptual landscape" of evolution: A possible instrument for training of earth and life sciences teachers. In C. Fortin & J. Gobert (Eds.), *The evolution of life: Teaching, learning and training, new approaches on current research in the didactics of evolution* (pp. 169–193). International Science and Technology Editions.
Piaget, J. (1977). *Recherches sur l'abstraction réfléchissante, tome 1: L'abstraction des logico-arithmétiques*. Presses Universitaires de France.
Pigliucci, M. (2013). The nature of evolutionary biology: At the borderlands between historical and experimental science. In K. Kampourakis (Ed.), *The philosophy of biology: A companion for educators* (pp. 87–100). Springer.

Powell, R. (2012). Convergent evolution and the limits of natural selection. *The European Journal of Philosophy of Science, 2*, 355–373.
Resta-Schweitzer, M., & Weil-Barais, A. (2009). Initiation scientifique et développement intellectuel de l'enfant à l'âge pré-scolaire. *Les Dossiers des Sciences De L'éducation, 21*(1), 101–113. https://doi.org/10.3406/dsedu.2009.1157
Ricœur, P. (1975). *La métaphore vive*. Editions du Seuil.
Ricœur, P. (1984). *Temps et récit Tome 2, La configuration dans le récit de fiction*. Editions du Seuil.
de Ricqlès, A., & Padian, K. (2009). Quelques apports à la théorie de l'évolution, de la « synthèse orthodoxe » à la « super synthèse évo-dévo » 1970–2009: Un point de vue. *Comptes Rendus Palevol, 8*(2–3), 341–364.
Sanchez, E. (2019). « *Game-Based Learning* ». In A. Tatnall (dir.), *Encyclopedia of education and information technologies* (pp. 1–9). Éd. Springer International Publishing.
Taylor, C., & Dewsbury, B. M. (2018). On the problem and promise of metaphor use in science and science communication. *Journal of Microbiology & Biology Education, 19*(1), 10–1128.
Simpson, G. (1944). *Tempo and mode in evolution*. Columbia University Press.
Solinski, B. (2015). *Ludologie: jeu, discours, complexité* (Phd thesis). Université de Lorraine.
Wright, S. (1931). Evolution in Mendelian population. *Genetics, 16*, 97–159.
Zabel, J., & Gropengießer, H. (2015). What can narrative contribute to students' understanding of scientific concepts, e.g. evolution theory? *Journal of the European Teacher Education Network, 10*, 136–146.
Zabel, J., Bruguière C., Paulin, F., & Zdunek, J. (2021, August). The narrative framework applied to evolution learning. Epistemological and didactic implications. In *ESERA*.

Chapter 22
Biology, Scientific Practices, and Everyday Life Situations Within an Entertaining Mesozoic Era Setting for Young Children

Marida Ergazaki and Iro Bardoutsou

22.1 Introduction

Edutainment refers to entertainment that creates opportunities for informal learning by providing information that is educational and entertaining at the same time (Okan, 2012). The idea is not new. Already in the 1950's, Disney produced movies and programs presenting wildlife to the general audience, or introducing science, math and history to children in ways that were supposed to be entertaining to them. Edutainment may come in many forms. It can be offered through, e.g., comics, books, stories, songs, games, hobbies, museum visits, nature walks, videos, computer games, software, virtual and augmented reality, animated cartoons on TV, computers, tablets or mobile phones. Research shows that animated cartoons may help children improve their cognitive and behavioral skills, enhance their motivation for learning, reduce their stress or boredom and also familiarize with scientific concepts and problem-solving in scientific ways (Fisch, 2000; Habib & Soliman, 2015; Hanif, 2020).

Many animated series present young children with science, engineering, and math concepts and practices. 'Sesame Street', 'The Cat in the Hat Knows a Lot About That!', 'Curious George', 'Sid the Science Kid', 'Elinor Wonders Why', 'Plum Landing', 'Dinosaur Train,' 'Octonauts', and 'Wild Kratts' are just a few. Our study is concerned with 'Dinosaur Train' in particular. Why is that? 'Dinosaur Train' is a top-rated, PBS animated series that aired worldwide over 100 episodes across five seasons. Being so popular with its young audience for over a decade possibly indicates a very good balance between education and entertainment (Cesiri, 2019).

M. Ergazaki (✉) · I. Bardoutsou
University of Patras, Patras, Greece
e-mail: ergazaki@upatras.gr

O. Morin et al. (eds.), *Challenges of a Changing World in Biology Education*,
Contributions from Biology Education Research,
https://doi.org/10.1007/978-3-032-05346-6_22

'Dinosaur Train' was developed by a multidisciplinary team of scientists, educators and communicators. It uses a well- informed Mesozoic era setting, with a train traveling in the Triassic, Jurassic, and Cretaceous through a time tunnel, to entertain preschoolers while introducing them to scientific concepts and practices blended with concepts of everyday life (About Dinosaur Train, PBS Shows, PBS KIDS for Parents, n.d). Moreover, being concerned with the history of life on Earth, 'Dinosaur Train' focuses primarily on biological concepts and discusses the nature of scientific research by highlighting scientific practices that might enhance children's thinking skills.

Our study aims at exploring the scientific concepts and practices, as well as the concepts of everyday life introduced in the over 100 episodes aired across the series' five seasons. Here, we present our analysis on the thirty episodes we started with, to address the research questions of (a) which biological concepts, (b) which scientific practices, and (c) which concepts of everyday life are presented to young children through the 'Dinosaur Train' stories and how.

22.2 Methods

22.2.1 The Overview of the Study

This chapter concerns the content analysis of thirty 'Dinosaur Train' episodes from seasons one, three, four and five. The episodes (Table 22.1) were selected conveniently as they were freely accessible. Each episode is about half an hour long. It includes two 11-min stories, as well as one or two short live-action segments moderated by 'Dr Scott, the palaeontologist', a real palaeontologist (Dr Scott Sampson) who was also involved in the development of the series (About Dinosaur Train, PBS Shows, PBS KIDS for Parents, n.d). We analyzed the stories of all thirty episodes and the live-action segments of thirteen (Season 1: episodes 1–4, 17, 27, 31, 34–35, 37–38; Season 3: episode 5; Season 5: episode 6) because of accessibility problems.

22.2.2 The Overview of the Data

22.2.2.1 The Leading Character

The leading character of the series is Buddy. He is a Tyrannosaurus rex but he doesn't know it, since he hatched in a Pteranodon nest with three little pteranodons, Tiny, Shiny, and Don. Buddy was adopted by the pteranodon family and grows with his three pteranodon siblings, all with different personalities and talents. He is eager to find out his own species and so he is very curious about different species of dinosaurs.

Table 22.1 The thirty episodes

Season 1	
1.1.1	Valley of the Stygimolochs
1.1.2	Tiny loves fish
1.2.1	The call of the wild Corythosaurus
1.2.2	Triceratops for Lunch
1.3.1	Beating the heat
1.3.2	Flowers for Mom
1.4.1	I'm a T. rex!
1.4.2	Ned the Quadruped
1.11.1	Derek the Deinonychus
1.11.2	Don's Dragonfly
1.14.1	The Theropod Club
1.14.2	Hatching Party
1.15.1	The Old Spinosaurus & the Sea
1.15.2	A Spiky Tail Tale
1.16.1	Night Train
1.16.2	Fossil Fred
1.17.1	Dinosaurs in the snow
1.17.2	Cretaceous Conifers
1.24.1	Erma Eoraptor
1.24.2	Under the Volcano
1.27.1	Iggy Iguanotation
1.27.2	Shiny Can't Sleep
1.28.1	Triassic Turtle
1.28.2	Tank's Baby Brother
1.31.1	The Wing Kings
1.31.2	The Big Mud Pit
1.34.1	Elmer Elasmosaurus
1.34.2	Dinosaur Block Party
1.35.1	Carla Cretoxyrhin
1.35.2	Train Trouble
1.37.1	The Amazing Michelinoceras Bros
1.37.2	Dad's Day Out
1.38.1	Great Big Stomping Dinosaur Feet!
1.38.2	Hornucopia!
1.39.1	The Good Mom
1.39.2	Diamond Anniversary
Season 3	

(continued)

Table 22.1 (continued)

Season 1	
3.5.1	Classic in Jurassic: Air, Water, Land
3.5.2	Desert Day and Night
3.7.1	Solar Train
3.7.2	Bird watching
3.9.1	One Big Frog
3.9.2	Caving with Vlad
3.10.1	Tiny's Fishing Friend
3.10.2	Butterflies
Season 4	
4.6.1	Junior Conductor Jamboree
4.6.2	Troodon Train Day
4.9.1	What's at the Center of Earth? Layers!
4.9.2	What's at the Center of Earth? Fossils!
Season 5	
5.1.1	The Tiny-saur Train
5.1.2	How Many Horns?
5.2.1	Mom Was a Kid Once, Part 1
5.2.2	Mom Was a Kid Once, Part 2
5.4.1	A Brand New Species, part 1
5.4.2	A Brand New Species, part 2
5.6.1	Love day
5.6.2	A new leaf
5.7.1	Under water Race
5.7.2	Buddy Wants to Fly
5.8.1	King and Crystal Play Red Rock
5.8.2	Nick of Time

He is also very curious about nature in general. Moreover, he loves having fun with his family and friends.

22.2.2.2 The Episodes Structure

The stories of the episodes have similar structure. At the beginning of the story, the members of the pteranodon family are curious about something that may concern another Dinosaur species or some other creature. So, they have a question, and in order to find the answer, they board the 'Dinosaur Train'. This travels through the 'time tunnel' of the Mesozoic era and takes them to different periods where they can meet the other species they are curious to learn about. On board, they talk with Mr

Conductor, a very clever and wise Troodon, about their question. When they arrive at their destination (Triassic, Jurassic, and earlier Cretaceous), they meet other species. They discuss with them, play, explore and learn together, while having fun by joking, playing and singing. On their way back home, they discuss their day and what they have learned in their trip.

22.2.3 The Data Analysis

The analysis of the stories of the thirty episodes (N = 60) and the live action segments of thirteen of them was performed with the qualitative data analysis software, NVivo. We first coded pieces from the stories and the live-action segments to our three overarching categories: (a) biological concepts (e.g. ecosystem, fossils, growth, adaptations, carnivores-herbivores-omnivores, biodiversity, evergreen-deciduous, reproduction) (b) scientific practices (e.g. asking questions, making observations, testing hypotheses, be open-minded), and (c) everyday life concepts (e.g., emotions, diversity, friendship, fair play). Then, we elaborated further on the biological concepts category. More specifically, we removed the references that concerned mere information about dinosaurs or other animals and plants (their traits, period, group or habitat) rather than concepts related to them. For instance, cases like the following in which there is plenty of information about a specific animal (Castrocauda), whereas there is no elaboration on concepts related to this information (mammal, oviparous animal, carnivorous animal) were not included in our concept-centered analysis: "It is a wonderful creature, an early mammal from the Jurassic. It has fur and webbed feet and lays eggs. It lives both on land and in water and loves fish. It is called Castrocauda". Next, the remaining contents of the category were organized into four broader subcategories to make better sense of our data: (a) interaction-related concepts (e.g., ecosystem, adaptations), evolution-related concepts (e.g., biodiversity, fossils), organisms' distinction-related concepts (e.g., carnivores-herbivores-omnivores, evergreen-deciduous), and biological functions-related concepts (e.g. growth, reproduction). Some episodes (1/3) were coded independently by each author, disagreements were discussed, and a satisfactory consensus was reached.

22.3 Findings

Our findings showed that 'Dinosaur Train' and specifically the 30 episodes we analyzed introduce (a) biological concepts essential for children's understanding about nature ($N_{references} = 287$), (b) scientific practices familiarizing children with how knowledge about nature can be built ($N_{references} = 368$), and (c) everyday life concepts valuable for a fulfilling life ($N_{references} = 194$) (Fig. 22.1).

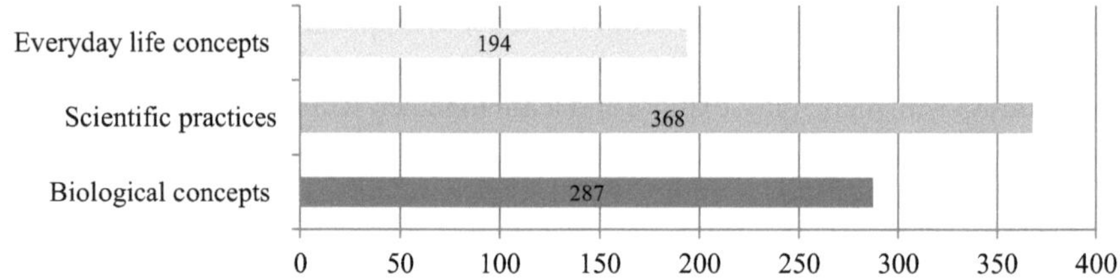

Fig. 22.1 Findings' overview

22.3.1 *Biological Concepts*

The biological concepts were categorized as (a) interaction-related ($N_{references} = 98$), (b) evolution-related ($N_{references} = 97$), (c) organisms' distinction-related ($N_{references} = 65$), and (d) biological functions-related concepts ($N_{references} = 27$) (Fig. 22.2). Each subcategory is presented below in more detail.

22.3.1.1 Interaction-Related Concepts

As shown in Table 22.2, the biological concepts coded as interaction-related ($N_{references} = 98$) were (a) adaptations, (b) ecosystem, and (c) pollination.

The ecosystem, particularly its biotic part, is presented as a cycle of life shaped by the interactions of the organisms in the context of their food relationships. For instance, Dr Scott, the paleontologist, points out that "An ecosystem is made up of plants, animals and other organisms that live together and make up a life cycle.

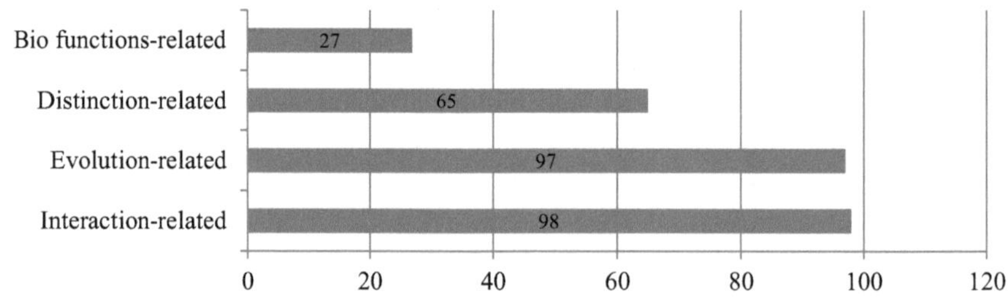

Fig. 22.2 Biological concepts ($N_{references} = 287$)

Table 22.2 Interaction-related concepts

Interaction-related concepts	References
Adaptations	89
Ecosystem	7
Pollination	2

The cycle starts with plants, which get their energy from the sun. Herbivores eat the plants and then carnivores eat the herbivores. The remains of all these animals then become food for insects and other small creatures that make food for more plants. This was happening in dinosaur times and it's still happening now. Everything in an ecosystem is connected to everything else through a life cycle. In an ecosystem everything becomes food for something else".

Moreover, being presented as a mutually beneficial interaction between insects and plants, the concept of pollination sheds light on another type of the interaction within the ecosystem's biotic part: "At some point, they see a bee on the red flowers, and Buddy says that the bee seems to be drinking something from the flower. Mr. Pteranodon, his dad, says that the bee drinks from each flower something called nectar, and while sucking the nectar, it sticks pollen on its legs, and the bee then carries it from flower to flower. Buddy asks if the pollen helps make more flowers, and his dad says yes".

On the other hand, the idea of interaction between the biotic and abiotic part of the ecosystem is highlighted through an analogy between 'members' of the ecosystem (biotic and biotic), and members of a family, which is a familiar entity for children:

- At some point it starts to rain, and Mr Pteranodon, says, "Look at the flowers; they need sunlight, soil, insects, even this rain. All these things together make an ecosystem". "Right", says his daughter Shiny. "An ecosystem is all the elements of nature working together as a family?" "Yes", says Mr Pteranodon. "The different parts of an ecosystem are like the different members of a family". Shiny doesn't like rain but then she remembers that rain is part of the ecosystem and helps the flowers grow".

Finally, adaptations, the idea with the most references in this subcategory, shed light on both biotic and biotic-abiotic interactions without any explicit connections with evolution. That is why we coded them in the interaction-related concepts rather than the evolution-related ones. The adaptations discussed concern efficient utilization of food, protection from other organisms (bodily appendages, behaviours), resistance to extreme environmental conditions (drought, heat, and cold), predation, efficient movement, and communication. The series is not limited in just describing adaptations; in fact, this is very rare ($N_{references} = 2$). On the contrary, it explains adaptations (a) mostly as functions made possible by specific structures or behaviours, and (b) rarely as serving a purpose, i.e. through teleology. For instance:

(a) "Stegosaurus plates [structure] must have helped them stay cool [function] in the summer heat", "Beelzebufo has a hard, plate-like armor on its body [structure] that helps him keep away predators [function]", "Archeopteryx sleeps upright [behaviour] and so it can be alert for predators [function]"
(b) "Cassie Castrocauda, an early mammal from the Jurassic, tells Buddy they both have lots of sharp teeth. She says hers are for catching fish and holding them in her mouth [teleology]".

Since teleology is a persistent bias in children's thinking about nature, the series' choice to mainly avoid it and provide the 'structure–function' (or behaviour-function) relationship as an appealing alternative, seems to be a good one.

22.3.1.2 Evolution-Related Concepts

As shown in Table 22.3, the biological concepts coded as evolution-related ($N_{references} = 97$) were (a) biodiversity, (b) natural history of Mesozoic Era, (c) fossils, and (d) the concept of evolution by natural selection, just once.

Biodiversity, the idea with the most references here, is highlighted through differences between (a) species, (b) species of the same group, and (c) related species of different periods. (To be more precise, the series presents genera rather than species as they are called to make it easier for young children). For instance:

(a) Tank (Triceratops) gives his palm and salutes with a "High five!" (Five-digit forelimbs). Tiny (Pteranodon) goes next: "High three!" (Three-digit forelimbs). Buddy (Tyrannosaurus) is the last one: "High two!" (Two-digit forelimbs).
(b) Buddy says, "You Stacey (Styracosaurus) and you Tank (Triceratops) are both ceratopsians. But your horns are different. Yours are thicker and more than his and have a different shape".
(c) Effie (a Triassic crocodile) takes part in a race, she gets last out of the water, and everyone is surprised because she runs on two legs. "I thought crocodiles only ran on four", says Tiny. "What's happening in the Triassic?" "I'm a different species of crocodile. I'm bipedal", says Effie.

The natural history of the Mesozoic Era includes descriptions of (a) flora, (b) fauna, and (c) climate, as concrete examples of the idea of change over time. For instance:

(a) "Did you know that flowers started growing in the Cretaceous? Before that, there were no flowers".
(b) "Mr. Conductor tells them that to see a turtle like Proganochelys they must go back to the Triassic when the first turtles lived".
(c) "He tells them to visit the Jurassic, which had a warm and humid climate and many jungles and forests".

Fossils are presented as the main evidence of change over time, and the subject of study by specialized scientists called paleontologists.

Table 22.3 Evolution-related concepts

Evolution-related concepts	References
Biodiversity	36
Natural history of Mesozoic Era	34
Fossils	26
Evolution	1

- "This was an ocean floor that became mountains and cliffs, which have fossils of marine creatures from the Triassic".
- "Buddy and Don are looking for trilobite fossils but a small digging mammal that is around is sure they won't find them there, because all the present fossils come from a more recent time period than trilobites: "I've learned after many years of digging that you don't find fossils from different time periods mixed together". "We just have to drill down a bit more to the fossils of an older time period".
- Dr. Scott asks children what they think a paleontologist actually does, and starts discussing fossils. "The paleontologists dig up the remains of dinosaurs and other animals that lived a long time ago. And which parts do we usually find?". "Yeah, [bones/teeth as children said] and shells, plants and even poop can sometimes become fossils", says Dr. Scott and shows a fossil. "Paleontologists spend a lot of time looking for fossils and when we find one, we carefully uncover it, wrap it up in a protective jacket and take it back to the museum. After all the bone fragments are glued back together, we make comparisons with the fossils of other animals to find out what we've found. It's trying solve a mystery!"

Finally, evolution of a new species is presented just once through a descriptive example provided by Mr Conductor, the wise Troodon who is in charge of the Dinosaur Train. According to him, the body traits of a species may change over very long periods of time and this may lead to a new species:

- "Creatures with different numbers of horns are called ceratopsians. They live in the Cretaceous and belong to the same family. Some ceratopsians that lived earlier began to grow horns. Most had bony frills on the back of their heads. After millions of years, the frills and horns got bigger, and we got a huge species like Kosmoceratops". The mechanism of natural selection is absent.

22.3.1.3 Organisms' Distinction-Related Concepts

Organisms are allocated to different groups based on criteria such as (a) food type, (b) activity time, (c) leg number, (d) leaf shedding. So, the biological concepts coded as organisms' distinction-related ($N_{references} = 65$) were (a) carnivores - herbivores - omnivores, (b) diurnal - nocturnal, (c) bipeds-quadrupeds, and (d) evergreen-deciduous (Table 22.4).

These concepts are descriptive and they are presented with concrete examples. For instance:

Table 22.4 Organisms' distinction-related concepts

Distinction-related concepts	References
Carnivores - Herbivores - Omnivores	48
Diurnal – Nocturnal	7
Bipeds - Quadrupeds	3
Evergreen – Deciduous	7

Table 22.5 Biological function-related concepts

Biological functions-related concepts	References
Growth	15
Reproduction	12

- "Mrs Pteranodon says it is snack time and offers all children fish. Mrs Maiasaura (mom of two) says that they're herbivores and don't eat meat or fish. "I brought them these leaves", she says.
- Mr Conductor says that nocturnal animals are more active at night and usually they can see well in the dark. "So when you sleep, a lot of nocturnal animals are out there looking for food."
- "Both were bipeds, which means they walked on two legs". "Quadruped meand four legs".
- "Most are evergreen. This means they are green and tasty all year round".

22.3.1.4 Biological Function-Related Concepts

As shown in Table 22.5, the biological concepts coded in this subcategory ($N_{references}$ = 27) were (a) growth, and (b) reproduction.

Growth is discussed in terms of (a) increasing size, (b) required factors, and (c) life cycles with metamorphosis. Reproduction is discussed in terms of (d) egg-laying. For instance:

(a) Mr Pteranodon says his wife and he were little pteranodons but now they are big. Mrs Pteranodon says that Buddy started out as a little tyrannosaurus, but one day he will be huge. The mom of Mr Conductor says that her son was once a little troodon, and then he grew up and tall like a conifer!
(b) Mr Pteranodon tells Tiny that rain, sun, and soil, are all necessary for the flowers' growth.
(c) Mr Conductor says "Dragon fly moms lay eggs in the water and the babies hatched are called nymphs and they look like little worms. When they get bigger, they sprout claws and soon after wings". "Yes, like Howard", says Tiny, "and fly around. I can't wait to fly everywhere with Howard and explore…".
(d) Tiny bets that other creatures, like insects or birds, hatch from eggs like dinosaurs, and her mom agrees with her. Mr Conductor says they're both right: "All dinosaurs hatch from eggs and some other creatures, too".

22.3.2 Scientific Practices

As shown in Fig. 22.3, the scientific practices recorded in our data ($N_{references}$ = 368) were (a) asking questions, (b) making comparisons, (c) making observations, (d) testing hypotheses, (e) drawing conclusions, (f) explaining how, (g) being open-minded or open to reconsidering, and (h) categorizing.

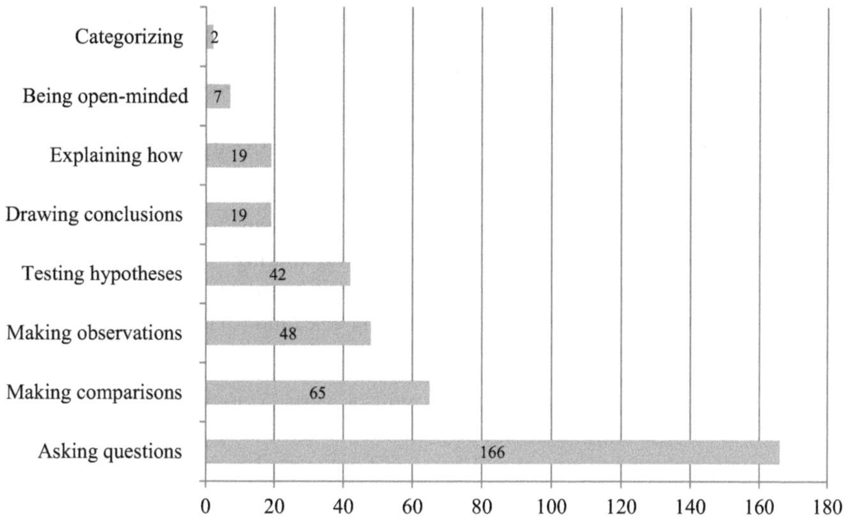

Fig. 22.3 Scientific practices ($N_{references} = 368$)

The most frequent scientific practice we recorded is asking questions about nature. This practice is obviously very important since it may trigger meaningful exploration and knowledge construction. It was used by the series characters within the episodes' stories and by Dr Scott within the short live-action segments. For instance:

- "What do you think Corey (Corythosaurus) is doing with these sounds?"
- "Why are there different kinds of leaves? Why do they fall?"
- "How did the fossils from the sea get here?"
- "Are the eggs of the different species different?"
- "Can you think of any animals with feathers or fur that live today in cold climates", asks Dr Scott.

The scientific practice of making comparisons was recorded quite often, too. The characters compare traits or behaviours of different species. Moreover, Dr Scott asks children for comparisons of extinct and extant animals. For instance:

- Tiny says that the claw of Deinonychus is bigger than the one of Troodons.
- Buddy and Tiny find more flowers, which look different from the previous ones. These have a longer stem and a smaller flower. Tiny sees a bee that seems to smell it. The bee leaves, they follow the bee and see some different flowers. They say that these have 6 petals in a circle and pink lines in the middle.

Dr Scott asks kids if they can compare Triceratops to any other animal today. And they think of rhino" and he says that both are herbivores with horns:

- Making observations is performed either as part of making hypotheses or independently. For instance:

- Buddy tells Iggy (Iguanodon) that he noticed he has a lot of sharp teeth and asks him if he uses them to eat meat.
- Buddy makes the observation that the mouth of Spikey (Stygimoloch) looks more like a beak. He asks Spikey if he can have a look in his mouth. His friend doesn't mind, so Buddy observes the open mouth, and says "strong jaws, no teeth".

Testing the hypotheses they make is one of the characters' favorite practice, especially Buddy's. "Buddy likes the word hypothesis. It means an idea you can test!" Whenever he has a hypothesis, he is very keen on sharing it with the others and searching with them ways to test it. So, this scientific practice is performed by the characters by (a) observing, (b) asking somebody else, and (c) trying out something. For instance:

(a) Tiny says that Cory (Corythosaurus) will be very excited because she'll have a new brother. Buddy says that he's anxious to see the eggs and wonders what they will look like. Tiny says "I have a hypothesis. Maybe the eggs will have crests like Cory's". To test the hypothesis, they go to the nest and observe the Corythosaurus eggs. There is no crest, so they conclude the hypothesis is wrong.
(b) They want to know about Cory's crest and they have several hypotheses: "Do you use it to breathe underwater?"; "Is it a crown? Are you a princess?"; "Has it something to do with your music?". Cory answers directly: the crest is related to the music she makes.
(c) Buddy has the hypothesis that the front legs of Ned (Brachiosaurus) may help him climb trees. To test the hypothesis, they ask Ned to climb the tree and eat the leaves. The hypothesis was correct: Ned was using his front legs to climb.

Children have opportunities to watch the characters engaging with the scientific practice of drawing conclusions. These derive from reasoning strands based on the empirical data at hand. For instance:

- Buddy and his siblings play detectives. They find something and Don says it looks like a tooth. Buddy says it has the right shape and it is pointy. So, they draw the conclusion that it may be the tooth of a carnivore.
- The legs of Castrocauda remind them of fish fins. So, they draw the conclusion that this animal is made to swim.

The scientific practice of explaining how, is connected with mechanistic reasoning which is very important for understanding biology (Shtechman et al., 2022, 2023). It seems that young children can have the opportunity to start familiarizing with it through the episodes we analyzed. For instance:

- Cory explains how her crest works: She breathes in air and the air travels through the crest and then comes out and makes sounds.
- Mr Conductor explains how the sea fossils ended up there: The earth plates float on top of a molten hot material. They slowly move and fall on top of each other, pushing huge land masses from the sea floor to the top. The ocean floor becomes mountains and cliffs and that's how fossils of marine creatures can be found there.

Open-mildness in science, which is connected with the provisional nature of scientific knowledge, is introduced by Dr Scott, who presents paleontologists as not having all the answers and as being ready to reconsider what they believe about dinosaurs. For instance, Dr Scott says that:

- Although they are not sure, paleontologists think that Styracosaurus rather used their horns to show them to other Styracosaurus.
- For a long time scientists were not sure how many babies hatched from a typical dinosaur nest. Now they know for most dinosaurs the number is 20–30 eggs.
- When paleontologists found a bone with a strange edge near Iguanodon fossils, they thought it was a nose horn but then it turned out to be a large thumb spike.

The scientific practice of categorizing was the one with the fewest references in the episodes we analyzed. However, it is valuable tool both in science and everyday life thinking (Ergazaki et al., 2018). The instances of categorizing are the following:

- Buddy and his friends observed that all leaves are not the same. They may have different shapes or colours. Some can be thin and pointy and they never fall. They found four different types of leaves and they put them in four different piles.
- Shiny looks for shells and separates them depending on how shiny they are.

22.3.3 *Everyday Life Concepts*

The everyday life concepts we recorded ($N_{references} = 194$) are shown in Fig. 22.4.

Concepts like these concern children's well being and meaningful interactions with others. So, giving them the opportunity to think about them through the stories of others with whom they can relate, may be beneficial to them. For instance: Annie tells Buddy that it must be weird to be the only T. rex in his family. Buddy says that some kids look like their parents and some don't. "It doesn't matter: my family loves me" (family); "You have horns, we have wings. Everyone is unique and that's what makes us interesting" (diversity); Tiny apologizes to Buddy for the disagreement they had about the Stegosaurus tail and tells him that he is her best friend (friendship).

22.4 Discussion

The thirty episodes we analyzed introduce biological concepts related (a) to essential characteristics of living organisms, such as interaction, evolution, and biological functions, and (b) to organism's distinction by a unique criterion each time. Although the biological functions of growth and reproduction are discussed and plants are portrayed as living, the overarching 'living-non-living' distinction on the basis of such functions is absent. Children are introduced to growth and reproduction per se and not as criteria for distinguishing between living and non living things.

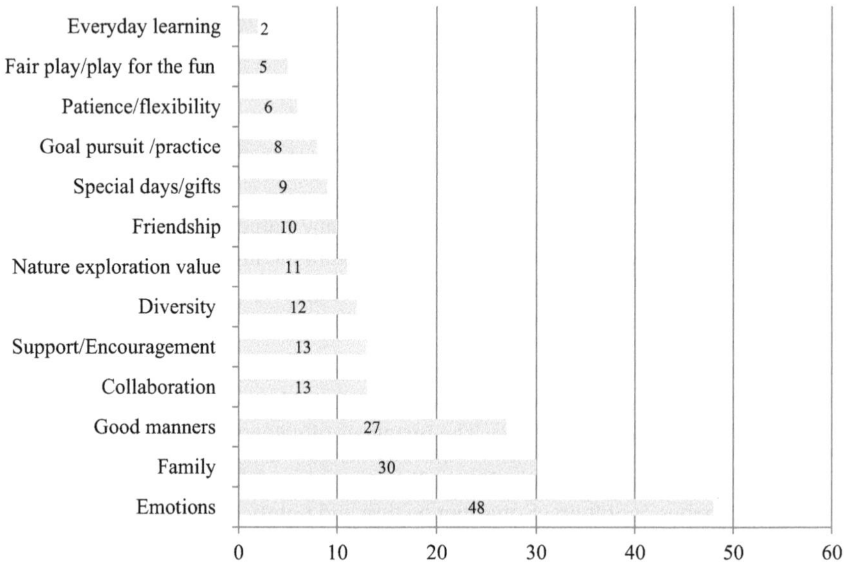

Fig. 22.4 Everyday life concepts ($N_{references} = 194$)

The concept of species is used interchangeably with genus; it is based on morphology and serves as a conceptual peg for the whole picture. This is rather expected considering how young the target group of the series is. The idea that life on earth changes over long periods of time emerges through the natural history of the Mesozoic Era and its concrete examples. Inter-species diversity is thoroughly discussed. The series presents young children with the idea that even species of the same taxonomic group (e.g., ceratopsians) or species related to each other (e.g., different turtle species) can have different body traits. However, intra-species diversity, which is a prerequisite for natural selection, is left out of focus.

The same goes for the natural selection itself. Namely, the concept of evolution on the species level is presented through a descriptive example about how a new "species" of ceratopsians, Kosmoceratops, came to be. The underlying mechanism remains hidden, although research suggests that young children could possibly get involved with it (Emmons et al., 2017). Similarly, the concept of adaptation is not dealt with as a process but as a product. The series introduces concrete examples of animal and plant adaptations and "explains" them through the cross-cutting concept of the "structure–function" relationship rather than teleologically. So, it emphasizes how the adaptation benefits the organism concerning its interactions with its biotic or abiotic environment.

The series introduces the important scientific practices as knowledge construction tools, and encourages children to use them in building their own understanding about nature. It familiarizes children with the nature of scientific research and the nature of scientific knowledge which is empirically-based and provisional. Finally, the series attempts to better prepare children for everyday life emotions, relationships, and behaviours. It employs the idea of biodiversity to emphasize that diversity in daily life should not be considered as an obstacle to building meaningful relationships based on support, encouragement, and collaboration. The examples of the Pteranodon family and various friendships that function very well despite including highly diverse members illustrate how inherent diversity is in everyday interactions and how important it is to embrace it.

In sum, according to our findings, Dinosaur Train seems to be a valuable informal resource for (a) engaging young children in biology learning, (b) familiarizing them with scientific practices as tools for expanding their knowledge about nature, and (c) having them reflect on everyday life emotions, relationships and behaviours. Moreover, according to Cesiri (2019) and her visual and verbal analysis on knowledge dissemination in Dinosaur Train, the series employs well-tested popularizing strategies to communicate demanding content to preschool children. So, it seems worth investigating how Dinosaur Train may be combined with preschool curricula, and how effective it may be in terms of measurable learning outcomes as well.

References

About Dinosaur Train, PBS Shows, PBS KIDS for Parents (n.d.). Retrieved June 27, 2022, from http://www.pbs.org/parents/dinosaurtrain/about

Cesiri, D. (2019). Knowledge dissemination in paleontology. A case study from the animated series dinosaur train. In S. M. Maci & S. Michele (Eds.), *Representing and redefining specialised knowledge: Variety in LSP* (pp. 223–244). CERLIS Series. http://hdl.handle.net/10446/155776

Emmons, N., Lees, K., & Kelemen, D. (2017). Young children's near and far transfer of the basic theory of natural selection: An analogical storybook intervention. *Journal of Research in Science Teaching, 55*, 321–347. https://doi.org/10.1002/tea.21421

Ergazaki, M., & Andriotou, E. (2010). From 'forest fires' and 'hunting' to disturbing habitats' and 'food chains': Do young children come up with any ecological interpretations of human interventions within a forest? *Research in Science Education, 40*, 187–201. https://doi.org/10.1007/s11165-008-9109-6

Ergazaki, M., Gasparatou, R., & Valanidou, E. (2018). Young children's reasoning when sorting pictures and objects. In N. Gericke & M. Grace (Eds.), *Challenges in biology education research* (pp. 199–215). ERIDOB. https://www.diva-portal.org/smash/get/diva2:1230716/FULLTEXT01.pdf

Fisch, S. M. (2000). A capacity model of children's comprehension of educational content on television. *Media Psychology, 2*(1), 63–91. https://doi.org/10.1207/S1532785XMEP0201_4

Habib, K., & Soliman, T. (2015). Cartoons' effect in changing children mental response and behavior. *Open Journal of Social Sciences, 3*(09), 248–264. https://doi.org/10.4236/jss.2015.39033

Hanif, M. (2020). The development and effectiveness of motion graphic animation videos to improve primary school students' sciences learning outcomes. *International Journal of Instruction, 13*(3), 247–266. https://www.e-iji.net/dosyalar/iji_2020_4_16.pdf

Okan, Z. (2012). Edutainment and Learning. In N. M. Seel, (Ed.) *Encyclopedia of the Sciences of Learning* (pp. 1080–1082). Springer. https://doi.org/10.1007/978-1-4419-1428-6_1938

Shtechman, Y., Ergazaki, M., & Haskel-Ittach, M. (2022, August-September). *Elementary school students' evaluation of explanations to biological phenomena.* Paper presentation at the European Researchers in Didactics of Biology (ERIDOB) Conference 2022, Nicosia, Cyprus.
Shtechman, Y., Ergazaki, M., & Haskel-Ittach, M. (2023, April). *Exploring how students evaluate explanations about biological phenomena in different grades of elementary school.* Paper presentation at the National Association for Research in Science Teaching (NARST) Conference 2023, Chicago, US. https://narst.org/sites/default/files/2023-04/NARST%202023%20Abstract%20Book%20FINAL.pdf

MIX
Papier aus verantwortungsvollen Quellen
Paper from responsible sources
FSC® C105338

If you have any concerns about our products,
you can contact us on
ProductSafety@springernature.com

In case Publisher is established outside the EU,
the EU authorized representative is:
Springer Nature Customer Service Center GmbH
Europaplatz 3, 69115 Heidelberg, Germany

Printed by Libri Plureos GmbH
in Hamburg, Germany